A guide to task analysis

A guide to task analysis

Edited by

B. Kirwan
(University of Birmingham)

and

L.K. Ainsworth
(Synergy)

Taylor & Francis
London • Washington, DC

UK Taylor & Francis Ltd., 4 John Street, London WC1N 2ET

USA Taylor & Francis Inc., 1900 Frost Road, Suite 101, Bristol, PA 19007

First published 1992. Reprinted 1993.

British Library Cataloguing in Publication Data
A catalogue record for this book is available from the British Library

ISBN 0 7484 0057 5 (cloth)
ISBN 0 7484 0058 3 (paper)

Library of Congress Cataloging in Publication Data is available

Cover design by Amanda Barragry

Printed in Great Britain by Burgess Science Press, Basingstoke on paper which has a specified pH value on final paper manufacture of not less than 7.5 and therefore 'acid free'.

CONTENTS

Executive Summary

Task Analysis is the name given in this *Guide* to any process that identifies and examines the tasks that must be performed by users when they interact with systems. It is a fundamental approach which assists in achieving higher safety, productivity and availability standards.

This document describes the task analysis approach, and gathers together and comprehensively describes the application of 25 major task analysis techniques, organizing them into a practical user guide.

The *Guide* is aimed at industrial practitioners such as design engineers, safety assessors, ergonomics practitioners, trainers, procedures writers, project managers, etc., with or without previous task analysis experience, who have a need to establish task requirements or to carry out task assessments for one or more of the following topics:

- Allocation of function (between human and machine)
- Person specification
- Staffing and job organization
- Task and interface design
- Training
- Human reliability assessment
- Safety management
- Performance checking
- Problem investigation

Guidance is provided on which techniques to use to achieve human–system performance goals (e.g. efficient and effective allocation of function) in the most cost-effective manner. Application of this *Guide* will therefore assist organizations in choosing and applying the most appropriate technique(s) to meet their particular needs.

Examples from some of these topic areas of when and where Task Analysis could usefully be applied are the following:

- Where safety is important (e.g. accidents or near misses have occurred; the technology is vulnerable to human error; or operator safety actions are claimed as part of the safety case)
- Where it is necessary to provide assurance that task allocation and feasibility are appropriate
- When operational feedback raises concerns over a particular human–machine interface
- When the requirements for productivity have changed
- Where personnel characteristics are different
- Where different crewing or organizational structure is intended

- Where a significantly different technology is being used
- Where the same task has to be performed in a different environment

This *Guide to Task Analysis* shows how and when various task analysis techniques can be used to address the above issues, and also gives practical insights into the use of these techniques. Additionally, the *Guide* assists in planning task analysis applications during the life cycle of a system, at any stage from early design onwards, to achieve cost-effective results. Finally, a number of real case studies demonstrate how task analysis has actually been used in practice, and the impact that task analysis has had in these industrial applications.

G. Pendlebury
Chairman of Task Analysis
Working Group

Preface

This *Guide* was produced because of a perceived need to transmit knowledge of techniques from practising ergonomists to others in industry who would benefit from the application of formal task analysis methods and strategies. Task analysis is fundamental to the analysis and enhancement of human–machine systems, yet few documents exist which cover more than a small number of the large range of tools available, and of these fewer still contain practical guidance. A working group of interested members with appropriate expertise from a range of UK industries, universities and consultancies, was therefore formed. This working group originated from the Human Factors in Reliability Group which had already developed two other guides, one on human reliability assessment and one on human factors in the process industries. The Task Analysis Working Group (TAWG) was then formed and worked as an independent group attempting to explain the task analysis process and how it worked in industrial systems design, development and assessment (the task analysis process, Part I of the *Guide*). TAWG then defined a representative set of task analysis techniques (25 in total, documented in Part II of the *Guide*), and supplemented this information via a series of documented case studies from a range of industries, exemplifying the different uses for which task analysis could be implemented.

This *Guide* is not a theoretical treatise on the nature of task analysis. Rather its emphasis is on practical applications and insights that have been gained from applying the techniques documented, whether aimed at solving a single aspect of a human–machine-system problem, or a programme of task analysis and human factors aimed at maximizing successful human performance in systems. Whatever purpose the reader has in mind for task analysis, the developers of the *Guide* believe it will be useful in achieving those goals, as this is fundamentally what task analysis is all about.

Disclaimer

The contents of this report, including any opinions and/or conclusions expressed are those of the authors, contributors and reviewers, and do not necessarily reflect the views or policy of the parent organizations.

Acknowledgements

The need for a *Guide to Task Analysis* was identified by the Human Factors Reliability Group (HFRG), who set up a Task Analysis Working Group (TAWG). The TAWG then developed the specification for the *Guide* and subsequently produced it.

The developers of the *Guide* thank five organizations in particular who have sponsored the *Guide*, listed here in chronological order:

British Nuclear Fuels plc, Risley, UK — for sponsoring the development of Part I
National Power — for sponsoring Part II
Nuclear Electric, Barnwood — for sponsoring Part III
ICI Engineering — for sponsoring the final but critical stages of the *Guide's* editorship and production
Synergy — for lending support in the word processing and final production stages of the *Guide*

In addition the producers of the *Guide* wish to acknowledge the major contractors involved in the development of the *Guide*:

Part I — Dr S.P. Whalley (*R.M. Consultants Ltd*) and Dr A. Shepherd (*Loughborough University*)

Part II — Dr L.K. Ainsworth (*Synergy*) (technique write-ups — Dr L.K. Ainsworth, Mr B. Kirwan, Mr A. Livingston, Mr D. Mullany, Dr A. Shepherd, Dr R.B. Stammers, Mr D. Visick, Mr T. Waters and Dr S.P. Whalley)

Part III — Ms J. Penington (*W.S. Atkins Ltd*)

Principal editorship/integration of the *Guide* — Mr B. Kirwan (*Birmingham University*)

The *Task Analysis Guide* was produced from approximately 1989 to 1991. The work was orchestrated by a main group who contributed to the *Guide* and developed its concept, scope and over-saw its content; and a steering group (Task Analysis Steering Group) who made executive decisions and over-saw the contractual obligations during the *Guide's* development. Both groups were chaired by Mr G. Pendlebury (*Nuclear Electric*, Knutsford), and the secretary to Task Analysis Working Group was Miss J. Penington (*W.S. Atkins*). The members of the two groups are listed alphabetically below. Where a person's involvement spanned more than one organization, both organizations are listed:

Task Analysis Steering Group

Derek Heckle	*ICI Engineering*
Barry Kirwan	*British Nuclear Fuels plc, Risley/Birmingham University*
Geoff Pendlebury	*Nuclear Electric*
Keith Rea	*NNC/British Nuclear Fuels plc, Risley*
Jerry Williams	*Nuclear Electric/DNV Technica Ltd*

Task Analysis Working Group

Primary working group members

Dr Les Ainsworth	*Synergy*
Derek Heckle	*ICI Engineering*
Barry Kirwan	*British Nuclear Fuels plc, Risley/Birmingham University*
Allan Lewis	*Glaxochem plc*
Geoff Pendlebury	*Nuclear Electric*
Johanne Penington	*W.S. Atkins*
Keith Rea	*NNC/British Nuclear Fuels plc, Risley*
Dr Andrew Shepherd	*Loughborough University*
Dr Rob Stammers	*University of Aston*
Ian Watson	*NCSR, UKAEA*
Dr Susan Whalley	*Lihou Loss Prevention Services/R M Consultants Ltd*
Jerry Williams	*Nuclear Electric/DNV Technica Ltd*

Other contributing members

Dr Jane Astley	*Technica Ltd./Four Elements Ltd.*
Andrew Livingston	*HRA Ltd.*
Keith Mitchell	*NNC*
David Mullany	*Vickers Shipbuilding and Engineering Ltd*
Julie Reed	*Electrowatt/British Nuclear Fuels plc, Risley*
Dr Donald Ridley	*Nuclear Electric/Hatfield Polytechnic*
Brian Sherwood-Jones	*YARD Ltd*
Peter Shields	*Associated Octel Ltd*
Dr Ian Umbers	*Nuclear Electric*
David Visick	*Rank Xerox*
Keith Webley	*Rolls Royce and Associates Ltd*
David Whitfield	*Nuclear Installations Inspectorate*
Gareth Williams	*CERL (National Power)*

Finally, various people who were not part of the TAWG contributed towards the contents of Parts II or III:

Francis Brown	*British Nuclear Fuels plc*, Sellafield (Part III)
Mike Joy	*Diam Performance Associates* (Part III)
Dr Nigel Leckey	*British Nuclear Fuels plc*, Sellafield (Part III)
Dr Ron McLeod	*YARD Ltd.* (Part III)
Helen Rycraft	*British Nuclear Fuels plc*, Sellafield (Part III)
Trevor Waters	*British Nuclear Fuels plc*, Risley (Part II)

Introduction

This section introduces the basic concepts, purposes and application areas of task analysis. It does this by answering fundamental questions on what task analysis is and why it should be used, as well as when it should be used and by whom. It also serves as an overview to the structure of this *Guide*, outlining the nature of its three principal sections, which cover the task analysis process, task analysis techniques, and case studies of actual applications of task analysis in industry.

What is task analysis?

Task analysis covers a range of techniques used by ergonomists, designers, operators and assessors to describe, and in some cases evaluate, the human–machine and human–human interactions in systems. Task analysis can be defined as the study of what an operator (or team of operators) is required to do, in terms of actions and/or cognitive processes, to achieve a system goal. Task analysis methods can also document the information and control facilities used to carry out the task.

Task analysis is therefore a methodology which is supported by a number of specific techniques to help the analyst collect information, organize it, and then use it to make various judgements or design decisions. The application of task analysis methods provides the user with a 'blueprint' of human involvements in a system, building a detailed picture of that system from the human perspective. Such structured information can then be used to ensure that there is compatibility between system goals and human capabilities and organization, so that the system goals will be achieved.

Why use task analysis?

It can be argued by managers, engineers and others involved in design that the human element within a system is already implicitly included in system designs. While this is to a large extent true, unless this is done systematically in an open manner which can be subject to careful scrutiny, it is unlikely that the human element will be optimized, or that the potential for error will be minimized. Usage of explicit task analysis approaches should therefore lead to more efficient and effective integration of the human element into system design and operations, in three principal areas:

Safety

Systems must be safe in terms of staff and public safety, system integrity, and the impact on the environment. Task analysis can impact on safety in four major ways. Firstly it can be used to identify hazards to the operator in the workplace, and as an input to defining safe systems of work. Secondly, task analysis aims to achieve a general level of system safety through the achievement of good design for human operation. Thirdly, it can form the basis for the analysis of human errors in systems, or human reliability assessments which can feed into quantified risk assessments of systems. Fourthly, task analysis can be used in incident or accident investigations, to define what went wrong and to help identify remedial measures.

Productivity

Task analysis can help in decisions about where to automate processes, how to determine staffing requirements, and how to train staff and ensure efficiency. The identification and reduction of error potential will also enhance efficiency.

Availability

Systems must be adequately maintained and run to keep downtime within acceptable limits. Task analysis can be used to identify maintenance demands and to define the need for maintenance support tools and systems of work. Optimal work design should also reduce errors that lead to unscheduled downtime.

The importance of a formal and systematic approach such as task analysis becomes even more evident when considering the complexity of some human operations within systems, and the co-ordination that will be required to integrate various design decisions on human tasks at different stages during the design process.

The task analysis techniques can either be used as a process (as described in detail in Part I of the *Guide*) to maximize successful performance in system design, operation and maintenance, or can be used to address specific issues. These issues fall into six general categories, which together comprise the full impact which the task analysis process can have on system performance (for full description see Part I):

Allocation of function

Allocating functions between personnel and machines, and defining the extent of operator involvement in the control of the system.

Person specification

Defining the characteristics and capability requirements of personnel to enable them to carry out the task effectively.

Staffing and job organization

Defining the number of staff required, the organization of team members, communications requirements, and the allocation of responsibility, etc.

Task and interface design

Ensuring adequate availability and design of information displays, controls and tools to enable the operator(s) to carry out the task adequately, whether in normal or abnormal operations.

Skills and knowledge acquisition

Training and procedures design.

Performance Assurance

Assessment of performance predictively via human reliability assessment, retrospectively via incident investigation or analysis, or concurrently via problem investigations.

These areas or human factors issues are described more fully in Part I of this *Guide*, and are exemplified in the case studies in Part III.

When should task analysis be used?

Task analysis is used when designing a system, evaluating a system design, or if a particular human–machine system performance problem has been 'targetted' to be analysed and resolved. These three modes of use of task analysis are briefly discussed below.

Task analysis in system design

All system designs tend to pass through similar phases: from the initial concept of the system, through its preliminary and then detailed design phases, to the system's construction, commissioning and operation, and ultimately its decommissioning (shut-down and storage or demolition). This process is known as the *System Life Cycle*, and any task analysis studies must occur within one or more of these stages.

Task analysis can be undertaken at any stage in the life cycle of a system, provided that the information requirements of individual task analysis technique can be met. Most task analysis techniques can in fact be applied during the design stages (i.e. before the plant goes operational). However, as the design becomes fixed, changes identified as desirable during the task analysis will become expensive. Consequently, the earlier that task analysis is applied in the *system design life cycle*, the more cost-effective the process is likely to be.

If task analysis is utilized early in the design process, it is also possible to establish a two-way flow of information, with knowledge about human requirements and limitations feeding into the design process, and design preferences and constraints feeding into the task analysis. This can avoid the situation in which equipment design dictates personnel requirements, which may lead to sub-optimal systems and the necessity of retro-fit design solutions later on in the project life cycle.

The timing of task analysis is thus a critical determinant of the usefulness of its results, especially if it is being used to assist optimization of system design. If task analysis is undertaken at the initial conceptual phase of a system, there is the opportunity to address all human factors issues in the most cost-effective manner. It must be stressed, however, that task analysis is rarely a 'one-off' process; instead it usually requires one or more iterations as more detailed information about the system becomes established and the roles of various personnel within it become clearer. As the design progresses, the task analysis will become more detailed.

In Part I of the *Guide* the six human factors issues are therefore related to the different stages in the system life cycle, from its early concept until the system's decommissioning. Part I also discusses the timing of the application of task analysis techniques to achieve cost-effective impact.

Task analysis for system evaluation

Safety assessors can use task analysis as a basis upon which to evaluate design proposals or constructed/operational plant, either because a problem has been identified, new equipment has been added, or as part of a periodic review. Such evaluations may be in the form of internal company audits, or more formal evaluations such as quantified risk assessments. In either case task analysis can be used to consider the adequacy of the design of the task, and hence the likely human performance in the system.

Task analysis can therefore be used successfully for assessment purposes on an existing plant as well as on one being designed or commissioned. Access to information for carrying out task analysis (e.g. by observation, interviews, procedures, incident records, etc.) will usually be far easier with an existing plant than with one at the design stage. However, this benefit is sometimes offset by the implications of retro-fit design, and the re-training of operators etc., should these needs arise.

Targetting task analysis

Task analysis can also be used to focus upon specific issues rather than examining the system as a whole. It can be used when:

- Safety is especially important
- Technology is vulnerable to human error
- System changes have created a high level of uncertainty about system integrity
- There are productivity/availability problems, or a particularly high quality of product is required which depends on human performance

It is therefore possible to use task analysis to look at particular areas of concern, for specific benefits, as well as applying more comprehensive task analysis programmes.

Who should carry out task analysis?

Anyone can carry out task analysis procedures, but users and interested parties usually fall into one of the following categories:

Designers: Those responsible for designing the system and deciding which functions should be achieved by machine and which by operator, and those responsible for designing equipment layout including displays and controls.

Operations departments: Those similarly responsible for allocation of functions, but principally involved in the staffing of their plants, and the training and provision of support (e.g. procedures) and organization for those staff.

Assessors: Safety assessors involved in risk assessments, ergonomists involved in evaluations, or others who must perform an evaluative function.

Managers: Those who are responsible for the Project, who may not necessarily be users of task analysis, but who are likely to be the ones who sanction such work if it appears useful or necessary for reasons of profit, safety or availability.

What type of task analysis techniques exist?

There are many techniques currently available in ergonomics and other disciplines which deal with the description and analysis of human involvements in systems. The development of this *Guide* included the consideration of a large range of such task analysis methods, of which 25 were selected as being representative of task analysis approaches, and were also considered relevant to process control and related systems. In Part II of the *Guide*, these techniques have been divided into five broad categories:

Task data collection methods	Collecting data on actual or proposed task performance
Task description methods	Representing such data in a pre-specified format
Task simulation methods	Creating simulations of the task
Task behaviour assessment methods	Assessing what can go wrong in task performance
Task requirements evaluation methods	Assessing the adequacy of the task environment and existing facilities available to carry out the task.

As noted in Part II however, many of the techniques are not easy to classify as above, because they could fit into more than one category. Nevertheless, if the reader is looking for a means, for example, of collecting data, then this classification will be a useful starting point, and will serve to identify appropriate techniques in most cases. This is therefore not a definitive categorization system, but it is aimed at aiding the reader by structuring Part II and in selecting techniques for particular applications.

The actual mechanics of the techniques vary tremendously, ranging from straightforward observation to the use of sophisticated computer programs. Each technique is detailed in Part II according to the following format:

- *Overview of the technique*: Its main function, rationale, and genesis.
- *Application*: Its major application areas in terms of the purposes for which it can be used.
- *Description of the technique*: The basic mechanics of the technique, how to apply it, calculational formulae, etc.
- *Practical advice*: Practitioner's warnings, and potential pitfalls, where relevant.
- *Resources required*: A statement of the approximate utilization of resources and special requirements (e.g. computing facilities and packages).
- *Links with other techniques*: Other techniques with which this technique or its results can be usefully combined.
- *Advantages and disadvantages*: Its major theoretical and practical strengths and weaknesses.
- *References and further reading*: Principal source references and other useful and accessible published material.

Many of the techniques have been the subject of long chapters and even books, and so this *Guide* cannot always explain the techniques in comprehensive detail. However, the intention is to give the reader sufficient information to evaluate the technique's utility for a particular application or problem area. The references at the end of each section in Part II will enable the reader to follow up the technique in more depth where this is desirable. Furthermore, it should be noted that some of the techniques are particularly useful and hence these have become more popular and used than others (e.g. hierarchical task analysis). The reader may gain an appreciation of this from Part II and from the case studies in Part III.

What has been achieved with task analysis?

Ten case studies of actual applications of task analysis in a range of industries are documented in Part III of the *Guide*. These are intended more to serve as examples of what can be achieved via Task Analysis, rather than how to implement task analysis methods. The case studies which are documented are as follows:

1 Balancing automation and human action through task analysis

This case study from the nuclear power plant domain deals with a new approach to Functional Analysis called FAST (functional analysis system technique) integrated with hierarchical and timeline task analyses. It shows the usage of task analysis early in the design life cycle.

2 Preliminary communications system assessment

This case study from the offshore sector, deals with the analysis of communications, primarily between drilling and production personnel, and makes use of several techniques such as interviewing, the critical incident technique and task decomposition.

3 Plant local panel review

This case study is one of three from the design phase of a nuclear chemical plant. It deals with the analysis and design change of local panel interfaces, using a range of methods such as hierarchical task analysis task and decomposition.

4 Staffing assessment for a local control room

This case study from a nuclear chemical plant illustrates an attempt to define staff numbers for a local control room early in the design life cycle. It shows techniques such as hierarchical task analysis and a simplified form of timeline analysis.

5 Simulation to predict operator workload in a command system

This case study from a naval application predicts operator workload of a naval crew in a mission scenario, using a computerized workload assessment system called Micro-SAINT.

6 Operator safety actions analysis

This study for a nuclear power plant (still at the design phase) entails the operational assessment of required procedural actions following the onset of abnormal events. It uses decomposition task analysis and aims to determine the adequacy of the task and interface design.

7 Maintenance training

This case study is in the general area of process control and concerns the development of a scheme for supervising the site training for mechanical fitters. It predominantly uses hierarchical task analysis, and takes place on an existing plant.

8 A method for quantifying ultrasonic inspection effectiveness

This study in the ultrasonic inspection (non-destructive testing) area is a performance assurance study. It is aimed at assessing human reliability by identifying human errors, and makes use of fault and event tree analysis techniques.

9 Operational safety review of a solid waste storage plant

This case study occurs on an existing nuclear chemical plant which had an identified operational problem to resolve. The main method used is hierarchical task analysis.

10 Task analysis programme for a large nuclear chemical plant

This case study is of an integrated task analysis process implemented during the detailed design of a plant. It therefore deals with the inter-relationships between a relatively large number of task analysis methods, showing the benefits and costs of such a process-oriented approach.

These case studies were selected because they were process-control related, and covered most of the major purposes of using task analysis, and also because they occurred at different stages in the design life cycle. Hierarchical task analysis may appear to be over-represented, but this is because it is one of the more popular, useful, and fundamental approaches. Furthermore, although most of the case studies are from the nuclear power/chemical field, this does not detract from the representativeness of the tasks and problems analysed for most process control situations. Many of the case studies are also equally relevant to the non-process control industries such as the service sector, transport sector, etc. Indeed, these industries have made use of task analysis methods and have benefitted from such usage.

The above has briefly introduced the concepts, purposes, and applications of task analysis. The two following sub-sections briefly outline how different readers may wish to use the *Guide*, and several aspects of the *Guide's* scope are explained. The remainder of the *Guide* then deals with the task analysis process, the major techniques which can be used in process control and other areas, and the case studies.

How to use this Guide

The *Guide* is divided into three parts:

Part I – The task analysis process: This part describes the human factors application areas in which task analysis can be used, and describes the system life cycle. It then shows when different issues are best addressed, and outlines which task analysis techniques are appropriate for addressing those issues.

Part II – Task Analysis Techniques: 25 techniques of task analysis are described in Part II, as discussed earlier. The techniques are all defined in similar format to enhance comparison of their attributes. In particular, this section has attempted to give practical insights into the techniques' usage.

Part III – Task Analysis Case Studies: Ten case studies of real applications in industry are documented in Part III. These also follow a consistent format, dealing not only with the context of the problem and what techniques were utilized, but also estimating the resources used in the study, and the benefits as perceived by the organization that hosted the study.

To summarize, Part I details what task analysis is and does and when to use it, and defines the task analysis process. Part II documents the major techniques and defines in more detail how and where to use each technique. Part III is a set of case studies which show what impact can be achieved by task analysis and also what costs can be incurred.

Guide for the reader

For the reader who has little or no previous knowledge of task analysis, it is advised that all of Part I be read first. If that reader has a particular industrial background or interest, then a related case study in Part III could be useful in demonstrating one role of task analysis in that context.

For the reader with a particular problem to solve, (e.g. workload assessment or training needs analysis), Part I will serve as a means for identifying which approaches are relevant, and when they are ideally applied. Then these methods can be reviewed in Part II, and if available, a relevant case study can be examined to see how a similar problem was solved in a real situation.

For the reader familiar with task analysis and with a problem to solve, or looking for more information on a particular technique, Part II, and if relevant Part III, can be accessed directly without reference to Part I. However, even this reader may wish to skim Part I, as the potential usage of the task analysis techniques, and the theory of the task analysis process, are themselves open to interpretation. It may be useful therefore, to check that the *Guide's* suggested usage of the technique agrees with the experienced reader's own expectations.

At the back of the *Guide* there is a glossary and a bibliography. Two companion documents listed in the reference section and bibliography are worthy of note. The first is the *Human Reliability Assessors Guide* (Humphreys, 1988), which describes and exemplifies contemporary human reliability assessment techniques. The second is the *Guide to Reducing Human Error in Process Operation* (Ball, 1991), which documents practical ergonomics advice on the design of systems. Both of these guides refer to and make use of the approach of task analysis, and for particular applications, serve as a useful supplement to this *Guide*.

Scope of the Guide

This document is not a theoretical treatise on task analysis technology. Instead it attempts to provide clear descriptions of how task analysis can be performed, and to define the ideal times within the system life cycle to achieve successful results. It is aimed at providing industry with practical guidance and knowledge on task analysis techniques, so that problems can be solved more effectively. It should also promote a higher degree of consistency amongst task analysis practitioners of various disciplines, enabling better communication of lessons learned through usage of the techniques, and hence ultimately more robust and efficient methods.

Task analysis application areas are diverse, and so in practice analysts will often tailor the technique(s) to fit the purpose of the evaluation. Because of this, the *Guide* does not specify definitively how to apply a particular technique. Some would go further and state that task analysis is still more of an art than a science, to the extent that there is no single correct way to carry out task analysis. While this is an over-generalization (e.g. hierarchical task analysis has become formalized in its approach), it would be inappropriate to constrain the potential user of task analysis techniques by over-specifying their recommended usage. In many cases, it is doubtful whether such rigid guidance could be given.

A further aspect of the practical use of task analysis techniques is the organization that is required to select, implement, and orchestrate task analysis techniques to deal with a particular application. Such organization differs for virtually every application of task analysis, and so is not dealt with in this *Guide*. However, the analyst must be able to fit task analysis techniques together to achieve the desired analysis of the system, according to the objectives of the application, in a resources-effective way. This is considered to a limited extent in Part I.

Certain techniques are useful for addressing a wide range of human factors issues, whereas some are highly specialized. It will be more cost-effective in certain situations for companies to consider using a technique which has multiple uses, which can then be used, refined, and re-used as the system life cycle advances, rather than use a series of discrete specialised techniques. Decisions and strategies such as this are not addressed in this *Guide* document, but there is in most cases sufficient information on the task analysis process in Part I, and on resources and uses of techniques in Part II, to make such decisions practicable for project managers.

Therefore, task analysis is a highly adaptable and flexible approach, but this *Guide* is limited in the extent to which it can provide specifications of how to carry out task analysis for each analysis situation. However, given the practical orientation of the *Guide*, the reader should be forearmed with a range of techniques and knowledge of their interactions, and forewarned regarding their respective pitfalls and shortcomings. This knowledge, supplemented with experience gained from using task analysis techniques, should enable the reader to gain the full benefits of this fundamental approach to dealing with human behaviour in systems.

Related guides

Ball, P. (ed.) (1991) *The Guide to Reducing Human Error in Process Operations.* Rept. No. SRDA-R3. Warrington: Human Factors in Reliability Group, SRD Association.

Humphreys, P. (ed.) (1988) *Human Reliability Assessors Guide.* Rept. No. RTS 88/95Q. Warrington: UK Atomic Energy Authority.

PART I

The task analysis process

Chapter 1
The task analysis process

Introduction

Task analysis involves the study of what an operator (or team of operators) is required to do to achieve a system goal. The primary purpose of task analysis is to compare the demands of the system on the operator with the capabilities of the operator, and if necessary, to alter those demands, thereby reducing error and achieving successful performance. This process usually involves data collection of the task demands, and representation of those data in such a way that a meaningful comparison can be made between the demands and the operator's capabilities. This allows 'analysis' of the task and, where necessary, the specification of ways to improve human performance. Sometimes this representation and analysis process may require simulation of the task in order to review more flexibly or dynamically the interactions between the system and the operator under a range of conditions. The analysis of the demands may also involve comparison of the task demands with a database on human performance limitations, or a checklist of recommended (ergonomic) ways of designing tasks and/or interfaces. This process of data collection, representation (and/or simulation) and analysis is termed the '*task analysis process*'.

To ensure a system is as effective and reliable as possible (i.e. that it is operable and maintainable), the human element must be considered in parallel with the equipment at all stages in the 'system life cycle'. This consideration of the human element (i.e. explicitly via the task analysis process), should be undertaken as an integral part of the system life cycle, making use of the different task analysis techniques at appropriate stages of design or operation.

This part of the *Guide to task analysis* therefore considers how the task analysis process can be used explicitly in an attempt to ensure that a system can be operated and maintained effectively from a human performance perspective. It firstly outlines the system life cycle stages, and then defines more explicitly the relevant human factors issues as key areas of task analysis application. The issues and their respective task analysis techniques are then placed in the context of the system life cycle, enabling a statement to be made of which techniques can be used for each human factors area, and of when these techniques are best applied within the system life cycle. An annotated summary listing of task analysis techniques is given in the Appendix, and 25 of these are then examined in more detail in Part II of the *Guide*.

The system life cycle

The system life cycle is the term used to describe all the stages through which a system passes from the initial idea of the system to its final dismantling and removal. The specific labels used to refer to these stages vary from industry to industry, but the basic ideas are quite general. The following list of life cycle stages therefore provides a generalized representation of the life cycle and includes a description for each label, so that the user who is not familiar with these specific labels can identify the different stages.

CONCEPT	A product or a service has been identified with the market requirements known. The basic inputs, outputs and principles are defined, but no details exist
FLOW SHEETING	The process or system units are defined, linked and associated with a system sub-goal (unit operation). An estimate has also been provided of the time required to fulfil the sub-goal
PRELIMINARY DESIGN	Specific methods for materials transfers and communication structures (e.g. pipework and instrumentation diagrams) are produced
DETAILED DESIGN	Equipment/system detail is established, drawn to scale and positioned
CONSTRUCTION	Building the system: in a large system this often occurs in stages, with those sections that have been designed in detail constructed while other sections' designs are still being finalized
COMMISSIONING	The system is built, and its operability tested and checked, usually with the need for minor on-line modifications to instrumentation and control. It is ensured that the system can function in the manner in which it was intended to function
OPERATING and MAINTAINING	The system is now running and fulfilling its role. Maintenance ensures that this situation continues for as long as the system is required

DECOMMISSIONING	The system is either no longer required or has reached its life expectancy and therefore must be stopped, dismantled, and removed or stored

For successful operations and maintenance, certain process stages function as useful check points, by which time particular human factors aspects should have been adequately dealt with. These check point stages, which will be referred to later, are:

1. Concept
2. Flow-sheeting
3. Detailed design
4. Commissioning
5. Operations and maintenance.

For systems addressed at the start of their life cycle, it is possible to consider all the human factors issues at the most economic and effective time using the most suitable task analysis methods and techniques. This will maximize compatibility between equipment and people within the system. The more mature the system, the less flexibility there is for ensuring that all human factors issues are successfully resolved, and the more likely it is that the cost of their resolution may rise. However, it is rarely too late to improve the system.

Before considering the integration of the task analysis approach into the system life cycle, it is necessary to expand upon the human factors issues raised earlier in the introduction. These application areas of task analysis are considered below.

Human factors issues: Task analysis application areas

There are six major human factors issues that influence system success, and their adequacy can be assessed and assisted by the use of task analysis. For each issue described, a table of related task analysis methods is presented. Each method in the tables is briefly described in the Appendix, and 25 are more fully described in Part II – see Appendix. However, in certain cases, earlier task analyses may be cited (e.g. task analyses from 'Person specification' are utilized for task and interface design). In a small number of cases where the available techniques are more numerous, a reference to a review document is given.

Allocation of function

Allocation of function (Table 1.1) takes place predominantly during the concept phase, and is concerned with the distribution of system functions between system equipment and human operators, as some tasks are best performed by machines and others by people. For example, a task requiring rapid, highly precise and repetitive actions will generally be best carried out by machines, whereas one involving the ability to deal with the unexpected will currently be best served by a human operator. Failure to consider the human element properly during the allocation of function stage may lead to the operator being asked to perform functions which are difficult to carry out reliably. These eventualities could lead to poor system operability and/or safety problems when the plant is commissioned and operated.

Table 1.1 Allocation of function

Human factors issue (HFI)	Description of HFI	Task analysis methods
Describe system functions	Describe clearly the functions of the system integrating all potential hardware requirements	*Functional analysis system technique* *Hierarchical task analysis* *Operational sequence diagrams* *Charting techniques* *Timeline analysis*
Allocate to either human operator or equipment	Consider the different requirements in relation to recommendations for allocating function	Fitts list (e.g. Singleton, 1974)

Person specification

A person specification (Table 1.2) details the characteristics needed by individual workers to achieve their tasks. It can also usefully specify which requirements must be fulfilled at the recruitment stage and which will entail training. These characteristics include physical and mental capabilities, qualifications, personality traits, and experience. A person specification meets the same type of need as a detailed specification for individual items of equipment (i.e. the designer needs to ensure that the system can function successfully).

Table 1.2 Person specification

Human factors issue (HFI)	Description of HFI	Task analysis methods
Identify potential recruitment pool	Identify potential recruitment pool and the general characteristics and capabilities of people within it	*Position analysis questionnaire*
Specify personnel requirements	Identify final sub-set of characteristics and capabilities of people who will be required to successfully complete tasks	*Position analysis questionnaire*

If it is known at the start of a new design that the workers will be chosen from the company's existing work force, it is possible to check the characteristics of this group relatively easily. These capabilities can then influence the design of the system so that tasks will demand capabilities that will be available, either through the basic work pool or through slight modifications to their training.

If the system is not to be limited by an existing work pool then the person specification can be frequently reviewed throughout the design process. Knowledge of the likely personnel available and the normal range of human capabilities can then contribute to decisions affecting job demands.

If workers and work requirements are considered together at an early stage in the process life cycle, incompatibilities between job demands and personnel capabilities are likely to be avoided, and the final personnel selection and initial training specifications will be more realistic.

Finally, when evaluating/auditing a system, a primary task will be to check that each person needed to run and maintain the system is suited to the job for which they are employed.

Staffing and job organization

After determining which tasks will be performed by people and which by machine, the numbers of people (referred to hereafter as staffing) and their organization (job organization) will need to be defined. This will be dependent upon the quantity and type of tasks, how long each task is likely to take, whether any tasks are required in parallel, and how much time there is to successfully perform tasks in terms of process requirements.

Staffing and job organization (Table 1.3) aim to ensure that individuals are not over-loaded or under-loaded in terms of the demands of the allocated tasks. Examining staffing and job organization will help identify whether the team size is suitable. In addition, there is the issue of distributing tasks between human operators. Some tasks are best performed by one team member, whilst other tasks are best distributed between several team members. Also, it has been shown that people work better when they can be made responsible for meaningful tasks rather than being required to undertake disjointed parts of tasks which have little apparent relationship to one another.

As part of job organization, communication requirements between team members must also be assessed, as well as co-ordination of their activities. When designing a system, task analysis will help to indicate the type of communication system(s) required. Note that as part of the manning and organization task checks it is particularly important to include an analysis of emergency tasks.

Issues of staffing and job organization can be broadly considered early on in the design cycle when little task detail is available. Decisions can be refined as design options become clearer. Later in the design of the system there must be a re-assessment of workload. The actual positioning of equipment and the ease with which instrumentation and controls can be used, as well as more realistic estimates of system process timings and delays will affect the time estimated to complete a task.

Table 1.3 Staffing and job organization

Human factors issue (HFI)	Description of HFI	Task analysis methods
Clarify task activities	Clarification and description of task activities and response to the process	*Hierarchical task analysis* *Operational sequence diagrams* *Observation* *Work study*
Establish system events and timings	System analysis in terms of patterns of events including time available for completion	*Walk-through/ talk-through*
Synthesize task times	Synthesis of task times compared with system timings. Check how long different task activities are likely to take and how frequently they will be required. Look for parallel task requirements	*Timeline analysis* *Computer modelling and simulation*
Check loadings on personnel	Human loading check both for physical and mental workload	*Ergonomics checklists* Person specification
Specify team organization	Team organization: on and between shifts, operations and maintenance, plus managerial decisions	*Management oversight risk tree*

Task and interface design

Task and interface design (Table 1.4) initially considers the types of information which personnel would need to understand the current system status and requirements (information and displays). In parallel it is necessary to identify the types of outputs that personnel will have to make to control the system (actions and controls). Once these have been identified it is then necessary to consider the specific ways that these will be provided.

In other words there are two sections to this task analysis process:

1. Assessing **What** will be needed to do the job
2. Determining **How** this will be provided

Table 1.4 Task and interface design

Human factors issue (HFI)	Description of HFI	Task analysis methods
	WHAT	
Record detailed task knowledge	Gather information on the task and represent the activities within the task	*Hierarchical task analysis* *Activity sampling*
Clarify information flows	An explicit statement of required information, actions and feedback	*Work study* *Task decomposition* *Decision/action diagrams*
Confirm adequacy of information flows	A check step to ensure that these provisions will be sufficient for successful completion of the task	*Table-top analysis* *Simulation* *Walk-through/ talk-through* *Operator modifications surveys* *Coding consistency surveys*
	HOW	
Identify links between attributes	Link attributes that must be used together to ensure success – includes a total system check	*Link analysis* *Petri nets* *Mock-ups* *Simulator trials*
Provide a detailed specification	Provide specific detailed recommendations; if an existing design, compare this with recommendations	Person specification *Ergonomics checklists*

Task design requires detailed information about tasks and will therefore need to be addressed continually during the system design process as fresh information is gained and design decisions are made. Most task analysis dealing with these issues takes place in the detailed design phase.

Skills and knowledge acquisition

Skills and knowledge acquisition (Table 1.5) ensure that people within the system are capable of undertaking the tasks required of them. However, this is only possible if the right types of people are selected, if adequate supporting information is provided, and if the personnel are adequately trained. If support is identified to be essential, then this should also be provided during training, and preferably identify task aids first then the type of training. This human factors issue therefore has three related components:

(a) A final check on the type of personnel required (person specification) and methods of identifying likely candidates;
(b) Checking the need for operator support and, if found necessary, the design of that support, (e.g. procedures, manuals, etc.);

(c) Determining training requirements and methods, to provide the necessary skills and system knowledge to effectively operate the system under a range of anticipated (and sometimes unanticipated) conditions.

Task analysis related to the area of skills and knowledge acquisition usually occurs later in the design process, when the design detail has been defined, and as commissioning is approaching.

Table 1.5 Skills and knowledge acquisition

Human factors issue (HFI)	Description of HFI	Task analysis methods
Check final person specification	Conclude final revision of Person Specification, ensure ability to recruit to the specification	*Structured interviews*
Generate a detailed task description (or use ones generated earlier in the system design cycle)		*Hierarchical task analysis* *Task analysis for knowledge description (TAKD)*
Establish the need for personnel support	Consider need for 'rules of thumb' and the frequency of operations (infrequent tasks require procedures) or the consequences of mal-operation (strict procedures or special tools for unreliable tasks)	*Ergonomics checklists*
Develop operator support	If task support is required then devise, develop and appraise resulting aids	
Design training	Determine decomposition level required to produce trainable units for effective learning (i.e. to ensure there is sufficient detail). Determine how skills are best acquired, and whether an on-the-job instructor would have control over range of tasks, situations and events that would have to be dealt with. Is there a useful knowledge base that people would need? Is simulation required? Identify training methods	*Hierarchical task analysis* *Verbal protocols* *Task decomposition* *Withheld information* *Use of simulators or mock-ups*

Performance assurance

This group of considerations is necessary to ensure that a system starts working correctly and continues to function as intended: safely and operably. Four human factors approaches are relevant to the achievement of this goals, as defined below, each with its related table of relevant techniques.

Reliability Assessment

This should commence during design and continue throughout the operational phases of the system life cycle and into decommissioning. It will involve considerations of human errors in normal and abnormal operations, and their impact on system risk and operability, and error reduction if required (Table 1.6 (a)).

Table 1.6 (a) Reliability assessment

Human factors issue (HFI)	Description of HFI	Task analysis methods
Define criteria	Consider the acceptable level of risk and errors to be avoided	The company and/or Government defines criteria – these are then interpreted through risk assessment to generate specific criteria for acceptable performance
Measure	Identify potential errors and their likelihood of occurrence	*Fault tree analysis* *Failure modes and effects analysis* *Hazard and operability studies* *Event trees* Human reliability assessment (Humphreys, 1988)
Investigate	Identify possible error causes and system influences	Human error identification techniques (Kirwan, 1992) *Failure modes and effects analysis*
Rectify problems	Consider how errors may be prevented or their likelihood reduced	*Ergonomics checklists*

Management safety structure assessment

This aims to ensure system operability and maintenance in its broadest sense, so that the system reaches its objectives in terms of safety and productivity. This will depend upon adequate management structures, functions within these structures, and interrelationships between these functions (e.g. if a system hazard has been detected by someone on the 'shop floor', how is this information transmitted upwards, so that the hazard receives attention in a timely fashion and protection is installed?). Management safety structure assessment will usually commence late in the design cycle e.g. during commissioning (see Table 1.6 (b).

Table 1.6 (b) Management safety structure assessment

Human factors issue (HFI)	Description of HFI	Task analysis methods
Define criteria	Management criteria cover a range of topics including supervision, communications, maintenance	*MORT*
Measure	Identify current management strategy and philosophy	*MORT* *Ergonomics checklists*
Investigate	Check implementation of management strategy and philosophy	*MORT* *Ergonomics checklists*
Rectify problems	Advise management of suitable changes	*MORT* *Ergonomics checklists*

Performance checking

This attempts to ensure that personnel continue to carry out their tasks as intended and trained (Table 1.6 (c)). This will also occur during the operational life of the system, and may be achieved by audits or by continuous monitoring of performance, 'near-misses' and incidents.

Table 1.6 (c) Performance checking

Human factors issue (HFI)	Description of HFI	Task analysis methods
Define criteria	Pre-operations a specification should have been produced detailing how the tasks should be performed	Earlier task analyses Procedures (skill and knowledge acquisition Staffing/job organization Person specification
Measure	Actual personnel performance is checked and compared with the original expectations	*Observation* *Walk-through/ talk-through* *Questionnaires* *Simulation*
Investigate	If performance differs, consider reasons for discrepancy (e.g. where original expectations valid	
Rectify problems	Solutions may involve re-training, providing job aids or even re-designing the workplace	*Ergonomics checklists* (e.g. Ball, 1991)

Problem investigation

This includes accident and incident investigations (Table 1.6 (d)). These will occur as and when required in the operational life of the system, but will be aimed at defining what went wrong and how to prevent its recurrence.

Table 1.6 (d) Problem investigations

Human factors issue (HFI)	Description of HFI	Task analysis methods
Define criteria	Define required system performance (e.g. product quality, throughput rate)	Earlier task analysis: Skills and knowledge acquisition Staffing/job organization
Measure	Check criteria against regular productivity reports and respond to incident reports	*Observation* *Walk-through/talk-through* *Questionnaires* *Simulation*
Investigate		*Management oversight risk trees* *Structured interviews* *Barrier and work safety analysis* *Critical incident technique*
Rectify problems	Remove causes of the problems rather than apportioning blame	Re-check task design *Ergonomics checklists*

The above six issues are the major human factors concerns in process control. The next section considers how task analysis techniques can be integrated into the system life cycle to enhance performance in the above six human factors areas, and also considers the issue of how techniques can be selected for particular applications. An example task analysis 'mini-programme' is also given, relating to a task analysis issue in decommissioning.

Integrating task analysis into the system life cycle

General integration of human factors issues

Table 1.7 indicates when each human factors issue (HFI) should ideally start to be addressed and the approximate period during which the analyst should maintain attention to the issues concerned. For example, performance assurance must be dealt with throughout the entire process. Task and interface design, on the other hand, cannot usually be usefully considered at the system concept stage, but requires consideration during the detailed design phase, and needs to be checked during commissioning.

Table 1.7 Life cycle stages

<table>
<tr><th rowspan="2">Human Factors Issue</th><th colspan="5">Preferred life cycle stage</th></tr>
<tr><th>Concept</th><th>Flow sheet</th><th>Detailed design</th><th>Commissioning</th><th>Operation and maintenance</th></tr>
<tr><td>Allocation of function</td><td>Best</td><td>Possible</td><td>Possible</td><td colspan="2">Check</td></tr>
<tr><td>Person specification</td><td colspan="4">Review</td><td>Check</td></tr>
<tr><td>Staffing</td><td>Too soon</td><td>Best</td><td>Possible</td><td colspan="2">Check</td></tr>
<tr><td>Task and interface design</td><td>Too soon</td><td>Best</td><td>Best</td><td colspan="2">Check</td></tr>
<tr><td>Skills acquisition</td><td>Too soon</td><td>Too soon</td><td>Best</td><td>Best</td><td>Check</td></tr>
<tr><td>Performance assurance</td><td colspan="5">Constantly</td></tr>
</table>

The types of tasks analysed and assessed should cover (ideally) the entire range of tasks that may be encountered, whether in normal operations, system abnormalities, or maintenance. They should thus include system start-up, shut-down, emergency tasks, communication, monitoring and supervision, administration, etc. Otherwise system performance will only be maximized for a sub-set of the likely system task requirements.

Therefore, to ensure that a system is successful throughout its life cycle, each human factors issue should be considered at its appropriate life cycle stage, whether to achieve safe and effective operating, maintenance, or decommissioning.

Task analysis could also be applied to a particular phase within the design cycle as if this phase was a system to be assessed, e.g construction, which is itself usually one of the most hazardous phases during the life cycle of the system. This is currently unusual, but in safety terms could be highly beneficial.

Detailed integration of task analysis into the system life cycle

Figure 1.1 illustrates in more depth when consideration of different HFIs using task analysis techniques should commence. There are identifiable steps within this task analysis process that must be completed and linked together, in order to ensure satisfactory solutions or answers to the different HFIs.

The links between the various steps are of two types:

- feed-forward of an increasingly detailed and re-iterated task description or 'profile', so that human factors issues can benefit from the earlier task analyses completed for issues already addressed;
- exchange of information between different task analysis functions relating to the different human factors issues requirements and the decisions taken to accommodate them. There can be 'trade-offs' between issues where a potential problem with respect to one human factors issue can be overcome by provisions relating to another. For example, a training objective which is proving difficult to achieve may be resolved by improved display format and special procedural support.

These steps within the task analysis process in Figure 1.1 differ for each HFI in the manner in which information needs to be collected and represented, and so differing task analysis techniques will be appropriate. These steps can be related back to the task analysis process tables.

As indicated during the introduction, there may be a number of reasons why not all the human factors issues need to be addressed or why only a part of the whole system will be 'task-analysed' (i.e. targetted). In recognition of this, the tables presented earlier address each human factors issue separately.

It should also be noted that Figure 1.1 and the associated tables represent an idealized task analysis process. Techniques can be applied later in the system life cycle than as represented in Figure 1.1, although such a task analysis application may be less cost-effective (e.g. due to having to implement retro-fit design). In

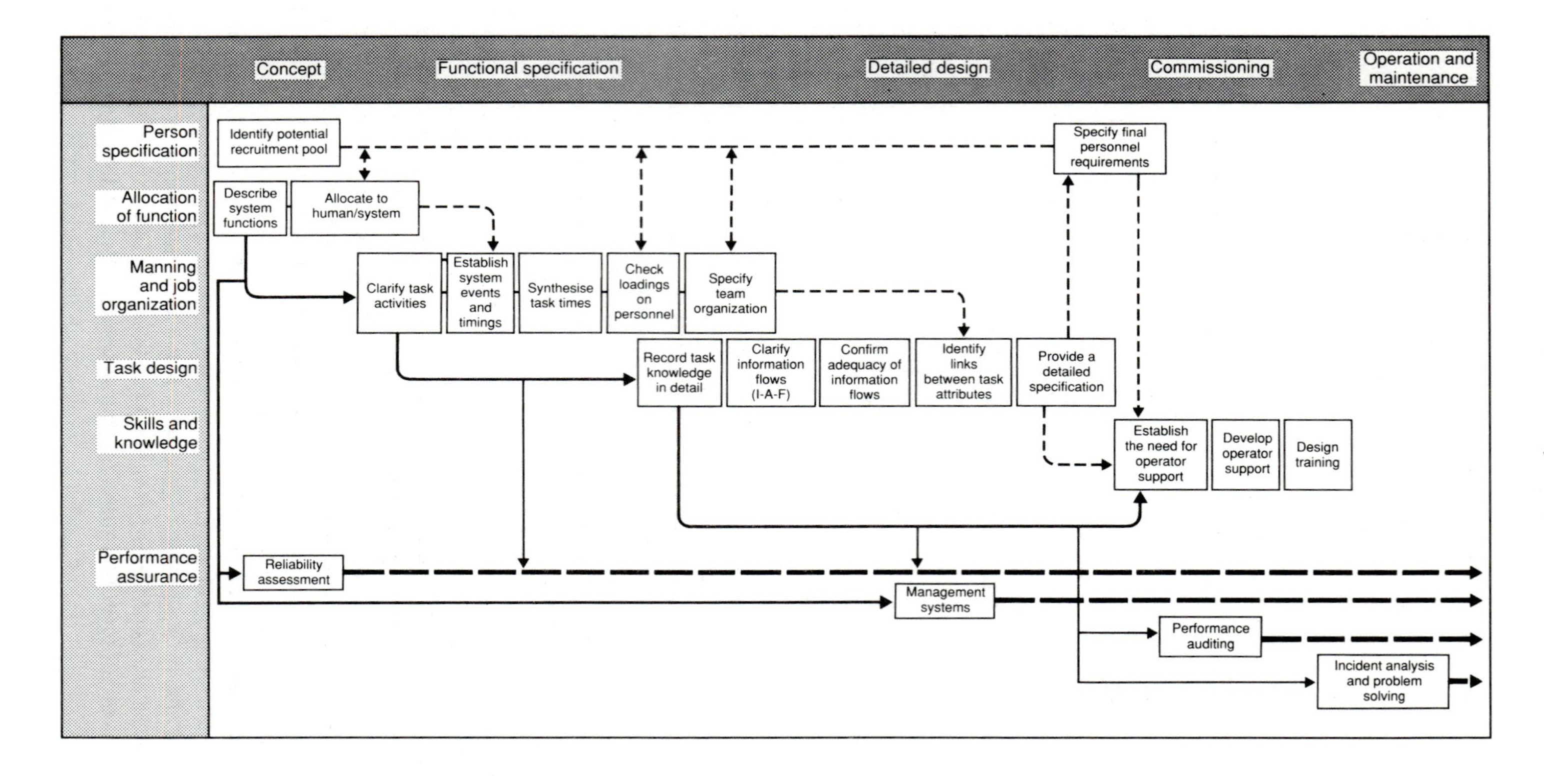

Figure 1.1 Optimizing human factors in system design

practice task analysis may often be applied later than its ideal timing, for various reasons. For existing systems there will be no option but to apply task analysis during the operational phase, targetting task analysis to achieve particular system performance goals. One advantage of this is that information and data collection on such tasks is usually far easier.

Selection of task analysis methods

To further aid the selection of task analysis techniques that might be used for a particular stage, there is a set of short summaries of task analysis techniques in the Appendix, with links to the more detailed descriptions in Part II, of 25 task analysis techniques. This summary listing, together with Figure 1.1 and Tables 1.1 – 1.7, will aid in the selection of a single technique or set of techniques, whether for a specific (targetted) study, or for a more broad-ranging system design support and/or evaluation study.

As an example, if a project manager is in the detailed design phase of a plant and wishes to determine what information needs to be presented on a VDU display in the central control room, this manager could first refer to Figure 1.1. This would suggest that this manager is in the right system design life cycle phase to 'clarify information flows.' Table 1.7 would suggest that such an undertaking could in fact have been started earlier, but nevertheless the detailed design phase is not a significantly late point at which to begin such an analysis. The manager could then consult Table 1.4 (task and interface design) and see a range of techniques listed. Two of these (GOMS and work study) are described only in the Appendix, and another technique (person specification) refers to earlier task analysis efforts which may or may not have been carried out in this particular project. The other techniques which are listed are summarized in the Appendix and documented in more detail in Part II. The project manager could then consult relevant Part II sections, and perhaps read the case study on interface design (Chapter 9) which also took place at the detailed design phase. The Appendix would be used to rule out some techniques which could not be applied at this design phase (e.g. if it was a new plant, operator modifications and coding consistency surveys could not be carried out, and walk-through and talk-through may be difficult). Ultimately the project manager would choose a small set of techniques (e.g. hierarchical task analysis, task decomposition, decision/action diagrams, table-top analysis, and ergonomics checklists) to carry out the analysis. Hierarchical task analysis results would also be useful later on for training purposes, and a table-top analysis may generate scenarios of interest for later human reliability assessment purposes.

Note that it is possible to find basic task analysis techniques that can be used in almost every circumstance (e.g. hierarchical task analysis). It may significantly reduce costs if one main technique is used, supplemented by specialist techniques for particular issues. The last case study (Chapter 16) in particular deals with a complete task analysis programme for a plant, both to support and to evaluate the design process. In that study, hierarchical task analysis played a particularly central and crucial role.

The reader, in practice, will probably find it most useful to return to Figure 1.1 and the foregoing tables, after having consulted Part II to consider candidate task analysis techniques for a particular application.

Decommissioning and task analysis

Decommissioning (or shut-down and storage/demolition) needs to be considered as a separate issue since, although it is the final stage of a system's life cycle, it can be potentially as hazardous as operating and maintaining, yet may often be overlooked in terms of task analysis requirements. Hence its associated human factors issues should preferably be considered at the pre-operational life cycle stages too. A system from the beginning can be designed to be easy to operate, maintain and decommission. If decommissioning is not considered during the design phase, which may be seen to be unrealistic in some industries, then all the associated aspects must at least be considered prior to commencing active decommissioning.

Using decommissioning as an example of a system problem for which task analysis could be helpful, and for which task analysis tools for each of the six HFIs must be selected, those responsible for decommissioning could approach the human factors of decommissioning as shown in Table 1.8.

Table 1.8 Decommissioning

Consider whether decommissioning relies upon human involvement	see	1. Allocation of function (Table 1.1)
Identify the type(s) of person needed to carry out the job	see	2. Person specification (Table 1.2)
Determine how many people will be required based in part on how quickly the system must be decommissioned	see	3. Staffing levels (Table 1.3)
Determine what information and facilities are required for successful completion	see	4. Task and interface design (Table 1.4) 5. Skills acquisition (Table 1.5)
Monitor the job	see	6. Quality assurance (Tables 1.6a – 1.6d)

By reviewing the sections and their associated tables mentioned in Table 1.8 (right-hand column), those responsible for decommissioning could identify a candidate set of task analysis techniques with which to achieve the goals specified in the left-hand column. Such techniques are noted in the Appendix, and will be described in detail in Part II. To complete the example however, the decommissioning team might select the basic approach of using *hierarchical task analysis* and its sub-technique, *function analysis system technique*, for allocation of function, person specification and skills acquisition; *timeline analysis* for workload assessment (staffing levels and job organization); *task decomposition analysis* for task and interface design; and *hazard and operability study*, *barrier analysis* and *work safety analysis* for quality assurance (QA). Other combinations of techniques could be constructed to achieve the same goals, but it would be up to the company concerned to decide which particular combination most suits their needs and resources. This type of decision therefore requires a knowledge of techniques mechanics, links with other techniques, and resources usage, and such factors are dealt with in Part II of the *Guide*.

Concluding comments on Part I of the Guide

Part I has argued that the maximization of successful human performance requires an orchestrated integration of task analysis techniques into the system life cycle, particularly during the design phase (i.e. prior to operation). It has further stated that task analysis techniques can deal with a range of problems or human factors issues, and that various techniques exist for dealing with each of these issues.

Part II gives much more detail on 25 individual task analysis techniques which have been shown in practice to be useful and effective in dealing with the human factors issues in a variety of contexts. To provide detail of some of the other accounts of task analysis that have been developed, the Appendix also includes short summaries of some further techniques which were mentioned earlier in this chapter. Part III shows some of the techniques as used in actual applications, which gives the user not only a view of a specific technique in action, but also a perspective on how various techniques were used together, what types of problem they have been used to solve, and how valuable their application was seen to be. These case studies, and the technique description sections in Part II on 'practical advice' and 'advantages/disadvantages', give the reader a practitioner's insight which should be helpful when actually selecting techniques and carrying out practical task analysis. It will therefore be useful, for the reader wishing to apply task analysis, to review Parts II and III and then return to the figures and tables of Part I, as well as the Appendix, before selecting techniques for a particular application.

References and further reading

Ball, P. (ed.) (1991) *The Guide to Reducing Human Error in Process Operations.* Rept. No. SRDA-R3. Warrington: Human Factors in Reliability Group, SRD Association.

Humphreys, P. (ed.) (1988) *Human Reliability Assessors Guide.* Rept. No. RTS 88/95Q. Warrington: UK Atomic Energy Authority.

Kirwan, B. (1992) An Overview of Human Error Identification Approaches for Use in Human Reliability Assessment. *Applied Ergonomics, (in press).*
Singleton, W.T. (1974) *Man–Machine Systems.* Harmondsworth: Penguin Books.

PART II

Task analysis techniques

Introduction

Part II of the *Guide* details 25 major techniques of task analysis. Part II cannot deal with all potential task analysis techniques, but these 25 were selected as being the most useful and representative ones in the field of process control. Some of these techniques are approaches rather than formal methodologies (e.g. for data collection). Nevertheless, since data collection is a necessary prerequisite of any form of task analysis, and since there are important practical insights for specific data collection approaches, these are dealt with as if they were techniques. The 25 task analysis techniques are grouped into five sub-sections which define their major or most popular role within the task analysis process, as detailed below.

Task data collection techniques

These are techniques which are primarily used for collecting data on human–system interactions, and which then feed into other techniques (e.g. observational techniques could be used as a basis for collecting data for a workload assessment).

These methods may be subject based (i.e. requiring verbal inputs from personnel familiar with the task), or observation based (i.e. using observation methods, such as watching personnel carry out the task). In all cases however, there is usually a prior or at least parallel phase of gathering and analysis of existing documentation, to find out what the task is about and its approximate (or specific) goals and task elements. Documentation analysis is not itself discussed as a technique in this *Guide*, although it is important, since if documentation analysis is omitted prior to task analysis, the analyst may well waste much time trying to obtain information which is already well-documented. It is therefore strongly advised that all task analysis studies commence with an initial familiarization/review of documentation. Documentation usually refers to existing or proposed procedures and training manuals, task and system specifications, relevant engineering drawings, shift logs if available, and any related problem or risk studies which focus on any aspect of the task under investigation. In some cases there may even be existing task analyses, possibly carried out for a different reason, but these may still be worthy of careful scrutiny prior to starting a new task analysis. Documentation analysis is therefore an implicit and necessary stage during the data collection phase.

The data collection techniques reviewed in Chapter 2 are as follows:

- Activity sampling (observation based)
- Critical incident technique (subject based)
- Observation (observation based)

- Questionnaires (subject based)
- Structured interviews (subject based)
- Verbal protocols (subject based)

Task description techniques

These are techniques which structure the information collected into a systematic format (e.g. the data from observation studies may be recorded by the use of flow charts, or operational sequence diagrams). Such formats may then serve either as reference material to enhance the understanding of the human–system involvement, or may be used more directly (e.g. the hierarchical task analysis technique can be used to identify training needs and content).

The task description techniques reviewed in Chapter 3 are as follows:

- Charting and network techniques
- Decomposition methods
- Hierarchical task analysis
- Link analysis
- Operational sequence diagrams
- Timeline analysis

Task simulation methods

These approaches are aimed at 'compiling' data on human involvements to create a more dynamic model of what actually happens during the execution of a task. Such techniques (e.g. the use of computer simulation), based on data from individual tasks, model tasks generically and can estimate factors such as time to complete the task.

The task simulation methods reviewed in Chapter 4 are as follows:

- Computer Modelling and simulation
- Simulators/mock-ups
- Table-top analysis
- Walk-throughs and Talk-throughs

Task behaviour assessment methods

These techniques, many of which are from the engineering risk assessment domain, are largely concerned with system performance evaluation, usually from a safety perspective. They primarily identify what can go wrong that can lead to system failure, and often deal with hardware, software and environmental events as well as human errors. Such techniques represent errors and other failure events, together in a logical framework, enabling risk assessment to take place, and/or the identification of remedial risk-reducing measures. The task behaviour assessment methods reviewed in Chapter 5 are as follows:

- Barrier and work safety analysis
- Event trees
- Failure modes and effects analysis
- Fault trees
- Hazard and operability analysis
- Influence diagrams
- Management oversight risk tree technique

Task requirement evaluation methods

These techniques are utilized to assess the adequacy of the facilities which the operator(s) have available to support the execution of the task, and directly describe and assess the interface (displays, controls, tools, etc.), and documentation such as procedures and instructions. These techniques are usually in the form of checklists, or special interface survey approaches.

The task requirement evaluation techniques reviewed in Chapter 6 are as follows:

- Ergonomics checklists
- Interface surveys

The categorizations above require qualification. Firstly, some of the techniques are difficult to classify, and appear to be able to fit into more than one category (e.g. Task decomposition analysis can be used as a task description method, a task behaviour assessment technique, or a task requirement evaluation technique). Secondly, some techniques offer the means to carry out a complete human factors process (e.g. the hazard and operability (HAZOP) technique collects and generates data, and identifies errors and mechanisms to overcome those errors). Such techniques are really frameworks in nature, and can actually incorporate other task analysis techniques (e.g. HAZOP can incorporate a task decomposition analysis approach or hierarchical task analysis). For these two main reasons, any classification of task analysis techniques must be treated flexibly. The classification in this *Guide* has been developed to add structure to Part II and to highlight the primary aim or most popular application of the technique. The reader must, however, remember that there will be other ways of classifying task analysis techniques, and so the classifications should not be seen as defining rigid boundaries between them.

Technique description formats for Part II

Each of the 25 techniques is described according to a common format. The sub-headings used throughout Part II are described below in the context of the reader wishing to select a technique or group of techniques for an application area.

Overview and application areas

The description begins with an overview of the nature of the technique, its genesis, and key references. The major application areas of the technique are then described, noting the purposes for which the technique can be used. This definition of major application areas is usually independent of type of industry, mainly because task analysis approaches are fairly generic in nature.

Description of technique

The description of the technique in some cases will be enough for the reader to apply the technique, but in other cases this would require a great deal of text, and such descriptions go beyond the scope of this *Guide*. In all cases however, there has been an attempt to describe at least the main concepts, inputs and outputs, and means of application of the technique, even if it has not been possible to describe it in full detail.

Practical advice and resources required

Practical advice and resources considerations are useful when the user has narrowed down the choice of techniques for a potential application to a small number. The practical advice comes from practitioners' experience, and ranges from helpful hints to potential pitfalls in the application of the techniques. When a technique or small number of techniques have been selected, then further more detailed reading may be required (see 'References').

Links with other techniques

The specified links with other techniques should not necessarily be seen as exhaustive, as there may be new links found with new applications. However, certain linkages as specified in this section have been found in practice to be particularly useful and effective.

Advantages and disadvantages

The disadvantages and advantages descriptions for each technique catalogue only the major ones, and are biased towards practical rather than theoretical insights, unless the latter are significant and detract from the technique's validity in particular areas of application.

References and further reading

The references given, where possible, are both the most useful and easily accessible. Many of these references are briefly annotated to describe the type of content to be expected in the reference (e.g. theory, case study, etc.). They are given at the end of each chapter.

Links to Other Parts of the Guide

Part II defines each technique independently, and although there is a section on links with other techniques, it is recommended that when selecting a technique the user should cross-refer to Part I to determine when to apply the technique and to review possible alternative techniques. If there is a relevant case study in Part III of the *Guide* (for a quick check, review the relevant technique description in the Appendix), this will also be worth reviewing to evaluate the technique in a practical application.

The remainder of Part II therefore documents the 25 techniques. Each technique is presented alphabetically within its sub-categorization.

Chapter 2
Task data collection methods

Activity sampling

Overview

Activity sampling is a method of data collection which provides information about the proportion of time that is spent on different activities. By sampling an operator's behaviour at intervals, a picture of the type and frequency of activities making up a task can be developed. The method was originally developed for use by work study practitioners, to help them determine exactly how time was utilized during industrial tasks (see Christensen, 1950).

Application

Not all tasks are suitable for activity sampling. A key requirement is that all, or most, of the task elements should be observable and distinguishable. If they are not observable, then normal sampling by observation will fail to detect them. If they are not describable, the analyst can never be sure precisely which task is being performed at any instant, and the results can be entirely misleading.

Another prerequisite is that the sampling should be able to keep pace with the task. Thus, if the task elements follow each other too rapidly, either an intermediate data recording technique must be considered (e.g. high-speed video), or activity sampling must be rejected. For example, activity sampling of manual circuit board assembly on a production line could be problematic because, although all the task elements are easily observable, for an experienced assembler they would take place so rapidly that an impracticably short sampling interval would be required. This sort of example would be a candidate for high-speed video recording. Another difficult case would be a task which involves significant periods of decision-making, monitoring, or other cognitive activities (e.g. supervision of an automated control room). In such situations, such a high proportion of activity would be unobservable that it would be impossible to construct an accurate profile of time usage.

A typical environment in which activity sampling would be ideal, would be the operation of a non-automated warehousing system. Here, all the essential operations – paperwork, search, distribution, stock intake, etc. – are observable. They also take place at a pace which would not defeat a manageable sampling interval (i.e. of minutes rather than seconds).

The intention of activity sampling is to produce a measure of what proportion of time is spent performing each of several identified activities. As such, it is capable of providing either very crude, or very refined results, depending on the skill of the analyst and the amount of care which goes into the design. Subsequently, such data can be used as the basis of efforts to reduce unnecessary tasks, or to understand which task elements constitute the greatest proportion of workload.

Description of technique

Activity sampling involves observing (sampling) task performance at predetermined times, and recording the activity which is then being undertaken. Much of the effort consumed by this technique should be expended on designing the study, rather than executing it. In order to design an activity sampling exercise effectively, there are four issues which must be considered, namely the classification of activities, the development of a sampling schedule, information collection and recording, and the actual analysis of activity samples. These are described below.

Categorization of activities

Any behaviour or system state which can be directly observed can be recorded by activity sampling. However, in order to ensure that useful activities are recorded, it is necessary for the analyst to have some impression of what activities would be expected during a particular task. Therefore the analyst should start by carefully observing the task, and if possible, supplementing this with a short pilot study.

The choice of activities to be recorded will generally be determined by the task itself, and any particular concerns that the analyst might have. As a rule of thumb, if the information is to be collected manually, no more than about 20 separate activities should be recorded. This latter limitation would not apply if the information were to be recorded on video for subsequent analysis, or if a form of direct computer logging were to be used.

It is also important that each of the activities can be clearly differentiated from each other. For this purpose, it may be necessary for the analyst to formulate very clear definitions of each activity, and to learn these thoroughly prior to the sampling. In writing such definitions, particular attention should be paid to defining the limits of an activity (i.e. when it begins and ends). If activity differentiation is difficult, and no recording aids are available, a longer sampling interval may be necessary.

Developing a sampling schedule

The most important issue related to sampling is to determine the sampling interval. The main criterion for this decision is to ensure that the shortest activities do not fail to be sampled. This is done according to the Nyquist criterion, which states that the interval between samples must be less than, or equal to, half the duration of the shortest activity which needs to be detected. However, an important constraint is that for manual data collection, it is unrealistic to expect the analyst to maintain an interval of less than 2.

The sampling interval can then be used as the basis for either a fixed sampling schedule, in which samples are taken at regular intervals, or a random sampling schedule. Random sampling will only give limited information about the relative frequency of activities, and so it is really only of use for tasks which are of very extended duration. Therefore, for most task analysis purposes, fixed interval sampling will be preferred. However, this sampling method cannot always be used if the task being studied is very regular and repetitive, because in such situations, fixed interval sampling will cause systematic sampling errors.

It is also necessary to decide for how long sampling will be continued (i.e. the total sampling period). This may be determined by the duration of the task, but for continuous tasks such as monitoring, sampling should be continued for sufficiently long to sample the full range of activities, and should yield in the region of 1000 sampling points for each session.

Information collection and recording

Prior to the sampling session, the operator undertaking the task should be told what is being done (and why it is being done), and be asked to undertake the task as normal. The analyst should then begin to record the task activities. If the recording is being done manually, rather than with a video or other recording device, the analyst must be given a reliable cue as to when to sample, such as a tone given through earphones.

The actual way that information is recorded will depend on the design of the information sheet being used, and there are two basic approaches:

Simple tally

This approach requires the analyst to make a simple tally of the number of times that each activity is observed. The information collection sheet for this purpose would be a list of all classified activities, and each time one of these is observed, the analyst should place a tally mark in the appropriate column. An example of a simple tally for the task of checking a word processor file, is shown in Figure

Activity	**Tally**	**Total**
File operations	IIII	4
Typing	~~IIII~~ ~~IIII~~ ~~IIII~~ II	17
Deleting	~~IIII~~ ~~IIII~~ II	12
Visual searches	~~IIII~~ ~~IIII~~ ~~IIII~~ ~~IIII~~ ~~IIII~~ ~~IIII~~ ~~IIII~~ II	37
Automated searches		0
Reading original copy	~~IIII~~ ~~IIII~~ I	11
Other operations		0

Figure 2.1 A simple tally data sheet for a word processing task

2.1. This is the simplest way of recording samples, but it is limited to providing only an estimate of the relative frequency of each activity, and gives no information about sequences of action, or how long each activity is sustained.

Sequential sampling

The alternative to a simple tally approach in activity sampling is to record the activities sequentially on an information sheet. This is a little more difficult for the analyst, but the resulting information can be used to determine the duration of each activity, and its relationship to other activities. If the sampling interval is relatively short, it will be advantageous to develop a simple coding or numbering system to facilitate the collection of this information. In Figure 2.2 a sequential tally is shown for the first 3 min of the same word processing task. This data sheet also illustrates how information can be distorted if the sampling interval is too large. For instance, although each correction had to be read (R), then located in the file (V), and finally it had to be amended (D and/or T), there were occasions when the correction was read between two sampling intervals (e.g. between 16 and 20 secs into the task), and so it appeared that the operator was making some corrections without first reading what had to be corrected.

A = Automated searches	D = Deleting	F = File operations	O = Other operations
R = Reading original copy	T = Typing	V = Visual searches	

Time	Seconds														
	0	4	8	12	16	20	24	28	32	36	40	44	48	52	56
0 min	F	F	R	V	D	V	V	D	T	R	V	V	R	V	D
1 min	R	V	V	V	T	T	R	V	V	V	R	V	R	R	V
2 min	R	V	D	T	R	V	D	T	R	R	V	D	D	R	D

Figure 2.2 A sequential tally data sheet for a word processing task

The analysis of activity samples

From any activity sample the relative frequencies of each task can be computed very easily, so that the analyst can discover where it is most effective to expend further effort. If a sequential sampling approach has been used, the analyst can also make estimates of the duration of each task element, and look at particular sequences of activity.

Practical advice

It is worth spending a significant amount of time becoming familiar with the task being studied, and checking that the behaviour categorizations are clear, exhaustive and mutually exclusive, such that each activity can only fit into one category. Errors at this preparatory stage can result in the entire study being of little benefit. To ensure success, it is recommended that a small pilot study is carried out and the results checked, before embarking on a full-scale study.

This data collection method has the potential for being one of the least intrusive, and consequently provides an opportunity for more objective data. To retain this advantage, the pilot study should also be used to check the smoothness of the data recording technique, to ensure that it can keep up with the requirements of the sampling interval and the environment with minimum interference.

It is useful to include a category such as 'other activity', to capture activities undetected in the pilot study, which cannot be identified as falling into one of the predetermined categories. If this category turns out to contain a large number of samples, the reason should be examined. It may be that the analyst has missed some significant task behaviours or has failed to define them exclusively, that non-task related activity is being observed, or that the experimenter has experienced difficulty identifying activities, perhaps due to an excessively short sampling period.

It is highly inadvisable to rely on memory, or clocks, as prompts for sampling intervals. A programmable beeper should be obtained, or, if possible, the sampling could be done from a video recording which has been time-stamped, or which can be played back slowly.

There is a danger of misinterpreting information, because frequency is not necessarily synonymous with importance. In one study, for example, users of a computer system had to interrogate it in response to customer enquiries. After only a few seconds on the primary display, they would often call up a display of background information. This second display was then usually left up until the next customer enquiry. Therefore, although the users only used this background information infrequently, its apparent frequency of use (based upon sampling which VDU display was being shown) suggested that it was much more important than the primary display. The analysts must therefore take care in interpreting the results.

Resources required

Low resources are a primary advantage of this technique. The major drain on resources is during the preparation phase, when great care and attention needs to be paid to the task and to behaviour categorization. However, after this phase, data collection sessions require only some means of indicating sampling intervals (a stop watch, or better still, a programmable beeper), an analyst/observer, and a supply of data sheets. For some types of task, a video recorder may be used. This may be either time-lapse, in which case recording takes place only at sampling times, or continuous and time stamped, in which case the behaviour can be *sampled* subsequently from the tape.

Links with other techniques

In determining which behaviours are of interest, it may be necessary for the analyst to employ another technique to aid understanding. This could well be a charting technique (pp. 81), because these can clearly reveal the interdependencies between task elements. The analyst can consequently determine which behaviours must be performed for successful task completion and which are peripheral.

Although the information generated by the technique is usually very limited, it can provide a very useful means of identifying the steps which have to be undertaken during a task. It can also be particularly useful in revealing whether people are doing what they are expected to do, or whether they are working outside their defined roles. Personnel may well have identified more effective ways of doing the job than those methods which were originally specified during system design.

It is worth noting that activity sampling data can be used to check the accuracy of other forms of analysis, because these will often imply that a particular range of behaviours are expected to take a particular amount of time. Comparison with activity sampling data can support, or refute, these expectations with relatively little effort.

Advantages

- The method is objective, in so far as the data recording does not rely on the operator or the analyst describing or interpreting what is going on.
- No active participation is required from operators.
- Although considerable skill may be required to set up the technique for a particular task, administering the technique is very simple and requires few resources.
- Data classification and tabulation is part of the process of setting up the technique, so there is little data interpretation required subsequent to recording sessions.
- The technique can reveal whether significant amounts of time are spent on unclassified and unanticipated activities. This can either increase the analyst's understanding of the task, or reveal activities which are not task-related.

Disadvantages

- Significant analytical skill is necessary if the task elements are to be properly described and categorized, prior to data collection. Otherwise, the data sheets produced will not be a full and accurate record of events.

- Complex tasks with many elements, particularly if they are lengthy, can cause the amount of time required for data collection to expand significantly.
- Activity sampling will shed little or no light on cognitive activities.
- There is a risk of misinterpreting the information from an activity sample, because the importance of any particular activity does not necessarily equate with the frequency with which it is reported.

References and further reading

Christensen, J.M. (1950) The Sampling Technique for use in Activity Analysis. *Personnel Psychol.*, 3, 361-368.

Heiland, R.E. and Richardson, W.J. (1957) *Work Sampling.* New York: McGraw Hill.

Critical Incident Technique

Overview

This technique sets out to collect data about incidents: these are events or features within a working environment which have a potentially important effect on system objectives. While the idea of negative effects usually springs to mind, it is also a way of identifying events that may have had a positive influence on the system.

The critical incident technique was first described in detail by Flanagan (1954). The technique is most widely associated with studies of human error and safety, but even in the Flanagan paper a much broader context for the technique in occupational psychology was envisaged. The wartime studies of Fitts and Jones (1947) remain the classic studies and somewhat influence the idea of the 'incident'.

The basic premise of the technique is that critical incidents will be inherently memorable to those working within a system. This means that they should be able to recall recent events that were deemed to be critical, or indeed, may be able to recall earlier incidents which happened to them or their colleagues. The technique specifically attempts to get at rare events, but may be used for those events occurring more frequently than the 'near-catastrophe'. The severity of the events which can be considered as 'critical' should be defined prior to eliciting responses from personnel.

Having generated a set of incidents, these can then be categorized and further analysed. It should then be possible to identify what changes can be made in the system to ameliorate problems, or to capitalize on positive instances of system success. The technique has made a major contribution to the reduction of human error, but, as mentioned above, can also be used to collect information about a range of other situations.

Application

It is most likely that the critical incident technique will be used in the early stages of a large-scale analysis, or at least as an initial 'trouble-shooting' technique. This is because its strength lies in the fairly rapid isolation of key problem areas within a system. Material collected in this way can then be used to form the basis for other more systematic and detailed information gathering techniques.

The critical incident technique is most likely to be applied to existing systems, because in such systems there will be a repository of information about events that have happened in the past. However, it could be the first stage in the production of a new system. Information gathered in this way can provide the basis for major, or minor, revisions in a system that is being developed.

Description of technique

Apart from Flanagan (1954), most reports of the technique give emphasis to different specific aspects of its usage. As such, it has come to take on a range of interpretations in different contexts. Flanagan paints a very broad picture of its application in personnel related areas, whereas other people have restricted its use to such task situations as the original Fitts and Jones (1947) study of cockpit activities. This can be seen as both a strength and a weakness. The weakness lies in not being able to pin-point exactly what the technique sets out to do. The strength lies in the flexibility afforded to the analyst by the technique and by its many modes of use. The technique can be seen as a very open ended one, where operators are simply asked to respond to questions of the kind put to them by Fitts and Jones (1947), (i.e. '*Describe in detail an error in the operation of a cockpit control... which was made by yourself, or by another person whom you were watching at the time*'). This apparently straightforward approach yielded a good deal of data, and the errors reported were classified into the categories shown in Table 2.1. From these data it was then possible to make recommendations about particular displays and controls, and also about the principles of their design both separately and within consoles (Singleton, 1974).

Alternatively, a much more systematic approach of interviewing operators on a daily basis, or getting them to fill in a daily, weekly, or monthly report on critical incidents, has been suggested. One recent development of the technique has been used in the aviation world, to solicit reports from aircrew, in an anonymous or confidential way, on incidents in aircraft operations. Such schemes are run in the USA and in the UK, and the data from them has led to changes in safety procedures and equipment (see Green, 1990).

Table 2.1 Error taxonomies used by Fitts and Jones (1947) for pilots (adapted from Singleton, 1974)

Controls		Displays	
	Substitution		Misreading
	Adjustment		Reversal errors
	Forgetting errors		Interpretation
	Reversal errors		Legibility
	Unintentional errors		Confusing instruments
	Unable to reach		Inoperative instruments
			Scale interpretation
			Illusions
			Forgetting

Information collection

Meister (1985) sets out a systematic approach for the application of the critical incident technique. He suggests that the technique does not have a set of established procedural steps, but rather, a set of guiding principles which should be modified to respond to the particular context in which the technique is applied. He identifies three major steps:

- Determining the general aims of the activity. By this is meant that the analyst should set out to determine what would be effective behaviour in the situation, and against which the critical incidents can be assessed. The more indeterminate the system is, and the less well-defined operators' activities are, then the more difficult it will be to set the general standards for this activity.
- Specifying the criteria of the effective/ineffective behaviours. There is a need to specify the situation in which the behaviour is to be studied, its relationship to the overall system aims and its importance in relationship to those objectives. In addition, questions arise as to who should do the observation. Whether this should be the analyst, the job incumbent, or some other person, will be a decision that needs to be made.
- Collecting the information. Various schemes have been worked out for information collection, in more, or less, systematic ways. A range of techniques have been used, such as personal interviews, group interviews, questionnaires, or various forms of recording on checklists etc. The interview has been deemed to be the most satisfactory way in which information can be collected. The use of standardized documentation may make the large-scale collecting of data more economic.

Sample size

The problem of sample size has not been fully resolved. The early advice of Flanagan (1959) was that if the task was fairly simple, about 50 to 100 incidents may suffice. However, in more complex systems, it is difficult to indicate how many incidents would be needed. Therefore, it is suggested that the cycle of collection and analysis should be an iterative one, in which information collection is ceased at a point where new incidents no longer extend, or alter, a category system of events that has to be created.

Analysis of information

Meister (1985) suggested that the analysis of the data themselves is probably the most important stage. This should begin with some determination of a taxonomy into which the collected material is to be organized. This will often be an arbitrary activity, and may well need to be revised in the light of experience. A second stage would be the classification of the items into various categories. Categories may well have to be similarly re-defined in the light of use, and broader, or more specific categories created in order to encompass all of the incidents which have been reported. The final stage requires the determination of the level of specificity, or generality, which is most appropriate for reporting on the information collected. This again will be determined by the specific purpose for which the analysis is directed.

Reliability and validity

Very little has been reported on the reliability and/or validity of the technique (see Andersson and Nilsson, 1964). One study suggested a fairly high predicted validity following a study of 4000 workers in 19 plants (Edwards and Hahn, 1980). This study showed that it was possible to predict, to some extent, the occurrence of accidents (correlation with critical incidents = 0.61) and disabling injuries (correlation with critical incidents = 0.55). This suggests that the technique is useful in being able to pin-point potentially hazardous situations.

Practical advice

From what has been said already it should be clear that it would be unwise to use the critical incident technique on its own as a way of gathering data about human operator activity within a system. Unless there is a very narrow objective in mind, the technique will need to be supplemented by other, more systematic, approaches that will cover the full range of activities and not just unusual, memorable, near-accidents. It is essential to reassure potential respondents that information which they provide will remain anonymous, and that they will not be identified with particular incidents. The analyst must also be wary of rumours, and consider that memory is eventually fallible, and that each person's account will be from one personal perspective. For incidents which are of particular interest to the analyst, therefore, corroboration of some kind (e.g. between independent interviewees or between an interviewee and logs, etc.) should be sought.

Resources required

A certain amount of training will be required for analysts. They will need preparation in order to be able to apply the technique with consistency, and to develop the tact to establish the necessary rapport with respondents and gain their belief in the confidentiality of the system. Analysts using this technique will need to gain the appropriate access to personnel, which may be done in both an on-job and off-job situation. The latter context is probably more amenable to a relaxed atmosphere, wherein people may be more prepared to report, and be more able to recall, critical incidents. In addition to this, the analysis may require the use of portable tape recording equipment and various forms of documentation.

Links with other techniques

There are clear links with other approaches to data collection. Much of the section on interviews is relevant in this context, because the interview is likely to remain a major source of data about critical incidents. Similarly, the design of questionnaires and checklists has relevance to this topic as they may be used to gather information. Observational techniques (pp. 53), while having relevance, are less likely to be of use here, unless a particular set of commonly occurring events have been categorized as critical incidents for the application of this methodology. The technique has similarities with verbal protocols (see pp. 71), in that it is asking people to articulate, albeit after the event, on things that happened to them.

Information generated by the critical incident technique may suggest that more detailed investigation is needed by the application of other techniques. For example, the identification of difficulties in particular operations could be studied further by structured interviews (pp. 66) and/or observation techniques. Critical incident techniques could serve as a first-pass approach to isolate areas of particular difficulty/criticality, which could benefit from the application of more systematic approaches. This technique can also be used to consider safety problems when carrying out HAZOP or MORT analyses (see pp. 194 and 208), or as inputs for fault and event tree analyses (see pp. 188 and 178).

Advantages

- The technique is helpful in highlighting the features that will make parts of a system particularly vulnerable, particularly from a human factors perspective. It will provide information on the type of errors that can occur, and when and how they are likely to happen.
- It is especially appropriate for the identification of rare events, which may not be picked up in other techniques that rely on operators reporting their everyday activities. The memorability of critical events is held to be the technique's strength.

- It also has the advantage of having high face validity, and may also have a particular appeal to system users who are interested in reporting on near-misses, etc.
- Depending upon the scale of activity, it can be fairly economical in terms of resources. If used in the early stages of an investigation, it can provide much useful information which can then be used to target the application of more costly time-consuming techniques that are likely to follow on from it.

Disadvantages

- A major problem arises from the nature of the reported incidents. The technique will rely on events being remembered by users and will also require the accurate and truthful reporting of them. The latter point is not meant to be critical of the honesty of individuals, but draws attention to the fact that they may be asked to report on potentially sanctionable incidents which they may not have reported in the past. Therefore, a degree of rapport must be built up between the analyst and the respondents, in order for them to feel confident that the material that they are reporting will be treated with confidentiality (which it must be). There may well be a desire on the part of operators not to open themselves, or their colleagues, to sanctions etc., by reporting events where they, or others, might be deemed responsible. In particular, problems may arise if only a few operators are involved, because the results of any analysis might be clearly attributable to specific individuals. Under such conditions it may be appropriate for the analyst to provide only the overall results of the analysis, rather than the actual content in terms of events etc.
- There is also the problem of fallibility in terms of memory, because many critical incidents may be forgotten, or not noticed at the time of occurrence. Such recollections as there are may well be distorted by events. Similarly, operators may be tempted to report on events reported to them, which may in turn take on the air of rumour, hearsay and anecdote. The analyst needs to be wary of such elaboration of critical incidents.
- A final point to emphasize is that the technique will, by its very nature, reveal rare and atypical events against a background of many years of normal working. The more everyday events will be missed by the technique. This reinforces the need to see it as one of a range of techniques used by an analyst in any large scale system examination.

References and other reading

American Institutes for Research (1973) *The Critical Incident Technique: A Bibliography.* Palo Alto, California, USA: American Institutes for Research. *A fairly recent collection of useful references to the technique.*

Andersson, B. and Nilsson, S. (1964) Studies in the Reliability and Validity of the Critical Incident Technique. *J. Applied Psychol.*, 48, 398-403. *One of the few studies which has reported on the reliability and validity of the technique.*

Edwards, D.S. and Hahn, C.P. (1980) A Chance to Happen. *J. Safety Research*, 12/2, 25-67. *A study showing how the technique predicted accidents.*

Fitts, P.M. and Jones, R.E. (1947) Analysis of Factors Contributing to 460 "Pilot Error" Experiences in Operating Aircraft Controls. Reprinted In *Selected Papers on Human Factors in the Design and Use of Control Systems*, Sinaiko, H.W. (ed.), 1961, pp. 332-358. New York: Dover Books, *The classic paper showing how information on the erroneous operations of controls could be collected.*

Fitzpatrick, R., Dysinger, D.W. and Hanson, V.L. (1968) *The Performance of Nuclear Reactor Operators*. Rept. No. NYO-3288-10 (AIR-481-9/68-FR). Pittsburgh, Philadelphia, USA: American Institutes for Research. *Report on the use of the technique in a nuclear power context.*

Flanagan, J.C. (1954) The critical incident technique. *Psychol. Bull.*, 51, 327-358. *The classic paper on the application of the technique and a review of early applications of it.*

Green, R. (1990) Human Error on the Flight Deck. In *Human Factors in Hazardous Situations*, Broadbent, D.E., Baddeley, A. and Reason, J.T., pp. 503-512. Oxford: Clarendon Press. *This chapter discusses the collection of aviation incident and near-miss data by the UK Civil Aviation Authority as part of the anonymous Confidential Human Incident Reporting Programme (CHIRP).*

Meister, D. (1985) *Behavioral Analysis and Measurement Methods*, New York: Wiley and Sons. *Useful overview of the use of the technique.*

Singleton, W.T. (1974) *Man–machine Systems*. Harmondsworth: Penguin Books.

Observational techniques

Overview

Observational techniques are a general class of techniques whose objective is to obtain data by directly observing the activity or behaviour under study (see Drury, 1990). This is in direct contrast to techniques which seek to measure some aspect of task performance or effect on the system. For instance, where a performance measure may directly record the time taken by an operator to respond to an emerging situation, an observational technique may be used to record the full overt sequence of actions, and possibly be supplemented by a verbal description from the operator of the decision processes taking place. The experimenter may subsequently choose to extract whatever information is of interest, usually from an audio-visual recording.

A wide range of observational techniques are in common use (e.g. direct visual observation, remote observation via closed-circuit television (CCTV) or video-recording, participant observation, time-lapse photography, etc.), and most can be combined or tailored to suit the particular requirements of a study. However, there is a strong tendency for people to react – whether favourably or unfavourably – to being observed and recorded, in a way that they would not if merely being 'measured' in an apparently more impersonal way. For this reason, one of the most important characteristics of an observational technique is the extent to which it intrudes, or appears to intrude, on the operator's sense of privacy.

Application

Observational techniques provide input for analysis. The objective is invariably to allow the data to represent 'natural' task performance in a relatively unconstrained environment. If the requirement is for precision performance data in a carefully controlled environment, observational techniques will almost certainly not be suitable.

Observational techniques are most appropriate when the information of primary interest is of a visual or audible form. Thus, observation methods can be particularly useful for recording physical task sequences, or verbal interactions between several people. In cases where a high proportion of the task involves covert mental processing, and very little value is added by overt behavioural information, it is probably not worthwhile recording visual data. However, it may still be very useful to record verbal reports (see verbal protocols, pp. 71) in addition to any objective performance measures which may be needed.

Because it is very difficult to achieve truly unobtrusive observation of a person at work, there are some situations in which observational techniques cannot be made to work. For example, where social interactions are a significant influence, the introduction of an observer, whether human or artificial, will always have an effect, although this may wear off after a time.

Description of technique

A good observational technique is therefore one which will capture all the significant visual and audible events, in a form which makes subsequent analysis reasonably easy, and which does not influence the performance of the task.

The requirement of minimum interference with the task can be modified somewhat if the objective is to introduce scenarios and monitor the reaction. If this is done, the technique falls into the general category of *participative* techniques. In the extreme case, the investigator will actually spend some time taking part in the activities of the group to be observed, and will influence the operators, or the environment, in such a way that the behaviours of interest in the study will occur. This is the antithesis of unobtrusive observation, and should only be practiced by trained and experienced investigators.

More commonly, a video or an audio record is made of a session in which it is known – or hoped – that a specific set of tasks will be performed in view of the camera. In such cases it is desirable for the recording process itself to be as unobtrusive as possible. The minimum requirement is that it does not get in the way. In longer term studies, it may be assumed that operators will largely 'forget' the presence of the recording equipment.

The different types of observational method can lead, in varying degrees, to a subjective feeling of 'intrusion' on the part of those being observed. Because this feeling can affect performance, and hence the validity of the data collected by observational methods, further discussion of this subject of intrusion is given below.

Observation and intrusion

Observation can occur at varying levels of 'intrusion'. Intrusion refers to the degree to which the observed personnel are aware of the physical presence of the observer. The personnel being observed should be told that they are to be observed, but the actual physical presence of the observer may affect the performance of the observed person or group. The first level of intrusion is *observer unobserved* (e.g. when observation is mediated via CCTV or video camera). The second is *observer observed*, in which case the observer is co-located with the operation being carried out, in that the personnel are aware of the observer's presence. This level is sometimes desirable (e.g. in certain verbal protocol situations in which the observer is recording verbal information, or adding prompts to the operator carrying out the task). The third level of intrusion is *observer participant*, in which the observer actually takes part in the tasks alongside other operators. This type of observation can be useful if aspects of team performance are being investigated to find out how the various team members are organized and carry out their tasks. It may also be a useful data collection approach if skilled performance is such that actions are 'semi- automatic', in which case operators may actually find it difficult to verbalize how they are achieving the task goals (i.e. because it is 'second nature' to them). In all observation studies it is worth considering the effects of observation on performance (and hence the validity of the information collected), and the analyst must then judge whether the appropriate level of observation has been selected.

Practical advice

Before conducting sessions relying on observation, it is essential to try and predict what information is expected to be extracted from the data. This is for two reasons: first, sophisticated recording techniques may be superfluous if it can be ascertained in advance that the events being looked for are obvious, infrequent (e.g. less than once a shift) and easily logged. Second, just pointing a camera at something does not guarantee that all important elements of the scene will be captured at all times. Problems posed by movement and interaction between individuals, and the difficulties of the video system capturing extremely detailed events (e.g. fine manipulations or parallel tasks by different team members, etc.), must all be considered in advance.

With any observational technique, a pilot session is invaluable, for assessing practical problems of data capture and the subsequent quality of the data. Unrepeatable data can easily be destroyed by the presence of something as simple as air conditioner noise.

The analyst needs to set clear rules regarding who (e.g. assistants) and what (e.g. cameras) should be unobtrusive during sessions, and define exactly what is meant by unobtrusive. Depending on the situation, a certain level of interruption or involvement from non-participants may be allowable. Similarly the existence of a camera and other recording devices may quickly be forgotten if they are placed discretely. However if this is not possible, operators will be continually aware of being observed. Some time should be spent on evaluating how sensitive the situation is, to determine whether or not the observation is truly unobtrusive.

Resources required

The resources required for effective observation, as described above, depend on the social and environmental conditions of the session and on the type of data required.

Assuming that audio-visual recording equipment is required, the sophistication and quality of it will affect the cost of the exercise. This is not just due to the equipment itself, but also because it may require technical support to set it up and keep it running smoothly.

The analyst should bear in mind that adequate sound recording is usually more difficult than adequate visual recording. It is worth experimenting with a variety of specialized throat microphones and unidirectional ones, and even a multi-band graphic equalizer in an effort to reduce background noise.

The resources required for subsequent analysis are similar to those described elsewhere (see verbal protocols and walk/talk-throughs pp. 71 and pp. 160). Basically, the data has to be reduced to a form which is useful to the analyst, and this usually involves first transcribing the audio part and annotating it with observed events. For this, transcriptional effort is required, and it must be assumed that on average one hour of a fairly relaxed session will take about 8 hours for the basic audio transcription alone. Logging events becomes increasingly time consuming as events become less easy to detect or identify, which is why it is so worthwhile ensuring that they are recorded properly in the first place.

Links with other techniques

However unobtrusive the observational technique itself is, it may be desirable that the investigator is involved with the operator under observation. Thus, video and audio records are typically made use of in sessions involving walk-throughs (pp. 160), verbal protocols (pp. 71), or structured interviews (pp. 66).

Observational techniques are thus simply a group of methods of recording data, and the data captured can be fed into a variety of representation and analysis techniques. Observational techniques do not have to use video, but if they do not, the opportunities for subsequent use of the data are potentially restricted. For instance, if the analyst decides to depend upon written notes, only the events which impressed the investigator at the time will be recorded, the data will be more sparse, and there will be no subsequent chance to focus on different events if required. The best approach, as with activity sampling (see pp. 41) is to carry out a pilot study first. During such a study it will become clear what type of information must be recorded. Written notes will be adequate in many situations, and require less analysis than video recording, and can sometimes focus on aspects which the video recording may fail to record in enough detail.

For more sophisticated analysis and potential re-analysis later, video is usually essential. It allows detailed descriptions of behaviours to be produced, and even very rapid and complex tasks can be traced and analysed thoroughly in precise order if the right equipment and technical support is provided. If the video is 'time-stamped' or otherwise referenced to external events, the observed behaviour can be related to system states or other logging systems already present on the plant. Consequently, video recordings can provide input for the application of hierarchical task analysis (pp. 104), and for a range of other representational techniques such as operational sequence diagrams (pp. 125). They may also provide a permanent record of tasks which may only be repeated rarely, if ever again. The best observational recording medium is therefore the combination of video recording and written notes.

Advantages

- Observational techniques reveal information which cannot be acquired in any other way. Detailed physical task performance data can be recorded, social interactions will be captured, and major environmental influences (noise, light, interruptions) can all be faithfully represented.
- Observation studies are ideal for pilot studies, as they can reveal potential behaviour patterns and influences which may not have been predicted, and can be subsequently analysed in more detail. They allow the analyst in an exploratory study to decide what to look for, after the data have been captured.
- They allow the investigator to become more familiar with the task, through being able to experience it in a natural medium rather than through a table of figures representing, for example, error scores.
- Observational methods can be used to identify and develop explanations of individual differences in task performance, which perhaps could only be guessed at otherwise. For instance a significant difference in performance between left and right handed operators would probably not be noticed, unless the investigator had enough insightful to predict it in advance.
- They provide objective information which can be compared with information collected by another observer, or by another method.

Disadvantages

- Observational data is the rawest possible form of data. Consequently the effort which must be expended on analysis is usually considerable. It is necessary to carry out the following tasks: identify and categorize the observed events; count them; relate them to the task and the system state at the time; produce a transcript of the audible content; etc. All this must usually be done while trying to avoid the natural human tendency to prejudge the behaviour of others. Observational data can be made impartial, but its interpretation cannot.

- Observation invariably has some effect on the observed party. Even if operators are being watched from behind a two-way mirror, they must be told of this and will be conscious of it. Most people eventually acclimatize to this, but waiting for this to happen is a considerable drain on resources. The analyst must therefore judge whether the type of observation selected will significantly alter the performance aspect under investigation.
- Situations which produce good or 'context-rich' observational data are rarely the ones which produce precise and controlled data. This is because observable behaviour is better produced in natural environments, rather than in the laboratory. Users of observational techniques must accept that data will tend to be incomplete and inconsistent, and certainly not usually compatible with automatic analysis.
- Observational techniques cannot provide information about underlying thought processes, and so they will be of little use for highly cognitive tasks.
- The equipment needed for producing high quality observational data can be expensive and difficult to set up. This will clearly depend on such factors as the mobility required, the lighting and noise levels, and the acceptable level of obtrusiveness. The resources for the analysis of the data collected can also be fairly high.

References and further reading

Drury, C.G. (1990) Methods for Direct Observation of Performance. In: *Evaluation of Human Work*, Wilson J.R. and Corlett, N.I. (eds.), pp. 35-57. London: Taylor and Francis.

See Chapter 15, which used observation as a task analysis data collection method.

Questionnaires

Overview

Questionnaires are sets of predetermined questions arranged on a form and typically answered in a fixed sequence (Cooper, 1986). The questionnaire is usually completed by the individual under study, but it is also possible for the questionnaire to be administered by a second party who will enter the necessary information. In this latter situation, it closely approximates a heavily-structured interview. The response required may be open-ended, or in terms of a 'Yes/No' indication, a choice of category of information, an indication of the extent of agreement/disagreement, or an indication of preference or attitude in terms of favourability towards an item. The questionnaire is typically a highly formalized activity although it is possible for open-ended questions to be included, where the individual may write sentences, or even paragraphs, in response to requests for information. Under these conditions it looks more like a normal structured interview. Questionnaires allow for a great deal of flexibility of administration. They can be filled in at any time by any number of individuals. They may be completed in the work place, or sent to a home environment. In a human factors context, questionnaires typically allow an investigator to either directly probe specific aspects of a task, or to examine the attitudes and feelings which may be expressed towards particular aspects of the task. The administration of a questionnaire is also very flexible, and this means that it is often relatively easy to gather the views of a variety of individuals who are involved with a particular task or system.

Application

Questionnaires can be used in a very broad range of applications. They can be used in order to isolate problems within a system by way of operator surveys, to evaluate specific aspects of a system, or to gauge opinions and attitudes towards aspects of a system. They can also be used for probing different perceptions and task knowledge of individuals with different backgrounds, either in terms of their use or experience of a system. For instance, differences between recommended procedures and actual operation could be investigated by comparing the responses of managers and operators.

Their main use would appear to be in the collection of specific items of information, and because of their economy in application, they can be used to sample responses from a large number of operators. Appropriate coding of the information can enable a fairly quick set of statistical tests to be run on the information which has been collected. They are probably the most widely used method for obtaining opinion and attitude data. Whilst questionnaires may remain subjective in their orientation, large amounts of data can be examined to determine the degree of consistency amongst operators in their perceptions of the system.

Description of technique

Four issues require discussion in describing the questionnaire approach, namely questionnaire items, types of questionnaire, their administration, and the use of published questionnaires. These are each described below.

Questionnaire items

The central issue in questionnaire design lies in ensuring that all respondents interpret the items in the way that an investigator intended. In order to overcome difficulties due to misinterpretation, it is necessary to pay a great deal of attention to the design and wording of questionnaires. In particular questionnaire items should:

- Use familiar words in short, simple sentences
- Avoid the use of negatives
- Avoid the use of technical terms and acronyms, unless the investigator can be certain that they are fully understood by all of the questionnaire respondents
- Cater as far as possible for all possible responses
- Avoid sensitive issues, unless anonymity has been ensured, or it is understood that information will be treated with confidentiality
- Avoid asking leading questions

Even after the most diligent efforts it is possible that there might be some questionnaire items which could give difficulties. Therefore, it is essential that a short pilot trial of any questionnaire should be undertaken prior to the main distribution.

Types of question

There are several types of question which can be used, and these are listed below.

Multiple Choice Items
Multiple choice items provide two or more specific responses from which respondents have to choose an item which is, for example, most representative of their understanding. Such items can be analysed relatively easily. These advantages can be negated on occasions because respondents may either see no difference between the two alternatives or else they may feel that neither alternative is preferable. This means that during the initial formulation of items, it is necessary to know the full range of significant alternatives, and attention must be paid to the wording. Even slight changes in wording may lead to misunderstandings.

Rating Scales
Rating scales that have been well constructed can be used to obtain subjective information from users which gives an indication of both the nature and magnitude of their opinions about certain aspects of a task. Provided that the items for a rating scale are carefully chosen, and enough respondents are used, it is relatively easy to analyse rating scale data statistically (see Brigham, 1975).

However, rating scales are particularly susceptible to subjective biases and distortions, so care must be exercised in designing them.

The items in rating scales should:

- Use a maximum of seven alternatives
- Use sets of response alternatives in which the intervals between responses are as nearly equal as possible
- Use standardized rating terms for all answers where possible such as: completely acceptable, reasonably acceptable, borderline, moderately unacceptable, extremely unacceptable

Paired associates (bipolar alternatives)
Paired associates can be regarded as a special type of rating method, where respondents have to decide which of two specific alternatives is most appropriate. When several response alternatives are paired together in a variety of ways, this method can yield highly reliable ratings.

Ranking
Ranking requires respondents to order items according to some specific criteria. It can be a very reliable technique, but its reliability falls rapidly when respondents are required to rank more than about ten possible alternatives. Unfortunately, rankings are of limited use, because they give no information about either the relative separation of adjacent ranks, or the actual effectiveness (or otherwise) of particular statements.

Open-ended
In this situation the respondents are required to write their own answer or comments on the question. This could be a few words or more, depending on the nature of the question and/or the fluency of the respondent. Such items can generate useful information, but it will be in a form which will require effort (from the analyst) to organize. Such an approach is more likely to be used in an exploratory study. Alternatively, many such items could be included, in order to provide the respondents with the opportunity to add details not covered by more restrictive items.

Administration of questionnaires

Before administering a questionnaire it is essential that the target population should have been identified. For instance, if determining how a complex task is actually performed, the most useful target population will probably be the users. However, if the analyst is mainly interested in training issues, or if the system is intended for general use by untrained users, it may be appropriate to administer the questionnaire to a representative sample of the intended users, directly after a user trial.

There are several ways that a questionnaire can be administered and this also influences the information which is collected. Important considerations are as follows:

- The presence of the investigator will provide respondents with an opportunity to obtain clarification, or explanation, where necessary, but may compromise the respondents' belief in the confidentiality of the replies.
- When respondents fill in questionnaires on their own, there is evidence that there is a tendency for them to rush through and to provide non-commital answers.
- Responses may be biased by talking to others who have filled in the same questionnaire.

The use of existing/published questionnaires

In some situations it will be possible to use existing questionnaires which have been widely used and validated. This will save the analyst much effort in questionnaire design, and it will facilitate comparison with other studies that have used the same questionnaire. There are questionnaires available for many purposes, but for task analysis a particularly useful set of questionnaires are those known as job component inventories. Probably the most widely known of these is the position analysis questionnaire, which is described below.

Position analysis questionnaire

The position analysis questionnaire (PAQ) is a method of gathering job information (McCormick *et al*. 1969). It attempts to identify general job characteristics, and has been useful primarily for personnel purpose, such as selection, job comparison/grading and career development, rather than for operational purposes such as work design and training. The technique and several similar approaches have limited application to operational and safety issues, since they deal with general statements of skill, rather than specific details of industrial processes.

The PAQ requires a job incumbent to identify which job elements (out of a total set of 194 elements) are present, or absent, in their occupation. It is organized into six divisions as follows:

- Information input
- Mental processes
- Work output
- Relationship with other workers
- Job context
- Other job characteristics

There are, as already noted, 194 job elements in all. The PAQ has been applied to a wide variety of industrial tasks, and an obvious advantage is that the technique is well established and is easily processed. A major problem with this kind of approach is that the questions may be misinterpreted and the ratings are subjective. Efforts have been made to improve the forms of the questions in order to reduce ambiguities, but it is extremely difficult to ensure that the rating scales will be used consistently by different respondents. Applying this technique in capital-intensive, low-labour- intensive industries is problematical, since it is not easy to collect data in sufficient quantities to eradicate response biases. A further major problem is that the outcome is somewhat general and cannot reflect the particular operating problems or skills that arise in specific complex industrial processes. Therefore, these techniques are limited to less specific personnel questions. But as such they can provide information that is useful in determining the skill content of tasks.

Practical advice

Anonymity of respondents is a significant consideration in the use of questionnaires for two reasons. Firstly, if the responses generated by the respondent could in any way lead to negative effects for the individual concerned, the questionnaires should be anonymous. Secondly (but related to the first point), people are more at ease with an anonymous questionnaire and it is more likely to be returned completed. Unlike interviews, the respondent for a questionnaire cannot ask questions for clarification or seek verbal assurance that the answers will not be quoted out of context or used for a purpose other than that stated on the front of the questionnaire.

It is essential that investigators fully appreciate what information they are seeking before they start formulating the questionnaire items. Also, it often helps to work out how the questionnaire information will be analysed and used subsequently. This will enable the questionnaire form to be appropriately designed and codified, for example, to facilitate quick data entry into computers. For these reasons it is important that a pilot exercise is always carried out.

A major feature of the use of questionnaires is the initial questionnaire design. Items will rarely be available off-the- shelf, and their development is not a minor activity. Standard texts describe how to determine questionnaire items, the process in detail being very context dependent (e.g. Cooper, 1986; Oppenheim, 1986). A good account from the ergonomics perspective is given by Brigham (1975).

The questionnaire can produce a wealth of information, and therefore planning for the analysis of this is important. Almost invariably the use of a questionnaire will fail to stand on its own as a one-shot single exercise. Therefore it should be used as a means of gathering initial information to be complemented by a follow-up questionnaire or additional interviews (a useful structure of this process is given by Sinclair, 1990).

It is always a good tactic to try to obtain questionnaires that have been used in other similar studies to see how they have approached the problem and to see what pro-forma they have used. The danger lies in uncritically accepting such items for use with a specific task. Also, there may well have been unreported shortcomings with a particular questionnaire. A number of standard questionnaires are now emerging for evaluation studies in the human–computer interaction area, and these could be of use for specific applications.

The analyst will need to determine an acceptable rate of response. This can be very high in a limited application of questionnaires in a system context, but low if a postal survey is used. Response rates can be improved by making the return of the questionnaire easy (e.g. by using stamped addressed return envelopes). It is also important to send reminders to those who have not returned their questionnaires.

Resources required

A major drain on resources from a questionnaire study is the amount of an analyst's time which is necessary to develop the initial questionnaire and run a pilot study. However, once an acceptable questionnaire has been developed, it will be very flexible to administer, and provided that it is not to be individually administered, it will demand relatively few resources at this point, other than the respondents' time, and the costs of printing and distribution.

The amount of effort which is required to analyse a questionnaire will depend upon its nature, and this must be carefully considered during design, so that it is easy to analyse. For most task analysis purposes, the analysis of questionnaires should be a relatively straightforward, clerical task, usually with some limited statistical analyses.

Links with other techniques

The most obvious links are with the interview technique (e.g. see Sinclair, 1990), which may use similar sources of information to the questionnaire and will generate material for questionnaire items. The questionnaire itself should always derive from an earlier study which has generated the material used in the items and thus is an hypothesis-testing, or further data gathering, exercise. Questionnaires may also be used to gather critical incident information (pp. 47).

Many of the information analysis techniques which will be used in the later stages of a task analysis may be dependent upon the information which has been gathered from a questionnaire. This means that the analyst must take particular care to ensure that all the necessary questions are included in any questionnaire.

Advantages

- The main advantage is in the area of administration. Questionnaires can be given simultaneously to a group of people. Respondents are often free to complete them at their own leisure and are able to add information which was previously forgotten or may have been omitted by the questionnaire designer. Because it is relatively easy to administer a large number of questionnaires, they are particularly useful for comparing the opinions, or understanding, of a large number of system users.

- Through the use of closed questions, questionnaires can prohibit tangential issues from being introduced. Although respondents may deviate from the main issue of the question, the next question should return them to the central theme.
- The correct use of statements and rating scales can produce answers that can be weighted for their reliability and consistency. This safeguards the researcher against acting on any hunches or insights gained from early, or particularly eloquent, respondents.

Disadvantages

- Respondents are not committed to give correct answers and may often be influenced by what they believe the analyst requires, or what they themselves wish to portray (while this is a valid criticism, it can be also levelled at data collected in face-to-face situations). Anonymity, motivation, forced-choice answers, and having the questionnaire completed remote from the analyst, and/or the work situation, can reduce any possible distortions.
- There is a real danger of loss of information or that subjective insights may be swamped by more trivial details, or inappropriate perceptions of what the question is asking. Correct choice of target population will minimize this danger.
- There is also the problem of subjective interpretation. If open-ended questions are incorporated, then they will always require interpretation on the part of the analyst.
- Requires a large, identifiable subject pool who are prepared to complete and return the questionnaire.

References and further reading

Brigham, F.R. (1975) Some Quantitative Considerations in Questionnaire Design and Analysis *Applied Ergonomics*, 6, 90-96. *A useful overview of the use of questionnaires in ergonomics.*

Cooper, M.A.R. (1986) *Fundamentals of Survey Measurement and Analysis.* London: Collins. *A useful general text.*

McCormick, E.J., Jenneret, P.R. and Machan, R.C. (1969) *A Study of Job Characteristics and Job Dimensions as Based on the Position Analysis Questionnaire.* Rept. No. 6, Occupational Research Center, Purdue University, West Lafayette, Indiana, USA. *This was the starting point for work on the PAQ.*

Oppenheim, A.N. (1986) *Questionnaire Design and Attitude Measurement.* Aldershot: Gower Publishing. *A useful general text.*

Sinclair, M. (1990) Subjective Assessment. In: *Evaluation of Human Work*, Wilson, J.R. and Corlett, E.N. (eds.), pp. 58-88. London: Taylor and Francis. *A useful chapter dealing with questionnaire layout and the relationship between questionnaires and interviews.*

Sparrow, J. (1989) The Utility of PAQ in Relating Job Behaviours to Traits. *J. Occup. Psychol.*, 62, 151-162. *Reports on the use of the PAQ.*

Sparrow, J. and Patrick, J. (1978) *Developments and Recommendations for the use of the Position Analysis Questionnaire in Selection and Training.* Applied Psychology Rept. 86, University of Aston. *Development work on the PAQ.*

Smith, P.C. and Kendall, L.M. (1963) Retranslation of Expectations. An Approach to the Construction of Unambiguous Anchors for Rating Scales. *J. Applied Psychol.*, 47, 149-155. *More detailed research on questionnaire design.*

Sudman, S. and Bradbum, N.M. (1982) *Asking Questions: A Practical Guide to Questionnaire Design.* New York: Jossey Bass. *A useful guide to questionnaire design.*

Youngman, M.M. (1982) *Designing and Analysing Questionnaires.* Maidenhead: TRL Rediguides. *A concise guide to questionnaire use.*

Structured interviews

Overview

The interview, whether structured, or unstructured, is probably the most commonly used approach for information gathering (Meister, 1985). It is widely used, not only in the human factors area, but as a general technique in personnel psychology and for knowledge elicitation. The term *structured* implies that the content of the interview, in terms of the questions and their sequence, is predetermined. The unstructured interview does have its role, particularly in the early stages of information collection. But the structured interview is more commonly used for the general collection of task-based information. It is suggested that the structuring of the interview offers the opportunity for more systematic collection of data. The interview will usually involve a single interviewer plus the interviewee, who will usually be someone with a detailed knowledge of the system, such as a subject matter expert, job incumbent, supervisor or manager of the system under study. However, on occasions, it may also be appropriate to use trainees, or naive users, in order to identify particular problems which such individuals might have.

Application

Interviews can be used at any stage during a comprehensive task analysis activity and may turn out to be the only form of data collection that is used, but they are more likely to be supplemented by other techniques. They can usefully be applied early on in an investigation for collecting basic information about the situation. In this context they are more likely to be unstructured in their format. In later stages of analysis a structured interview may play the main role in the data collection. Alternatively, it may be used at the end of an exercise for checking the accuracy of information collected by other methods.

The interview will be utilized because of its high face validity, and also because of it being probably the most economic approach that can be used. Due to limited resources, or alternatively, difficulties in measuring and observing in the job situation, it may turn out to be the only viable source of information. The technique is relatively efficient for dealing with large numbers of people, and it is possible in some situations to carry out quite large-scale survey interviews. Two main types of interviews relevant in this context have been described by Meister (1985). The first of these is the *informational interview*, which sets out to collect a wide range of information on a particular task situation. The second type, the *survey interview*, has a more specific objective, and sets out to collect data in a more systematic and codified way. In fact, extremes can be seen running from the completely unstructured open-ended interview, perhaps used early on in a study, to the more narrowly-defined survey interview, which can, in an extreme form, approximate to an administered questionnaire.

Description of technique

Preparing the interview

It is important that before the interview begins, effort should already have been applied to the derivation of the questions to be used in the interview. This presupposes that a process of determining the structure of the interview has already gone on, with the appropriate selection of questions and their sorting into a suitable sequence. This will depend upon a process of defining the objectives of the study itself, and some degree of preparation in terms of unstructured interviews and discussions with relevant parties (e.g. see MacFarlane Smith, 1972).

The structured interview can be fairly broad in its aims, with a large number of fairly open-ended questions, in the sense of determining information. Alternatively it can be highly structured, so that only a limited range of very specific responses, such as 'Yes' or 'No' answers, are required. In order to orientate the interviewee to what is to be asked, it is important to choose words carefully and to determine an optimum sequence of questions. The interview usually ends with an open-ended question such as 'Is there anything you think we should know about the task situation that we have not yet raised?', or 'Is there any more information you think we should have about this situation?'

Choosing a location

The structured interview is best carried out near to the respondent's place of work, but it is important that special facilities should be provided for the interview activity, so that the respondent will be at ease and relaxed by using a non-threatening, or familiar, situation. Therefore, some special room should be provided where there is privacy and a lack of distractions and interruptions.

Conducting an interview

Interviewers should be confident in their manner. They should be competent in putting the respondent at ease and generally establishing good rapport. It is wise to avoid overbearing approaches that will not yield effective communication from the respondent. Interviewing skills will need to be developed and assessed, before the interviewer can begin work in critical situations, where highly qualified and skilled people are giving of their time, etc. Difficulties can arise if these fairly basic assumptions are not taken seriously.

In terms of the interview itself, the interviewee should see the relevance of the activity and should not be made to feel insulted, or belittled, by the experience. Efforts should be made to establish good communication, but if this is done too well, there may be irrelevant or misleading material generated in an effort to be helpful. Therefore, the interviewer must maintain control of the situation.

The end result will be very much dependent upon the objectives set for the analysis and the conduct of the interview. It may be an end in itself, or it may well be a prelude to more detailed studies based on the information generated from the interviews.

The analysis of interview information

After the interview, the information which has been collected should be transcribed into a written form for more detailed analysis. This can be done by categorizing the information in some way (see pp. 42), or by using one of the other techniques which is described within this *Guide*.

Practical advice

As with questionnaires (and the critical incident technique), the need for preserving the anonymity of the interviewee should be considered by the analyst at the outset, i.e. prior to any interviews taking place. If there is any chance that the interviewee could suffer difficulties as a result of giving information during the interview, then responses should remain anonymous.

As it is an interpersonal situation, the interviewer needs a skilled approach to the interview situation. The need to establish rapport, to make the interviewee feel relaxed and not feel threatened or embarrassed by the situation is a primary concern. The more open-ended the interview situation, the more important these points will be. Interviewers need to be able to motivate respondents to give more information about the areas being discussed, without biasing the interviewee. Preferably the interviewer has knowledge of the topic area, of interview techniques, and is sensitive to people. As Sinclair (1990) notes, collectively such criteria in a single person are not easy to meet.

In the personnel selection context it has been shown that there can be low reliability between interviewers in interviewing the same person, and so if more than one person is going to be conducting the interviews, it is important to ensure that a consistent approach is taken. However, it must be stressed that the information collection role of an interview in this context is very different from that in which it may be used for job appraisal or job selection.

One practical question that does arise is the amount of time that it is reasonable to expect people to be interviewed for. While this may be determined by their availability, it is suggested that a 20 min period will be a useful minimum, and that 40 min should be a maximum. This does not preclude the possibility of a series of interviews with one individual. Repeated interviews may be used to pursue a fairly lengthy subject matter, or to gain increasing detail on a particular operation under study. The effort required on the part of the interviewer should also not be overlooked.

It may be useful where practicable to utilize two interviewers, one to conduct the interviews, and the other to record the replies. This can be more cost-effective than transcribing the tapes, and allows for a more rapid and fluent interview.

Questions of validity and reliability of information collecting need to be asked (see Meister, 1985), and while the data that are available are fairly positive, There is also sufficient variation to suggest that more systematic research is needed in this area.

Resources required

The resources needed for this technique have been mentioned previously, and usually revolve around a relatively isolated, comfortably furnished room where the respondent and the interviewer have privacy and feel at ease. A small tape recorder will usually suffice for sound recording. More extensive studies using videotape have been carried out, but this is usually in order to check on the nature of the interview process itself rather than to provide any extra detail. However, video recordings might be appropriate where the interviewee feels the need to demonstrate an aspect of the task using equipment, or use diagrams and drawings, etc. The recording techniques will obviously need to be adapted to the situation, but the resources required for the subsequent analysis of such tapes should not be forgotten. A highly structured interview using a pro-forma would require the derivation and appropriate try-out of such a pro-forma before its more widespread use.

The need for interviewers to be trained is clear and should be provided for. Experience from the personnel selection area has identified the importance of training interviewers to be consistent in their approach and to utilize a number of well-established techniques for getting the most out of the interviews. While open-ended questions can be used to generate information for subsequent studies, it is important that a great deal of effort should go into the preparation of questions for an interview, both for the specific role of the interview and for its data collection objectives. There is also a need to establish the ground rules for turning the interview into an effective interpersonal communication activity. This may well involve the interviewer gaining familiarity with the task situation vocabulary and the nature of the tasks. It may also involve the interviewer in establishing credibility with the interviewees. This can take place through the former's presence within the work situation, participation in training courses and familiarization with various aspects of the process.

Links with other techniques

As the structured interview forms the basis of so much information collection, it has obvious links with many other techniques, which use such information. Observation studies (pp. 53) might give rise to the need to explore particular areas in detail, and the interview is likely to be used for this detail. There may be a need to supplement the straightforward interview with other techniques such as the critical incident technique (pp. 47). It also has strong links with the use of questionnaires (pp. 58; see also Sinclair, 1990), in that an interview approach may precede the formulation of a questionnaire (although the converse can and does occur).

Advantages

- The main advantage of the technique is that it is familiar and would seem to most respondents to be a natural approach to adopt.
- The interviewer can deal with unexpected information, 'on-line', in a better way than a more structured and 'off-line' approach (such as a questionnaire).
- Usually it should be fairly easy to arrange for an interview to be carried out in the normal workplace. However, if the interview itself takes the interviewee away from the normal pattern of work, it has the advantage of bringing an element of variety, and can be seen as a 'perk'.
- The use of structuring within the interview offers advantages in that a consistent set of questions can be asked of each individual, and if appropriate, the response data can be treated statistically.
- There is also a flexibility in that other points can be made and particularly interesting lines of discussion can be followed up.

Disadvantages

- While a tape recording may offer certain advantages in terms of information collection, the time- consuming nature of the subsequent analysis should not be overlooked. Although the technique can be an economic one, it can also prove to be relatively costly in person-hours, and probably more expensive than the use of questionnaires, etc. However, it is perhaps not as expensive as detailed observation.
- Another disadvantage lies in focusing upon the more routine aspects of the task, which may mean that the interview should include the use of the critical incident technique in order to get at the more rare, but potentially disastrous, situations.

- The advantage of it being a social, interpersonal interaction situation can also carry with it disadvantages. The relationship established between the interviewer and interviewee may cause difficulties if there are questions of perceived work status, perceived social status, or a misconception about the purpose of the interview. There could also be a problem if the knowledge which is being requested is not accessible by the interviewee from memory, or it cannot easily be verbalized, as is sometimes the case with highly skilled operations. There is also the possibility of bias (interviewee or interviewer), or the interviewer may be told what is expected, rather than what actually happens.

References and further reading

Cordingley, E.S. (1989) Knowledge-elicitation techniques for knowledge-based systems. In *Knowledge Elicitation: Principles, Techniques and Applications.*, Diaper, D. (ed.), pp. 89- 175. Chichester: Ellis Harwood. *Discusses the use of interviews for knowledge elicitation.*

McCormick, E.J. (1979) *Job Analysis, Methods and Applications*, pp. 33-40. New York: AMACON. *Has a review of interview techniques for job analysis.*

MacFarlane Smith, J. (1972) *Interviewing in Market and Social Research.* London: Routledge and Kegan Paul.

Meister, D. (1985) *Behavioral Analysis and Measurement Methods.*, pp. 353-368. New York: Wiley and Sons. *Contains a useful section on the use of interviews in human factors work.*

Sinclair, M. (1990) Subjective Assessment. In: *Evaluation of Human Work*, Wilson J.R. and Corlett, E.N. (eds.), pp. 58-88. London: Taylor and Francis.

See Chapter 8 which utilized structured interviews.

Verbal protocols

Verbal protocols are verbalizations made by a person while they are carrying out a task, in the form of a commentary about their actions and their immediate perceptions of the reasons behind them. These self-commentaries are made while the task is being undertaken, to avoid the inevitable distortion, or forgetting, that could occur if the reporting were left until afterwards. Although the technique will provide information about simple control movements, the main aim is to obtain (or infer) information on the covert psychological processes which underlie performance, which are not directly observable, such as the knowledge requirements or mental processing involved. However, it is important to stress that not all mental processes can be verbalized directly, and for these cognitive activities, verbal protocols are wholly inappropriate. Furthermore, there is always the risk that the act of verbalizing about a task will interfere with the behaviour which is being investigated. Therefore, verbal protocols must be interpreted cautiously, and if possible they should be corroborated via other data.

There are several reports of verbal protocols being used in carefully controlled laboratory settings for very detailed examinations of complex problem solving tasks, but perhaps the most useful discussion of the technique in a practical process control environment is given in Ainsworth and Whitfield (1983).

Application

It is only possible for a person to verbalize about their ongoing mental activities if they can readily retrieve the appropriate information from their short term memory. This has two important ramifications:

- The protocols must be produced concurrently with task performance, because within a few seconds they will either have passed into long term memory, or have been forgotten. Therefore if verbalization is delayed the individual will first have to retrieve or infer the necessary information. In either case, this will reduce the accuracy and completeness of any verbalizations, and will also tend to interfere with performance on the basic task which is being investigated.
- Useful protocols can only be produced for information which is coded in a verbal form while it is in short-term memory. This will preclude the use of the technique for tasks which have processed information as visual images, etc., or for information which has not been heeded because it has either been deliberately ignored as trivial, or else has been otherwise overlooked. Also, parts of tasks which have become subconscious or 'automated' due to overpractice, will be inaccessible to this technique. The application of verbal protocols to such task types will, therefore, provide little useful information, and may even produce misleading reports. It is for this reason that so- called 'expert knowledge' is so difficult to extract.

Verbal protocols are a method of data collection, and are therefore used at the commencement of a task analysis to provide information input to other analysis techniques. The protocols obtained from a task should enable an analyst to derive a task description of the steps required to accomplish a particular task, but their main purpose is to provide some insights as to how the steps are accomplished. If an analyst only requires a task description, it is not necessary to record detailed verbal protocols, and therefore this technique should only be used for investigating decision-making tasks, or other tasks involving mental processing that can be verbalized.

The use to which the information will be put should determine who is asked to provide the verbal protocol. For example, skilled operators must be used if an analysis of skilled performance is the objective. However, the skilled operator's shortcuts will only be revealed by comparison with the more painstaking and proceduralized performance of the novice. Similarly, if the goal is to evaluate system usability for untrained users, then untrained users will be the only valid group of respondents.

Because verbal protocols will often pinpoint difficulties experienced by the operator of an existing system, the technique has been used successfully to identify training deficiencies, and to select alternative procedures for task execution. Verbal protocol analysis can also reveal instances of excessive demands being placed on individuals, either through sheer quantity of work in a given time, or through overly complex information processing requirements. However, for particularly demanding tasks the respondents may not be able adequately both to undertake the task and to verbalize about it, and so either the task performance will be unrepresentative, or else the protocols will be incidental, patchy and limited.

Description of technique

The technique can be considered in four stages, namely the preparation stage, recording the verbal information, supplementing the protocols, and analysing them. These four stages are considered in turn here.

Preparing to collect verbal information

To prevent the respondents from elaborating on, or rationalizing, their thought processes in any way (e.g. making them appear more coherent and structured than they are in reality), it is essential to encourage a continuous, flowing commentary. It is also very important that the process of providing a verbal commentary must not affect the way in which the task is carried out. For both of these aims to be achieved, reporting must be spontaneous, and respondents should be discouraged from making delayed and considered assessments after the event. Every effort must be made to remove any impediment to the free and uninterrupted generation of verbal commentaries.

The following guidelines should be followed to encourage task verbalization:

- The analyst should try to establish a rapport with the respondent, and should assure him or her of confidentiality, so that comments will be made directly, without any attempt by the respondent to filter out, or modify, responses which the respondent feels might not be acceptable to the analyst, or the respondent's superiors or peers.
- Many respondents will experience initial difficulty in verbally expressing their mental processes, and so an opportunity for practice should be included in the session, and constructive feedback should be supplied afterwards. Similarly, an explanation of the goals of a study often helps the respondents to understand what is required of them.
- Ensure that data recording is not intrusive, and does not increase workload for either the respondent or the investigator. For example, hand-held tape recorders should not be used because they may interfere with control actions. Similarly, the respondent should not be asked to undertake any additional duties, such as periodically reporting the time before undertaking particular actions. All such duties should be performed automatically, or if this is not possible, they should be undertaken by the investigator.

Recording the verbal information

It is essential to have a record of every utterance which is made during a verbal protocol session, and while the rates and amounts of information which could be given are highly variable, the analyst must be prepared for a rate of something in the region of 150–200 words per minute. This means that some form of audio-recording will be essential, and such a recording system should be checked for its clarity before commencing to record any protocols.

If verbal protocols are being recorded for teams, it will be necessary to arrange for each voice to be recorded separately, so that the individual sources can be readily distinguished. It will also be necessary to ensure that the different recordings can be correlated with each other during their analysis.

During the recording of the protocol, the analyst should avoid making any comments, because these could distract the respondent's attention, and either cause modified respondent behaviour, or else impede the free flow of the commentary. It will also be helpful if the analyst can link the protocol to any system behaviour occurring at that time. There are several ways that this can be achieved:

- By manually noting the time or the values of particular indicators, when particular utterances were made. This timing can be done either during the original recording or while listening to a playback
- By relating some form of computer logging to a timed transcript or play-back of the protocol
- By focusing a video-camera on a clock, and/or, particular displays, during the protocol recording session

Supplementing verbal protocols

It is unlikely that unconstrained verbal reports will provide all the information required, and it is therefore prudent to consider making some provisions for supplementing verbal protocols, either directly from the respondent or from other sources. There are three ways in which clarification can be sought from the respondent:

- The analyst can interrupt the respondent at appropriate points during the protocol, and ask for additional detail. However, there is a risk that such interruptions might disrupt performance on the basic task. If too many interruptions are given, it will not be possible to effectively study cognitive behaviour, and the protocol session will be reduced to being a walk-through, which will provide little valid information about the underlying mental processes.
- More specific instructions can be given to the respondent, explaining which aspects of the task are of particular interest, and then the respondent is likely to make greater efforts to report on these aspects of the task.

- After giving the protocol, the respondent can be asked to expand verbally on specific points. However, because verbal protocols are based upon transitory information which has only been held in short-term memory during task performance it will be necessary to provide some memory cues, such as a written transcript or a recording of the protocol. For this purpose video recording is particularly effective.

Each of these methods has risks associated with it, either because of the potential for disrupting the main task, or due to the inevitable fading, or distortion, of memory subsequent to the initial session. Therefore, it is common to supplement verbal protocols with some form of observational technique, such as video recording, or computer logging of user/system interactions. If it is still necessary to obtain some clarification from the respondent, this should be done as soon as possible after the initial protocol recording session.

The analysis of verbal protocols

The manner in which a verbal protocol is analysed will depend upon the complexity of that protocol and the requirements of the analyst, but generally it should include some, or all, of the following stages. The first stage in the analysis of verbal protocols is the direct transcription of the tape recordings into a written form. This initial transcript should be a faithful record of every word which was uttered, regardless of its grammatical correctness. It may be helpful to insert new lines into this transcript at points where there are obvious pauses, but otherwise this transcript should not incorporate any interpretation or amendment of the words which were spoken.

The analyst can then attempt to organize and to standardize the protocol by producing two further transcripts, as shown in Table 2.2. The second transcript should be an organized version of the protocol, which maintains the original words, but which breaks these down into separate activities by adding new lines to form separate segments at each point that a new activity is described. Also at this point, the analyst should add notes at the appropriate positions in the protocols to indicate the elapsed time, or critical system parameters. A second transcript should then be made which incorporates these amendments, with the notes added by the analyst being shown in additional columns.

The next stage of the analysis involves standardizing the presentation of the protocol information, so that it is easier to draw inferences from it. This will involve some, or all, of the following processes.

Extraneous vocabulary reduction

The analyst should search through the transcript for words which do not add anything to the explanation that is being given. However, many of these words could be indicative that the operator is having some problem, and so the analyst must be careful before any words are deleted. One way round this is to use a

'Comments' column to record things such as operator errors or difficulty in locating particular information.

Table 2.2 An example of how a verbal protocol can be progressively transcribed

min	kW	**Second transcript**	**Third transcript**
7.00	497	We'll also put this on auto by switching from manual to auto.	Put coal feeder on auto by switching from manual to auto
		That's okay now, it's also on auto	Confirm it is now on auto
		It'll take a few minutes now before the feeder settles down, so we'll just have to wait a bit	Wait for coal feeder to stabilize
8.50	495	This feeder has been pulled off too much,	This coal feeder rate has reduced too much
		so I'm bringing the PA flow back.	so I'm increasing the PA flow
9.10	495	We'll now take some bias off,	I'll have to reduce the bias on the mill rating
		which will increase the flow	which will increase the PA flow
		and speed up the feeder	and cause a speed up of the coal feeder (because it is on auto)
10.40	497	As the steam pressure is increasing so the load is going up,	The steam pressure is increasing and so the load is going up
		so we're pulling it back a bit, say half a division, on the mill rating.	so I'm reducing the mill rating by half a division
11.35	497	Table diff. is getting high	Table differential pressure is increasing towards its limit
		so increasing the PA flow is necessary now.	so I will increase the PA flow

Synonym reduction

Inevitably, different words and phrases will be used to define the same instruments, or types of activity, and so the analyst should reduce these synonyms to a more restricted set of common words and phrases. However, if the analyst has any reason to believe that such ambiguous references might have some particular significance, the analyst should retain the original words, and add others, enclosed within brackets, which resolve the ambiguity. For instance, it may be particularly significant that an operator refers to 'the top dial', because this could demonstrate that the grouping and layout are being used as important cues.

Resolution of ambiguities

During any protocol, the respondent may refer to particular instruments, or information, in a fairly ambiguous way. For instance, the respondent may state that 'The temperature is falling', when there are several temperature gauges, or he may simply report that 'It has reached 280 now'. In such situations, the analyst should try to establish the precise indication or information which is being reported, and then note this. Clues to assist in the resolution of such ambiguities can be obtained from elsewhere in the transcript, from records of the system state (such as video, or computer logs), or by questioning the respondent.

After all these processes have been completed for the protocol, a third version should be produced. This may be sufficient for the analyst to use as a basis for examining the content of the protocol, but some analysts may prefer to re-order the words in each phrase, so that they are all consistent. For instance, the analyst may wish to break the protocol down into a series of verb/noun clauses.

Practical advice

For many tasks, verbal protocols probably provide the most direct means of access to information about mental processes currently available. However, care must be taken to ensure that the processes being reported are appropriate for concurrent verbalization without distortion. Therefore, if possible, corroboration by other methods should be sought before inferences are made about how information is processed in a given situation. For the analysis of expert decision processes, there is a growing battery of 'knowledge acquisition' techniques available. These generally have less apparent or 'face' validity, but claim the advantage that they are not hampered by respondents' inability to express complex thought processes verbally.

The practical administration of verbal protocols may appear fairly straightforward, but care should be taken to avoid falling into some simple traps:

- Respondents often feel embarrassed at the prospect of speaking into a tape recorder, especially in front of colleagues. A useful suggestion is that where possible, an investigator should remain beside the respondent, so that he or she can feel that comments are being directed at a real person. This also provides an opportunity for the investigator to make additional data records, such as timings or critical display readings.
- Background noise can make data unusable, so it is necessary to use the best recording technology and environment, and to check this by conducting a short pilot recording.
- The analyst must not put words into the respondent's mouth and should restrict involvement to assisting the flow of commentary in a general way.

- It may take more time than anticipated to establish the required system state for commencing to give a particular protocol, and similarly, it may be necessary to undertake additional, non-planned actions on other systems during a protocol session. Therefore, the analyst should make allowances for this when planning any programme of protocol sessions.
- Significant wasting of resources, or loss of data, can be avoided by considering the purpose of the protocol before starting any recording, so that investigator and respondent are fully aware of what data are required and why. For instance time-tagging may be superfluous for determining skill requirements when developing a training programme, but could be essential when looking at work organization, or stress, and this could affect the instructions given to respondents.

Resources required

Verbal protocol recording alone can be achieved fairly easily by one investigator and a respondent, with a good quality tape recorder. However, if additional data are being recorded, it could be advisable to enlist some support. Otherwise, the session may appear unwieldy to both respondent and investigator, and there will be an increased likelihood of missing data, or interference with task performance. With experience verbal protocol sessions can be completed in little longer than it takes for the respondent to complete the task. However, it is possible that there could be delays in waiting to record protocols, because respondents are performing other duties, or because the system conditions are not appropriate for a specific task to be started. Such factors must be considered during the planning of a protocol recording programme.

If further clarification of the protocol is required, it should be obtained as soon as possible after the initial session. If secretarial assistance is available on site, to provide rapid transcription of results for discussion, this can be very beneficial, as can the use of video recordings, especially if they provide facilities for an additional dubbing track, so that the analyst can add comments such as timings.

The major resource drain with verbal protocols will be the effort required to transcribe and analyse the data. It is invariably more economical to enlist secretarial help to produce written transcripts of recordings, rather than trying to analyse the recordings themselves. One hour of tape will typically shrink to a few pages of text, which should be quite manageable. The amount of time needed to analyse the protocols will depend upon the detail which is required and the amount of ambiguous references which have to be resolved, but as a rule of thumb for most practical applications (as opposed to laboratory-based cognitive psychology experiments), 1 hr of protocol recording should take between 4 and 8 hrs to analyse.

In summary, verbal protocol analysis is relatively time-consuming, and it is recommended that in each case, simpler methods with a clearly defined output should be considered first if available.

Links with other techniques

In addition to being supplemented by the data collection methods described earlier (e.g. observation, pp. 53), verbal protocol data can be restated and analysed using several of the techniques described elsewhere in this report (e.g. as an input for task decomposition methods, pp. 95).

Advantages

- Verbal protocols, when used carefully, can give accurate and if required, detailed information about straightforward motor responses and simple decision-making.
- They can be assumed to provide a basis for investigating the underlying mental processes of complex tasks which cannot be studied in other ways.
- Although on occasion the actual validity of verbal protocol data may be questioned, the face validity of the technique is generally high. This means that the results of applying the technique will be credible and comprehensible to non-specialist observers or users.
- Data collection itself is typically quite rapid, because very few special arrangements need to be made on-site, and data analysis will usually be conducted off-site. The time spent recording protocols should be little longer than it takes the respondent to perform the task.

Disadvantages

- Language may be an inadequate descriptive technique for the process being scrutinized. Many mental processes rely on visual, auditory, or even mathematical imagery for their execution. Hence it may be very difficult, if not impossible, to effectively verbalize such events.
- Verbalizing may interfere with the basic task, changing either the speed, or the method of execution. This will be particularly true if the basic task demands excessive use of short term memory.
- Verbal protocols may reflect only the results of cognitive processes, rather than the processes themselves. Thus, the depth of knowledge which can be obtained is often limited. If further insights are required, it may be necessary to do some additional probing after the task has been completed.
- For some tasks, verbal protocols may be biased or inaccurate. Therefore, wherever feasible the protocols should be corroborated from other sources. Particular caution should be exercised in interpreting protocols if the analyst suspects that the respondent is rationalizing the protocols and not verbalizing directly from immediate memory. For instance, if responses are delayed, or if the information is held as visual images rather than in a verbal form, protocols may be a distorted report of the true mental processes.
- Individuals differ in their ability to verbalize their mental processes, although performance may be improved to some extent by practice. Quite commonly, people will only verbalize overt actions, which are observable

in any case. They may require some coaching before they spontaneously explain the decisions which underlie these actions.

- The quality of the description provided by a verbal protocol does not necessarily reflect the quality of task performance.
- The analysis of verbal data can be particularly time-consuming.

References and further reading

Ainsworth, L.K. (1984) Are Verbal Protocols the Answer or the Question? *Proc. Ergonomics Society Annual Conference*, 68-73. *Presents a useful taxonomy of verbal reports, in terms of reporting technique and nature of task. Helps the reader judge if the technique is appropriate.*

Ainsworth, L.K. and Whitfield, D. (1983) *The Use of Verbal Protocols for Analysing Power Station Control Skills.* Applied Psychology Rept. 114, University of Aston. *An evaluation of when verbal protocols are appropriate, with a practical example of the use of verbal protocols in a power station.*

Bainbridge, L. (1979) Verbal Reports as Evidence of the Process Operator's Knowledge. *Int. J. Man–machine Studies*, 11, 411-436. *Investigates the extent to which verbal reports reflect accurately mental processes. Results not clear, but suggests that there is a range of conditions in which usefully low distortion can result.*

Bainbridge, L. (1985) Inferring from Verbal Reports to Cognitive Processes. In *Uses of the Research Interview*, Brenner, M., Brown, J. and Canter, D. (eds.), pp. 201-215. London: Academic Press.

Ericsson, K.A. and Simon, H.A. (1980) Verbal Reports as Data. *Psychol. Rev.*, 87/3, 215-25. *An excellent discussion of the effectiveness of different kinds of verbal reporting for eliciting particular information.*

Ericsson, K.A. and Simon, H.A. (1984) *Protocol Analysis: Verbal Reports as Data.* Cambridge, MA: M.I.T. Press. *This is the accepted handbook for the analysis of verbal reports of highly structured cognitive psychology experiments. However, it is of limited use for less structured, practical applications, such as the task analysis of complex systems.*

Newell, A. and Simon, H.A. (1972) *Human Problem Solving.* Englewood Cliffs, New Jersey, USA.: Prentice-Hall. *This book presents a detailed study of complex decision making, which was based on verbal protocols of laboratory tasks. It is useful primarily for its examples of the ways in which verbal protocols can be used when studying mental processes. Also gives a good impression of the type of results that can be expected.*

Sanderson, P.M., Verhage, A.G. and Fuld, R.B. (1989) State-space and Verbal Protocol Methods for Studying the Human Operator in Process Control. *Ergonomics*, 32/11, 1343-1372.

Chapter 3
Task description methods

Charting and network techniques

Overview

Flow charts and networks are graphic descriptions produced from the analysis of a system which can be used to describe tasks within a system. Many charting and network techniques have been developed since process charts were first published (Gilbreth and Gilbreth, 1921), and this section can only present a selection of the more useful and influential methods available for use in process control analyses.

The principal aim of charting and network techniques is to use a formal graphical representation of the task which is either easier to understand or significantly more concise than a textual description, or else to emphasize particular aspects of the task.

Eight charting/network techniques are reviewed briefly in this section. These techniques range from the early, but still useful, development of the process chart, which can be highly useful for representing semi-manual tasks, to the Petri-net technique, which can be used to show relatively complex plans and control systems.

The techniques vary on one particularly important representational dimension, namely that of time. Charting techniques usually represent time more than network techniques do, the latter often being time-independent.

An important aspect of charting and network techniques is that many of them represent not only the human element, but the system elements also. Indeed some of the techniques, such as the signal flow-graph method, were developed to represent system phenomena before being adopted for task analysis purposes. Thus, whereas some task analysis techniques only focus on the human involvement in the system, charting and network techniques are far more 'integrative', displaying the human role firmly embedded within the system context.

Three examples of networks are provided: critical path analysis, Petri nets, and signal flow graphs.

Application

Both types of technique offer a flexible means of analysis, because each can be analysed to its own required level of detail. They provide a systematic approach to describing tasks and provide a formal graphical representation that is easily followed and understood.

The techniques have a range of applications, for example, the charting techniques are particularly useful for the derivation of branching or looping procedures, and may even be used as the final format for the procedures themselves. As the techniques tend to represent human and system functions together, they can also be used in functional allocation decisions (i.e. how to split functions between operator and machine).

Description of techniques

Input–output diagrams

Probably the simplest representation of a task is an input–output block diagram. The technique, developed by Singleton (1974), involves first selecting the task or step of interest, and then identifying all the inputs and outputs which are necessary to complete this task or step. Figure 3.1 shows a basic input–output diagram for the task of operating a boiler; several such blocks can be linked to form functional chains. Although this form of representation is relatively simple, functional chains soon become difficult to construct. Furthermore, although the outputs can generally be identified by the observation of skilled behaviour, it may be impossible to determine all the (input) sources of information which are being used. Conversely, the observation of a management task may detect all the inputs, but often the outputs will remain covert–lack of action is not necessarily lack of decision.

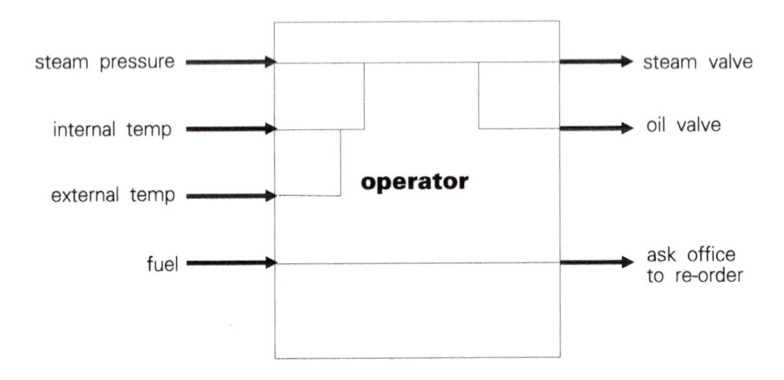

Figure 3.1 Input/output block diagram. Adapted with permission from Singleton, 1974

Input–output diagrams are nevertheless useful for understanding the role of the operator in systems, showing what information, control and communication/administrative facilities are required to keep the system/process going. If functional chains can be developed, then communication links can be highlighted between different personnel.

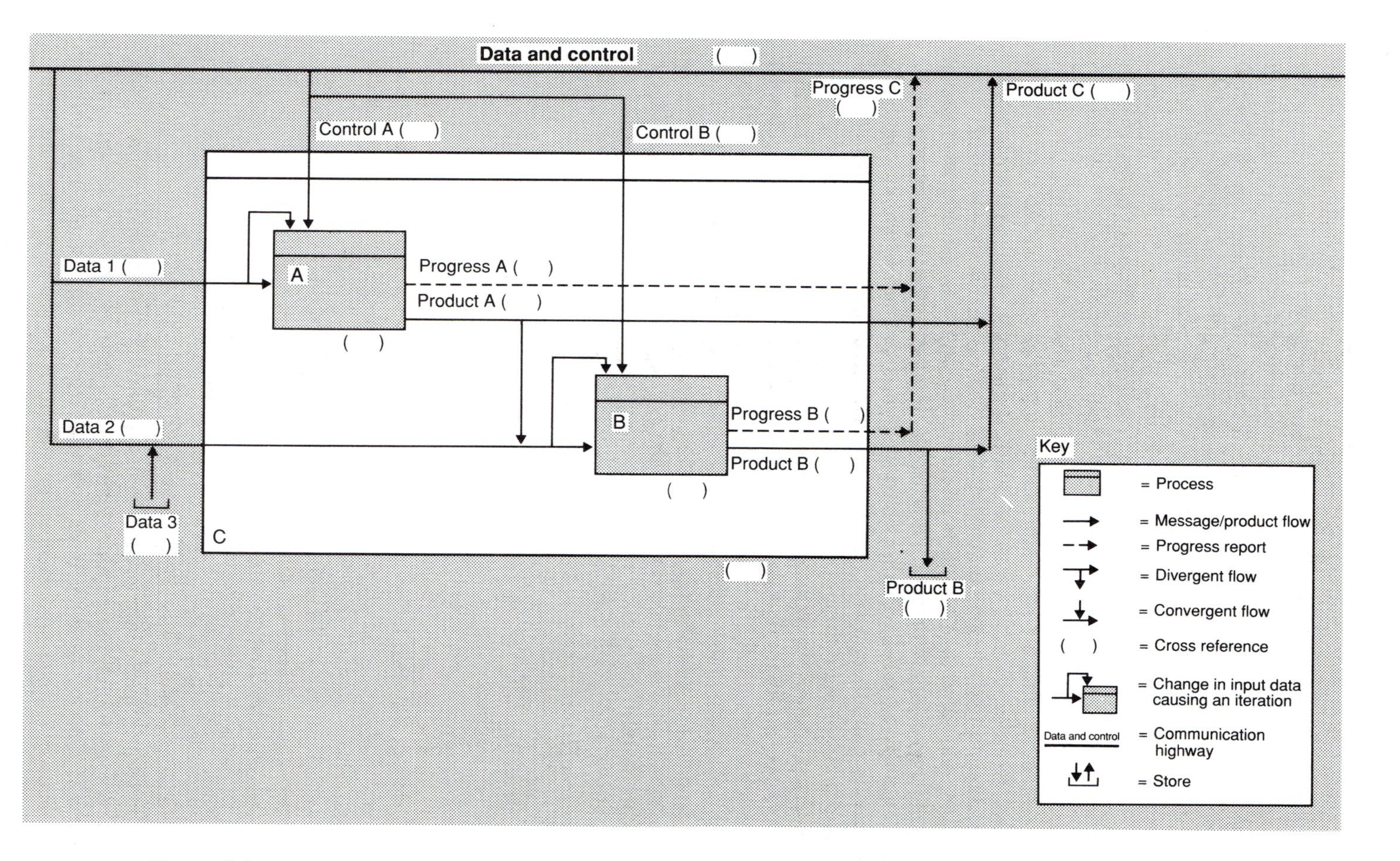

Figure 3.2 Action diagram – examples of basic notation. Reproduced from HUSAT with permission.

The extension of this technique through algorithms is currently being advocated (HUSAT, 1988) via reference to 'action diagrams'. These are defined as *'a record of System processes and message flows in the form of a 'structured flowchart' or 'algorithm' drawn to highlight the sequencing, nesting and control of processes and the converging and diverging of message and material flows. They are one of a class of similar notations derived from structured approaches to Software Engineering'*.

The complete notations used in these action diagrams cannot be described in detail here, but an example is provided in Figure 3.2.

Process charts

Probably the first attempt at charting a work process was the classic work of Gilbreth and Gilbreth (1921), who proposed the construction of process charts as visual representations of tasks. These charts were top-down flow diagrams of the task in which each behavioural element was classified and then represented by a particular symbol, termed a '*therblig*' (an anagram of Gilbreth). In their original proposal 53 of these therbligs were suggested (e.g. 'moved by conveyor', and 'inspected for quantity by weighing'), with a subset of 29 which were considered to be sufficient for most analytical purposes. Operations carried out by a person were represented as a circle which enclosed an operation number, and often a short description of the operation was given alongside. The American Society of Mechanical Engineers (1972) have reduced these symbols to five (see Figure 3.3) and this has now become the standard set of symbols which are used for process charts.

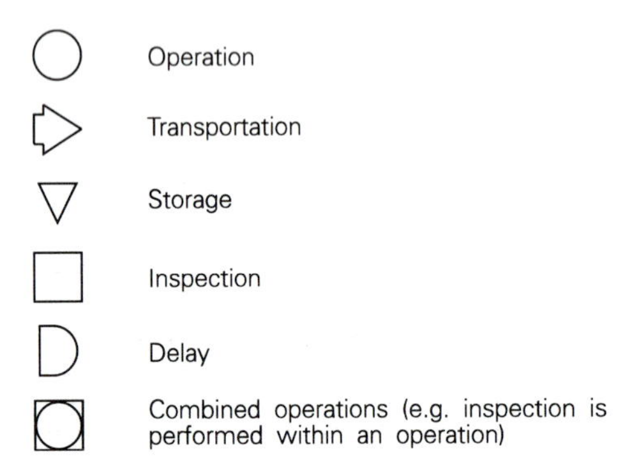

Figure 3.3 Basic set of process chart symbols

In their simplest form, process charts are a single, top-down flow line which links a sequence of activities or events. If it is appropriate, information can be recorded about the time taken for particular steps, or the distance a person or a product moves, and this can then be shown in a column to one side of the process chart. Figure 3.4 gives an example of a basic process chart. For some applications, distances moved or times taken, could be shown on the left.

If the analyst needs further information about the interactions between different individuals or systems who are simultaneously involved with the same task or process, it will probably be appropriate to adopt a multi-column format. In the additional columns, parallel process charts can be produced for different systems or individuals. One particularly useful version of this approach is the triple resource Chart, which has columns for the operator, the material and the equipment. Another version of this approach is the two-handed chart, which separately analyses and plots the movements of each hand (these are sometimes called Simo charts). However, it must be stressed that these two-handed charts are only really useful for short repetitive manual task sequences.

For activities which are different, but which must come together at some point, such as the production of a component which consists of several sub-assemblies, separate process charts can be produced for each activity, which only join when the activities directly link.

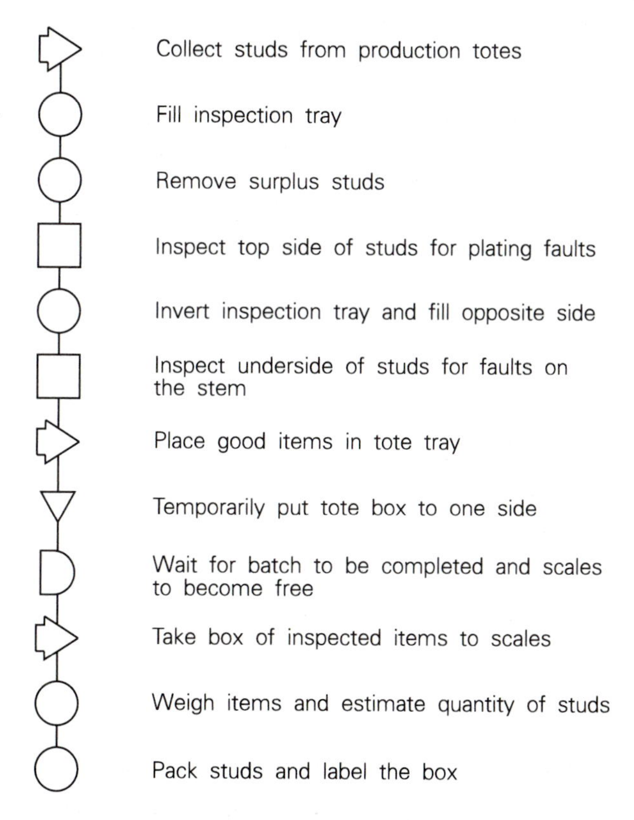

Figure 3.4 Process chart for a simple inspection task. Produced from data in Ainsworth, 1979.

For tasks with relatively short steps it can be difficult to accurately record and categorize the task steps, and so it will be particularly useful to prepare special data recording sheets. These present a column for each of the five activity classifications shown in Figure 3.3 so that the analyst merely has to mark the appropriate one. Columns are also set aside on this data collection form for other information which might be necessary.

Although process charts show the sequence of activities, they do not indicate graphically the relative time which is required for the different activities. However, it is possible to include a vertical time scale, in which case they become a vertical timeline. Further discussion of this application is given on pp. 137

Functional flow diagrams

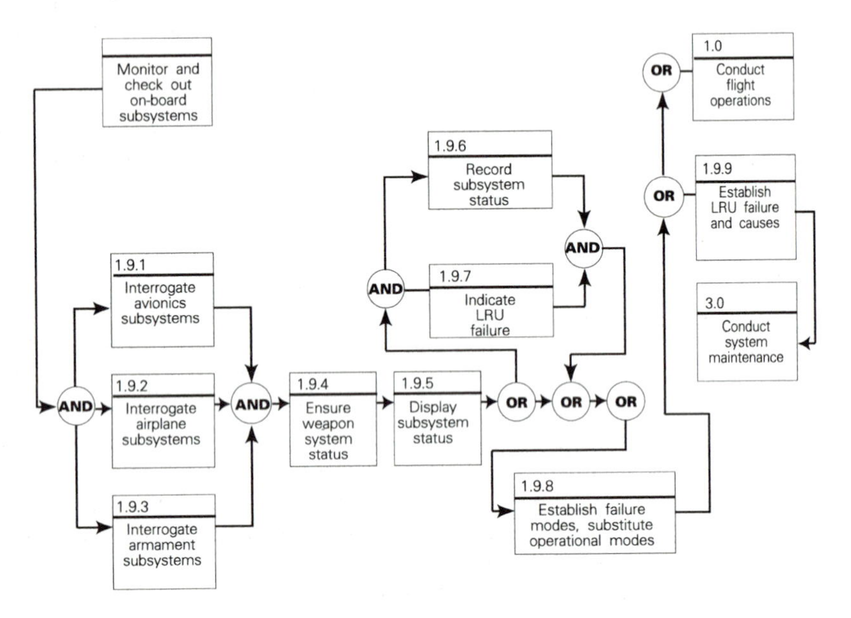

Figure 3.5 Functional flow diagram. Adapted from Geer, 1981.

Functional flow diagrams are block diagrams which illustrate the relationships between different functions. They are constructed by identifying the functions to be performed in the system, and then arranging these as a sequence of rectangular blocks which represent the interrelationships between the functions. It is usually most helpful to express these functions as verb–noun combinations and to identify each function by a numbering system which is presented at the top of each block. This numbering should provide some indication of the sequence of the functions. Where similar functions are repeated in different parts of a diagram (or linked series of diagrams), it is often helpful to use the same numbering for all of them.

Each of the blocks is linked by arrows, which indicate the sequence of activity. Where different functions have to be completed before the next function can be undertaken, these are linked through an **AND** gate. Where there are alternative courses of action, these pass through **OR** gates.

Functional flow diagrams can be drawn for individual tasks, or for complete systems. However, if they are being prepared for a complete system, it will usually be necessary to draw several diagrams at different levels, using a common numbering system throughout. For instance, high-level system functions could be shown on a first-level diagram, which could then be split into a series of second-level diagrams, etc. However, the way in which these functional levels are defined will depend upon the particular application, with more levels being required to get down to the same level of detail on more complex systems. An example of a functional flow diagram of part of a complex system is presented as Figure 3.5.

This technique is particularly useful for helping to determine how to allocate and order functions in a complex system, and to ensure that all the necessary functions are provided. However, the information in these diagrams is only of limited use for the detailed analysis of task steps.

Information flow charts

These are also known as decision–action diagrams and decision–action–information diagrams. They depict the progress through a system in terms of the decisions and actions which have to be performed. Decision elements are usually represented as diamonds and the possible outcomes are labelled on the exit lines. Although the questions are usually in the form of 'Yes/No' answers, this is not a specific requirement of the technique and multiple choice is permissible. An example of a decision–action diagram is presented as Figure 3.6. If it is accepted that the correct decision will not always be taken, it would seem natural to consider what outcomes might ensue.

Thus, for example, in the task of making coffee, an information flow chart may be used to ask if there is sufficient water in the kettle, and on the answer 'No' this would lead to 'add water'. An alternative approach would be to consider the consequences of heating the kettle without sufficient water, with sufficient water and with too much water. Tracing through the various stages produces a tree of possible outcomes, of which one or more will result in the successful completion of the task. Assigning probabilities to these courses produces event trees, which are described on pp. 178.

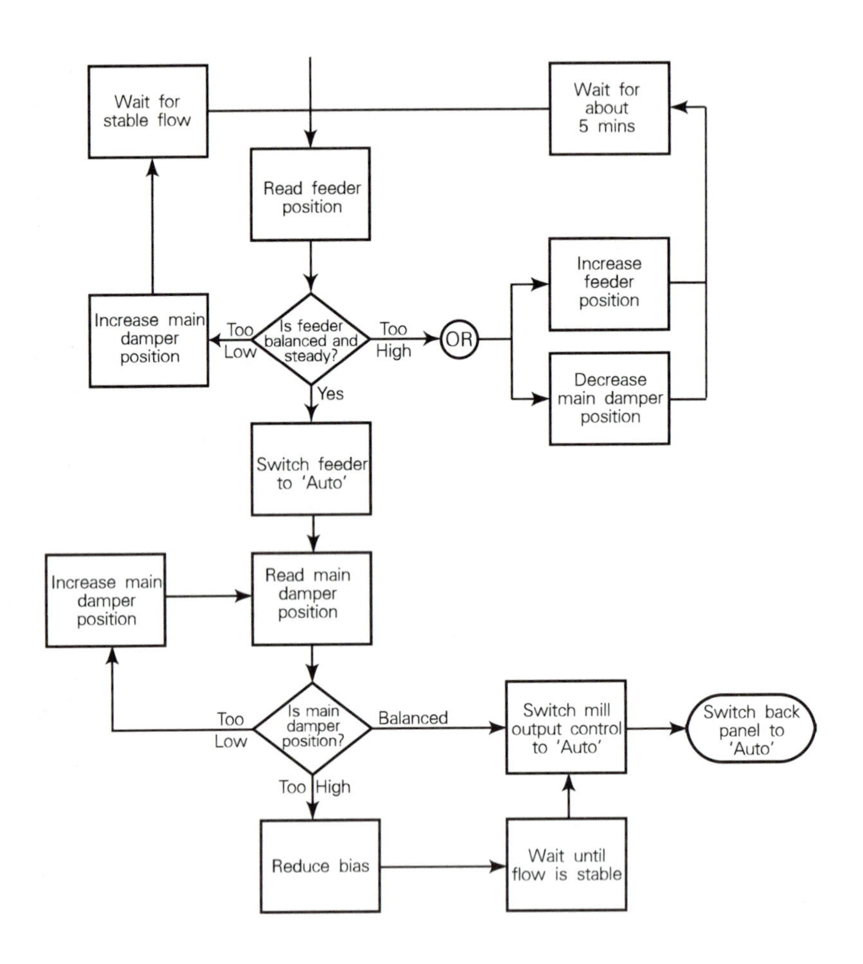

Figure 3.6 Decision-Action diagram. Adapted from Ainsworth and Whitfield, 1983

Murphy diagrams

When there is already a system and a catalogue of error events associated with it, an alternative approach can arise from retrospective examination of these errors. Such a method has been developed by Pew *et al.* (1981) and has been termed a Murphy diagram, based on the axiom of Murphy's Law, which states that 'if anything can go wrong, it will'. This method starts from a description of an accident (or other significant error sequence), and then an attempt is made to identify all the individual sources of error which occurred, using a standard set of eight Murphy diagrams to describe these errors. These Murphy diagrams define, at a general level, all the likely errors associated with the following decision processes:

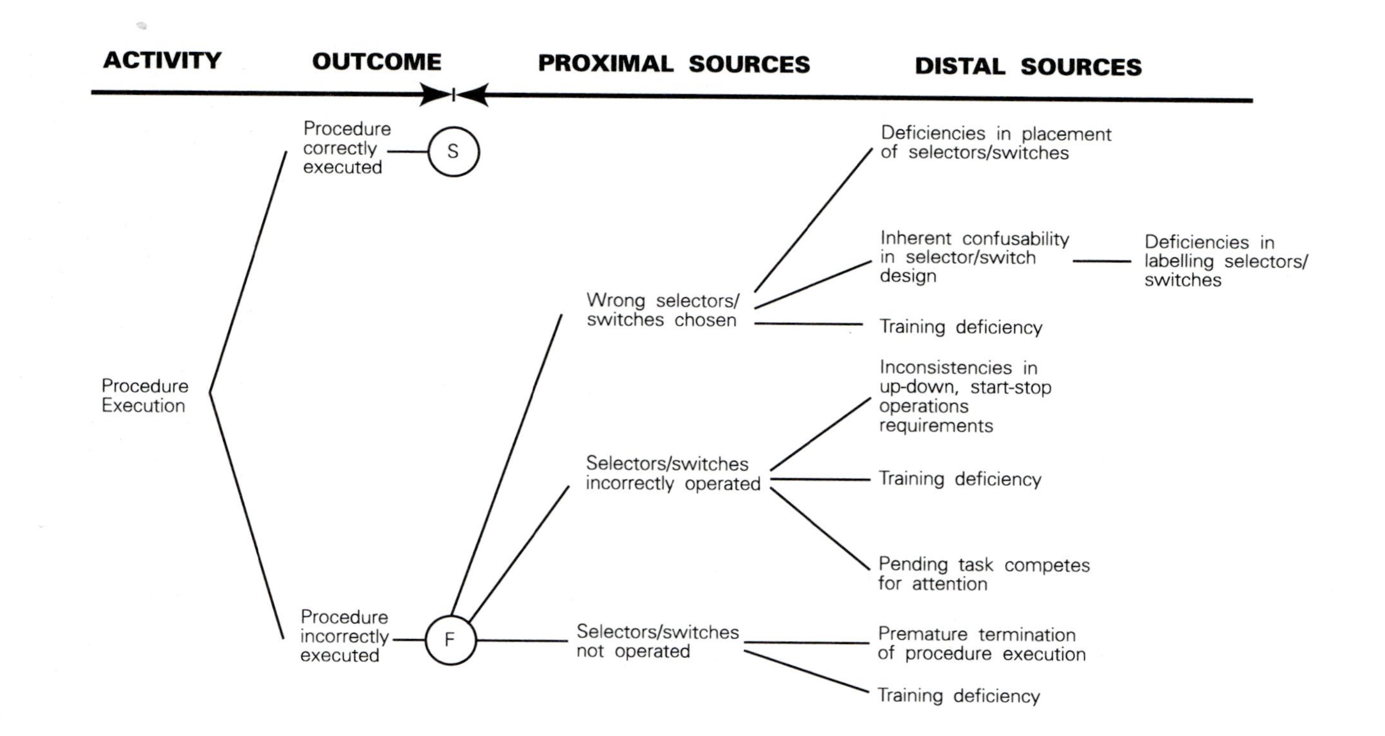

Figure 3.7 Murphy diagram for procedure execution. Adapted from Pew et al., 1981 and reprinted with permission. © 1981, Electric Power Research Institute, EPRI NP-1982. Evaluation of proposed control room improvements through the analysis of critical operator decisions.

- Activation/detection
- Observation/data collection
- Identification of system state
- Interpretation of situation
- Evaluation of alternatives
- Task definition/selection of goal state
- Procedure selection/formulation
- Procedure execution

Each Murphy diagram is a tree diagram, in which the initial decision process is successively split in three stages, according to the success of the outcome and then the immediacy of the source of error. The arrangement of a typical Murphy diagram for 'procedure execution' is shown as Figure 3.7. On each diagram, the first branch is always a dichotomy between successful and unsuccessful performance. Unsuccessful performance is then re-defined in terms of those sources of error which exert an immediate effect, and these are known as proximal sources of error. Each of these proximal errors is then further broken down into its underlying, or distal causes. Thus, for example, the proximal error of 'choosing the wrong selectors or switches', can be broken down into three distal problems, such as 'poor placement of those selectors or switches'. If appropriate, these distal sources of error can then be redefined so that they suggest remedies for the error.

The standard structure of Murphy diagrams enables an investigator to use structured interview (pp. 66) or table-top (pp. 155) techniques to rapidly identify and focus upon the specific sources of error which might have contributed to a particular human failure. If necessary, specific Murphy diagrams can then be produced, on which these sources of error can be clearly presented.

Pew *et al.* (1981) have stressed that the model upon which Murphy diagrams are based is neither predictive, nor prescriptive. However, despite these reservations, it must be concluded that it has certainly proved valuable in analysing events which have already occurred, and there appears to be no obvious reason why it could not also be effectively used in predictive analytical work.

Critical path analysis

Critical path analysis has been mostly applied in the area of management, in order to enable schedules to be designed with attention paid to the related events whose duration is largest. A similar analysis can be performed upon the different tasks which might have to be undertaken in a system in order to assess whether the non-completion of one task might affect the successful completion of another, and if so, what steps can be taken to avoid this. The tasks are represented as nodes, with links made between dependent tasks. The durations of these tasks can be annotated upon these links, and this should highlight any potential discrepancies in the system.

Petri nets

Petri nets were developed from state-transition diagrams (see Murata, 1989). The nodes represent a state or condition and they are connected by events which change the state of a system. Petri nets are particularly effective when the representation is not tied to a particular time base. They are also useful for demonstrating the relationships between the elements of a system. In mathematical terms they can be used to model information systems that are not deterministic, or else are considered to be parallel or concurrent, asynchronous, distributed, or stochastic. A simplified example is shown in Figure 3.8.

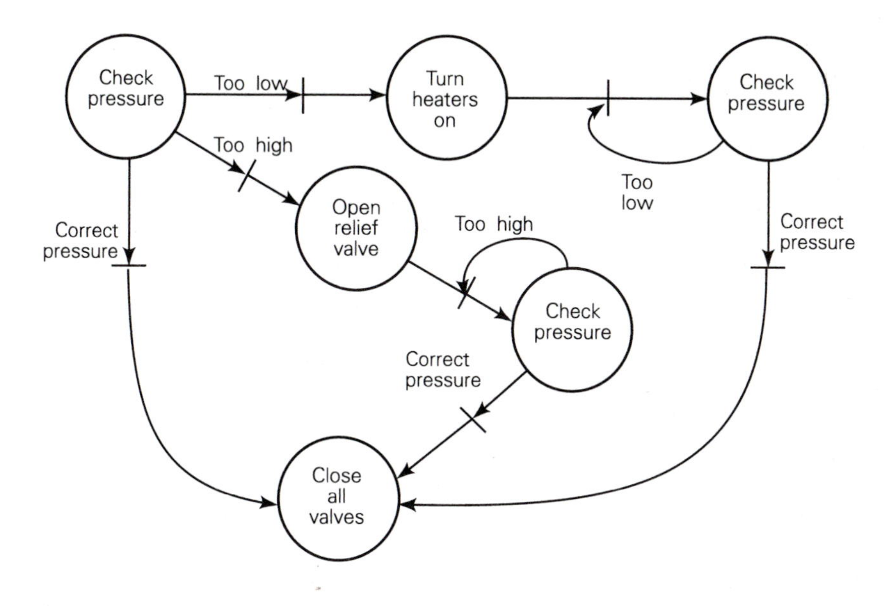

Figure 3.8 Petri net for the task of establishing a specific operating pressure within a vessel

Petri nets are a form of directed graph with arcs connecting two types of nodes called places and transitions (represented by codes and bars respectively), and usually representing states and events respectively. Places may be marked with tokens, and tokens determine the 'firing' of transitions within the Petri net. This representation of the system enables the modelling of different sequences of events depending upon initial system states and subsequent inputs to the system. More detail is presented by van Biljon (1988) in the context of modelling software interface dialogues.

Signal flow graphs

Although originally developed for the analysis of electrical networks, signal flow graphs have been applied effectively in operator studies. The method consists of identifying the important variables, and how they relate within the system. The analysis is conducted by selecting a system output variable, and then identifying all the variables which could influence this. The network presents the system variables as nodes connected by flow lines.

The time dependency of these graphs can be increased by annotating temporal tallies to the nodes and tracing these through the system. An example signal flow diagram is given in Figure 3.9 together with a block diagram of the same system (Figure 3.10). In this signal flow diagram, circles represent variables, and those with a 'V' on them are controlled variables, and those with an 'I' on them are input variables. Arrows with a plus or minus represent positive and negative dependency, respectively, and arrows with an 'I/S' type of notation are Laplace transforms. Dotted lines represent connections across the man–machine interface. Two arrows feeding into a variable (circle) via a node represent multiplicative dependency (e.g. valve orifice and water pressure in Figure 3.9), and otherwise (two arrows feeding directly into the circle) dependence is additive. For more detail see Beishon (1967) and Sinclair *et al.* (1966).

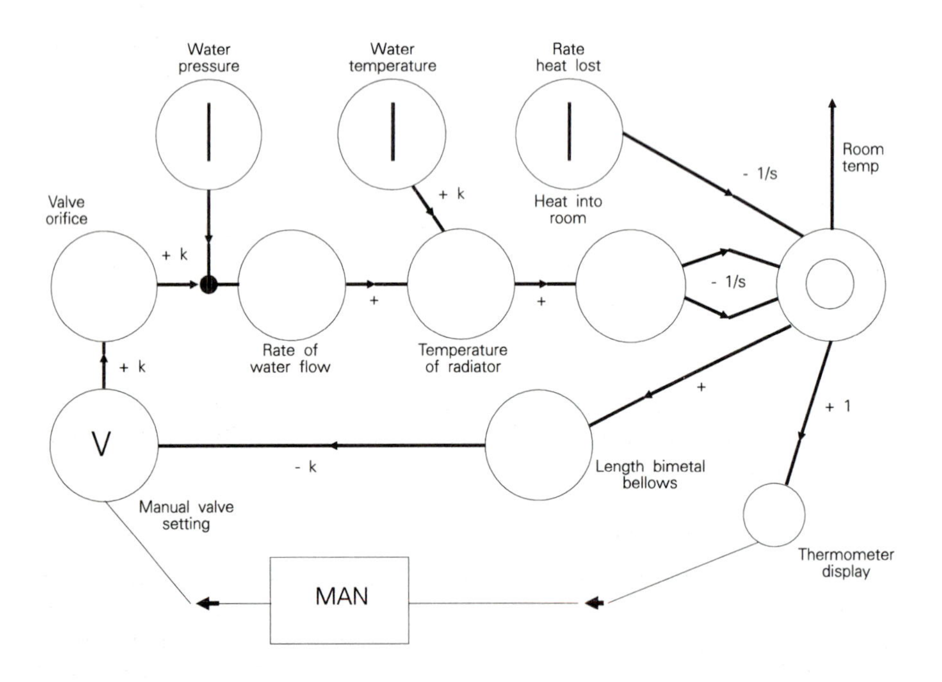

Figure 3.9 Signal flow diagram. Adapted from Beishon, 1967.

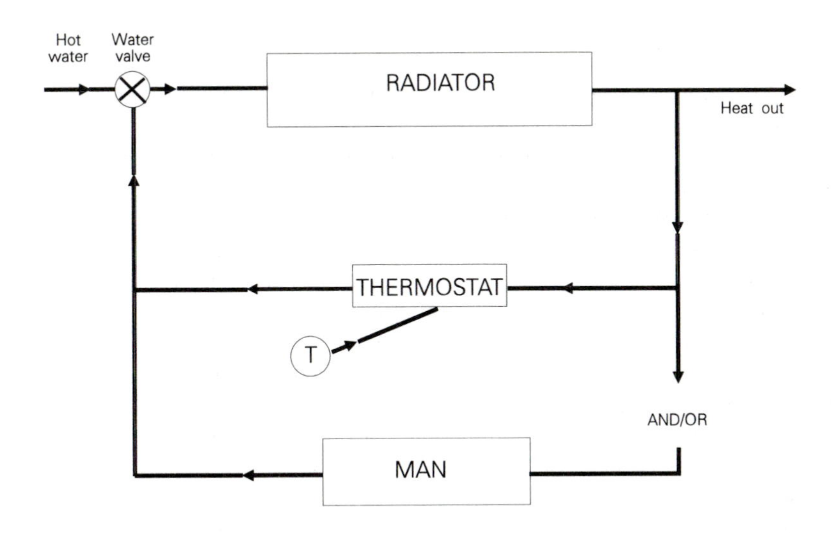

Figure 3.10 Block diagram. Adapted from Beishon, 1967.

Practical Advice

Flow charts are best applied to tasks of short or medium duration, particularly sequential, procedural tasks. For clarity and comprehension it is often useful to nest flow charts, so that on a master chart only the main tasks, or task elements, are shown, and then each of these is further broken down to create a more detailed picture on subsequent charts. Charting techniques can be particularly effective in the analysis of manual tasks, although they are also useful for other types of tasks.

Networks will usually prove more effective in the analysis of cognitive tasks, particularly where several strands of activity occur concurrently, or when the links between elements are complex. Networks also tend to become more appropriate as the task length increases.

Resources required

Few specialist resources are necessary for the analysis, other than a detailed knowledge of the technique's workings and access to data collected by methods outlined in Chapter 2.

Links with other techniques

Flow charts can also be useful for describing the action plans of hierarchical task analyses (pp. 104) and can feed into event tree analysis (pp. 178), timeline analysis (pp. 135), and operational sequence diagrams (pp. 125). Murphy diagrams can be a useful complementary analysis for most of the behavioural assessment methods (Chapter 5). Network techniques can be used to represent the output from verbal protocol analysis (pp. 71).

Advantages

- These techniques are capable of offering a very clear representation of a task, with the diagrammatic format facilitating comprehension. The form is often familiar, even to non-specialists of task analysis, and this may reduce training needs.
- The technique offers a breakdown of the tasks into clear units which allows the final representation of the analysis to be easily used.
- Flow charts and (usually to a lesser extent) network diagrams can incorporate the representation of the flow of time. Both are excellent at describing the conditions under which a task is carried out, and the requirements that need to be met to allow the successful completion of a task.
- The charts can be used to represent human tasks, system functions, material flows and equipment involved, so that many parts of the system can be represented in a common format. This can be useful, for example, when making decisions about the allocation of functions between man and machine.
- Flow charts can also provide information in a format that can be directly used for creating operational procedures. The step-by-step sequential flow of the diagrams makes them ideally suited for this.

Disadvantages

- As the cognitive (i.e. mental) content increases, chart representation becomes less satisfactory. In particular there may be many potential internal cognitive mechanisms and the real cognitive structures used by the operators may be 'opportunistic', rather than clear decisions made on strict criteria. Modelling such activity can be difficult.
- Only a limited amount of information can be assimilated from one chart but the temptation always exists, particularly with complex tasks, to depict an extra level of sophistication. This can lead to complicated diagrams, and it is preferable to show further detail on a nested chart.

References and further reading

Ainsworth, L.K. (1979) *The Utility of Signal Detection Theory as a Diagnostic Tool for the Analysis of Industrial Inspection Tasks.* Ph.D. Thesis, University of Nottingham.

Ainsworth, L.K. and Whitfield, D. (1983) *The Use of Verbal Protocols for Analysing Power Station Control Skills.* Applied Psychology Rept. 114, University of Aston.

American Society of Mechanical Engineers (1972) *ASME Standard – Operation and Flow Process Charts.* ANSI Y15.3-1974. New York: ASME. *A standard approach to process charting.*

Beishon, R.J. (1967) Problems of Task Description in Process Control. *Ergonomics*, 10/2, 177-186. *This paper gives a description of the signal flow-graph technique.*

van Biljon, W.R. (1988) Extending Petri Nets for Specifying Man–machine Dialogues. *Int. J. Man–Machine Studies*, 28/4, 437-458.

British Standards Institute (1987) *Specification for Data Processing Flow Chart Symbols, Rules and Conventions.* BS 4058, Milton Keynes: British Standards Institute. *A standard approach to process charting.*

Geer, C.W. (1981) *Human Engineering Procedure Guide.* Wright-Patterson Air Force Base, Ohio, USA: Rept. No. AFAMRL-TR-81-35.

Gilbreth, F.B. and Gilbreth, L.M. (1921) Process Charts. *Trans. ASME*, 43/1818, 1029-1050. *The definitive guide to the Gilbreths' original views on process charts.*

HUSAT (1988) *Human Factors Guidelines for the Design of Computer-based Systems, – Part 1*, HUSAT, Loughborough University, Loughborough. *The latest developments in the application of input – output models, which they term action diagrams.*

Kadota, T. (1982) Charting Methods in Methods Engineering, In *Handbook of Industrial Engineering*, Salvendy, G. (ed.), chp. 3.3. New York: Wiley and Sons. *A comprehensive survey of various charting techniques.*

Murata, T. (1989) Petri Nets: Properties, Analysis and Applications. IEEE Proc., 77/4, 541-580. *Applications and historic developments of the technique are discussed.*

Pew, R.W., Miller, D.C. and Feehrer, G.G. (1981) *Evaluation of Proposed Control Room, Improvements Through the Analysis of Critical Operator Decisions.* EPRI Rept. NP 1982. Palo Alto, California, USA: Electric Power Research Institute. *A description of a study which used Murphy diagrams.*

Sinclair, I.A.C., Sell, R.G., Beishon, R.J. and Bainbridge, E.A. (1966) Ergonomic Study of L.D. Waste-heat Boiler Control Room. *J. Iron and Steel Inst.*, 204, 434-442. *A range of methods are given for describing tasks, including an example of the signal flow-graph technique.*

Singleton, W.T. (1974) *Man–Machine Systems.* Harmondsworth: Penguin Books. *Chapter 3 provides a survey of many of the techniques mentioned in this section. Particular mention is made of input – output models, algorithms and charting techniques including flow and process charts.*

See Chapter 7, which deals with a form of functional flow diagram.

Decomposition methods

Overview

Task decomposition is a structured way of expanding the information from a task description into a series of more detailed statements about particular issues which are of interest to an analyst (Miller, 1953). It starts from a set of task descriptions which describe how each element within a particular task is (or should be) undertaken. For each of these task elements, the analyst then systematically gathers further information about specific aspects of the task (such as the time taken for that task element), or the facilities which are provided for it (such as the nature of the displays which are used). This information can then be presented for each task element using an appropriate set of sub-headings, so that the total information for each step is decomposed into a series of statements about limited aspects of the task.

The sub-headings which are used to decompose the task elements must be specifically selected by the analyst according to the purpose of the particular investigation, and so there are many different decomposition categories which could be used. One common use of task decomposition is to determine the control and information requirements of each step so that it is possible to ensure that all the necessary interfaces for a task will be provided.

It should be noted that some authors use the term 'tabular task analysis' to refer to the task decomposition process.

Application

Task decomposition is an information collection tool which is used to systematically expand upon the basic description of the activities which must be undertaken in each task element. Therefore, decomposition should be applied to task descriptions which have been produced by hierarchical task analysis or some other method. The technique can be applied to any task descriptions, regardless of whether they are superficial, or highly specific.

Description of technique

There are three steps which must be undertaken in any task decomposition.

Obtain a task description

The starting point for a task decomposition is a set of clear task descriptions for all the task elements which are associated with a particular task. Normally, these descriptions would be derived from a hierarchical task analysis, but for short task sequences, or certain highly proceduralized tasks, these task descriptions could be produced as a simple linear sequence of task elements, representing the expected, or observed, order in which they would be undertaken.

The actual format of the task descriptions does not matter, providing that they supply the analyst with sufficient information to ascertain what has to be done during each task element. In some situations, the analyst will only require tasks to be described very generally, while on other occasions much more detail will be required. Therefore, these task descriptions must be written at a level of detail which is appropriate for the analyst's purposes, and if any descriptions are not at that required level of detail these particular task elements must be broken down further, to produce new task descriptions.

Once the analyst is satisfied that all the task descriptions contain sufficient information, the task descriptions for each of the lowest level task elements should be rewritten on an information collection sheet.

Choice of decomposition categories

The originator of decomposition methods for task analysis was probably Miller (1953), who suggested that each task element should be decomposed into the following categories:

- Description
- Subtask
- Cues initiating action
- Controls used
- Decisions
- Typical errors
- Response
- Criterion of acceptable performance
- Feedback

The above decomposition approach still has relevance to the analysis of tasks today. However, this categorization does not cover all issues which might be of interest to an analyst, and so in order to address any other issues, it will be necessary to develop other decomposition categories. These can cover any aspect of the task, or the interface and job aids associated with it, but they must be chosen to answer some particular concern which is held by the analyst. In many cases, the categorizations will require information which can be obtained directly, by observing task performance, or by consulting the appropriate documentation. However, in many circumstances, behavioural information will be required which cannot be obtained directly, but must be inferred from a knowledge of the underlying psychological processes. A typical example of such a categorization would be 'skills required'.

It is convenient to consider three basic types of decomposition category (descriptive, organization-specific, and modelling), but it should be noted that this is only done for ease of description in this *Guide*, and for many applications, an analyst may find it necessary to use more than one type of category.

Descriptive categories

The most common type of decomposition category requires the analyst to elicit some form of description about a particular feature of the task, so that issues of concern to the analyst can be addressed. Any features of the task, or the resources and facilities (control, display, job aids etc.), which may be necessary to undertake it, can be used as the basis for such categorizations. However, it is essential to ensure that the analyst is clear as to the need for particular information, before establishing any particular decomposition category.

Although the decomposition categories should enable the analyst to address particular issues, they should not necessarily be restricted to one particular issue, but should be used to collate background information, which may or may not also permit other issues to be addressed. The analyst can then give each decomposition category a brief label, but the reason for collecting the information must always be kept in mind, because this will determine what is actually needed, with the label itself only serving as a cue to the analyst. For instance, the decomposition into 'controls used' could, according to the purpose of an analysis, require any of the following information:

- Name of control
- Location of control
- Type of control
- Grouping with other controls
- Label on control
- Direction of control movement
- Range of control movements
- Scale marking on control
- Accuracy of control movement

However, only a sub-set of this information would generally be required, dependent upon the issues of interest which, to continue with this example, could include the following questions:

- Can the operator adjust this parameter?
- Can it be controlled to the required tolerance?
- Can the operator use this control from his normal operating position?
- Could this control be confused with others?
- Could this control be inadvertently operated?
- Is it likely that this control action could be omitted?

As an aid to an analyst with the selection of descriptive decomposition categories, a taxonomy of such categories has been produced from a compilation of several studies, and this is shown as Table 3.1.

Organization-specific categories

These categories cover task information which is very specific to the organization for whom the task analysis is being carried out. They involve assessing whether some criterion, which is of particular importance to that organization, is being met. For instance, it may be necessary to assess whether any task elements violate a particular code of practice, or standard. Usually, such categories can be treated as a checklist item, which will allow the analyst to focus upon aspects of the task which do not meet some important criterion. Such criteria may not necessarily be clearly defined, as would be the case, for example, if a category such as 'potential plant licensing difficulties' was used.

These categories can also be used to indicate where further inputs are necessary from other personnel who are involved with the system development. For instance, by recording the 'safety classification of equipment', it would be readily apparent to safety personnel if a particular task element relied on equipment which was not classified to the necessary safety level.

Table 3.1 A taxonomy of descriptive decomposition categories which have been used in various studies

Description of task	Task difficulty
Description	Task criticality
Type of activity/behaviour	Amount of attention required
Task/action verb	**Performance on the task**
Function/purpose	Performance
Sequence of activity	Time taken/starting time
Requirements for undertaking task	Required speed
Initiating cue/event	Required accuracy
Information	Criterion of response adequacy
Skills/training required	**Other activities**
Personnel requirements/manning	Subtasks
Hardware features	Communications
Location	Co-ordination requirements
Controls used	Concurrent tasks
Displays used	**Outputs from the task**
Critical values	Output
Job aids required	Feedback
Nature of the task	**Consequences/problems**
Actions required	Likely/typical errors
Decisions required	Errors made/problems
Responses required	Error consequences
Complexity/task complexity	Adverse conditions/hazards

Modelling categories

If an analyst wishes to use some behavioural model to look at a particular task, it will probably be necessary to classify each task element into a structure which is consistent with the model. Before undertaking such a classification, the analyst must have a clear understanding of the model which is to be used, so that he or she can classify the task elements accurately and consistently. It may also be necessary to use a carefully controlled vocabulary to define behavioural elements, in order to ensure consistency and avoid the proliferation of too many items which are broadly similar.

Collection and presentation of decomposition information

Once an analyst has chosen the set of decomposition categories to be used, these should be listed on a set of information collection forms, using a set of clear sub-headings, and starting with the task description. Traditionally, these have been presented horizontally across the page as a series of narrow columns, with several task descriptions following below each other on the page (see Figure 3.11). Such a format permits a rapid overview of large chunks of a task, providing that each decomposition category can be presented very succinctly within a few words. However, if more text is required, such a format can give some practical problems, and so it may be preferable to adopt a less tabular presentation with each task element on a separate page, with the decomposition sub-headings positioned down the page, as in Figure 3.12.

Whichever information collection format is chosen, the analyst should then write down all the task descriptions in a clear, concise manner on the information collection forms. Then the analyst should systematically work through these forms, and gather the additional information which is required under each decomposition sub-heading. This will necessitate using many sources of information, such as:

- Task observations
- System documentation (e.g. system descriptions, piping and instrumentation diagrams, panel drawings)
- Procedures
- Training manuals
- Discussions with users, or technical experts

Eventually, when all the necessary information has been collected, the information collection forms can be used as a source of reference for further analysis or textual summaries can be prepared.

Practical advice

The partially-filled information collection forms can provide a useful context for technical experts who have to be consulted. Therefore, in order to assist them to gain a rapid overview of the context in which a particular question is being asked, the information collection forms should be kept as up-to-date as possible.

For some purposes it will be useful to collate all the information from a particular decomposition category into a structured list for subsequent analysis. For instance, if an 'information required' category is used, the analyst could generate an ordered list of information requirements, which could then be compared with a similar list of display sources in order to determine whether the system provided all the interfaces which were necessary for a particular task.

TASK ELEMENT	PURPOSE	CUE (for element)	DECISION	ACTION	DISPLAYS	CONTROLS	FEEDBACK	LIKELY ERRORS
3.1 Check primary circulators have stopped	Avoid pumping out water if the circuit is breached	Trip sheet	If any are still operating, stop them manually	Turn off manually after 20 secs.	Mimics at top of Panel F (should be positioned horizontally) Ammeters on desk E should indicate no current.	Key protected knobs at right side of desk E (not compatible with mimic grouping)	Mimics should change. Ammeters should show no current.	Step missed Mimics misread. Wrong dials read. Ammeters not checked.
3.2.1 Check de-aerator level.	If it drops too low,there will be insufficient pressure for normal feed to boilers.	Trip sheet.	If < 1.5ft, trip feed pumps.	Trip feed pump if necessary.	Analogue dial LI501 on Panel G, gives level.	Boiler pump discrepancy switch BFP1, or BFP2, on the 11kV supply.	Level on LI501 will increase.	Step missed. Wrong dial read. Dial misread. Feed pumps not stopped.
3.2.2 Check that one main D_2O main circ. is running	Cooling of D_2O to avoid circuit damage.	Trip sheet.	If none running, try to start one.	Try to start, if necessary.	Mimic at top of Panel E, and ammeters on D_2O pumps (analogue dials at bottom left of Panel E).	Two-position knob at top of Desk E, labelled 'No. 2' AND 'No.2 D_2O emergency circ pump'.	Mimics should change if successful.	Step missed. Mimics misread. Wrong knob turned. Wrong mimics read. Wrong dial read. Dial misread.

Figure 3.11 Example of a task decomposition

TASK ELEMENT
3.1 Check that primary circulators have stopped.

PURPOSE
To avoid pumping out any additional water if the circuit is breached.

CUE (for element)
Trip sheet entry.

DECISION
If any circulators are still running, try to stop them manually.

ACTION
After 20 secs, manually turn off the primary circulators.

DISPLAYS
Mimics at the top of Panel F (these are not grouped together well and would be better if they were positioned horizontally. Ammeters (labelled XI1, XI2, XI3 and XI4) on the desk section of Panel E, should provide a further indication of whether the primary circulators are drawing current.

CONTROLS
Key protected knobs at the right side of desk E (but these are not compatible with the grouping of the mimic displays.

FEEDBACK
If the primary circulators are stopped, the mimics should change and the ammeters should show no current.

LIKELY ERRORS
Step missed.
Mimics misread.
Wrong dials read.
Ammeters not checked.

TASK ELEMENT
3.2.1 Check de-aerator level.

PURPOSE
If it drops too low there will; be insufficient pressure for normal feed to the boilers.

CUE (for element)
Trip sheet entry.

DECISION
If < 1.5 ft, trip feed pumps.

ACTION
Trip feed pump if necessary.

DISPLAYS
Analogue dial LI501 on Panel G gives the level.

CONTROLS
Boiler discrepancy switch BFP1, or BFP2, on the 11 kV supply.

FEEDBACK
The level shown on LI501 will increase.

LIKELY ERRORS
Step missed.
Wrong dial read.
Dial misread.
Feed pumps not stopped.

Figure 3.12 A vertical format for a decomposition data sheet.

It must be appreciated that in large and complex systems, few if any individuals will have a full and accurate picture of the entire system. Therefore, the analyst should anticipate that there could be inconsistencies in the information received, and should guard against this by probing for inconsistencies and seeking corroboration wherever possible. If it is available, dates of each reference source should also be cited: and this will be particularly important for systems which are currently undergoing rapid development.

If any inconsistencies are discovered, it may be necessary to return to the technical experts with further questions. If these experts feel that their original statements are being challenged, they are unlikely to give their full co-operation, and therefore such approaches must be made sensitively. A useful approach is for the analysts to give a précis of their understanding of the system, and then to ask the technical expert to comment upon its accuracy, thereby relieving experts of the need to defend their previously-stated comments.

Resources required

Task decomposition can be a lengthy process, involving the use of many sources of information, and much cross-checking. In every case this will entail a considerable amount of the analyst's time. The other resources which are used will depend upon the particular study, but the most significant is likely to be the time spent with system users and technical experts.

Links with Other Techniques

Decomposition is a way of systematically gathering more information about a task, based upon a set of task descriptions, and so it should logically follow on from a hierarchical task analysis (pp. 104) or some other means of deriving task descriptions. However, it must be stressed that the boundaries between activities are not always clearly defined in task analysis, After starting a particular decomposition, for instance, it may well be found necessary to return to the task description stage, in order to further break down particular task descriptions.

The information within the decomposition categories will form the core of the background information for much of the subsequent information analysis of a task, and therefore at the time that the decomposition categories are formulated, it will be important to fully consider how the task information will be analysed.

Decomposition information is particularly useful for deriving control and information requirements lists. These define the control which a user must be able to exert upon a system, and the information which that user requires from it. These lists can then be compared to lists of the equipment which is available (i.e. control and display lists), which can be derived from interface surveys (pp. 223). It will then be possible to determine whether the system provides all the necessary user interfaces. An example of the use of a decomposition task analysis methodology to assess the adequacy of control/display provision is given in Kirwan and Reed (1989).

Advantages

- The structure of the decomposition should ensure that the issues of interest to the analyst are systematically considered for every task element.
- The format of the decomposition makes it relatively easy to determine when advice from subject experts is required, and the task descriptions should provide a convenient frame of reference for these subject experts, which facilitates responses from them.
- The method is particularly useful as a basis for the preparation of control and/or information requirements lists.

Disadvantages

- The main difficulty associated with task decompositions is that it may take a considerable amount of time and effort to obtain some of the necessary information, or to evaluate conflicting comments.

References

Kirwan, B., and Reed, J. (1989) A Task Analytical Approach for the Deviation and Justification of Ergonomics Improvements in the Detailed Design Phase. *Proc. Ergonomics Society Annual Conference*, 36-43.

Miller, R.B. (1953) *A Method for Man–machine Task Analysis.* Wright-Patterson Air Force Base, Ohio, USA: WADC Tech. Rept. No. 53, 137. A description of the first recorded use of task description.

See Chapters 9, 10 and 16.

Hierarchical task analysis

Overview

Hierarchical task analysis (HTA) is a broad approach to task analysis, which prompts the analyst to establish the conditions when various subtasks should be carried out in order to meet a system's goals. HTA produces a hierarchy of *operations* – these are basically the different things that people must do within a system – and *plans* – these are statements of the conditions which are necessary to undertake these operations. Since the resultant task description is hierarchical, the analysis can be developed in as little, or as much detail, as is necessary to deal with a particular task. HTA provides an effective means of stating how work should be organized in order to meet a system's goals. Thus, the activities of the human operator are linked directly to the requirements of the system. It can be used flexibly by the analyst as a framework for using other task analysis methods, or human factors expertise, to gain information, or to suggest system modifications. HTA was developed as a general method for representing a wide range of tasks, including those with significant cognitive aspects (Annett *et al.*, 1971). It was initially developed for training applications, but has since been developed and applied in a number of other contexts.

Application

HTA can be used to deal with specific issues, such as interface design, work organization, the development of operator manuals and job aids, training and human error analysis. It can be used systematically throughout the system life cycle to help designers articulate how tasks should be carried out. The analysis can then be developed and modified as design decisions are made.

Description of technique

HTA is a process of developing a description of a task in terms of operations – things which people do to attain goals – and plans – statements of conditions when each of a set of operations has to be carried out to attain an operating goal. The analyst begins by stating a goal that a person has to achieve. This is then redescribed into a set of sub-operations and a plan (or plans) governing when they are carried out. When the analyst is satisfied that this redescription is adequate, he or she must decide whether any of the sub-operations should be taken further. The reasons for stopping may vary in different situations as described below. If further redescription is warranted, then each operation needing further redescription is broken down into operations and plans as before. The analysis is represented in the form of a table, or a hierarchical diagram. An example of a hierarchical diagram is given in Figure 3.13, while a table representing the same task is included as Table 3.2.

Basic terms and concepts

To carry out HTA effectively, it is necessary to understand some basic concepts, such as *goals*, *tasks* and *operations*. These are used in a specific way within the technique, which may vary slightly with the ways that these ideas are used another task analysis methods, and so brief descriptions of HTA usage are given below.

Goals

Purposeful activities at work entail human beings trying to attain goals. Goals are desired states of systems under control or supervision (e.g.'*a plant operating at maximum efficiency and safety*', '*a shutdown executed safely and with minimum cost*', or, '*a quick pump changeover*').

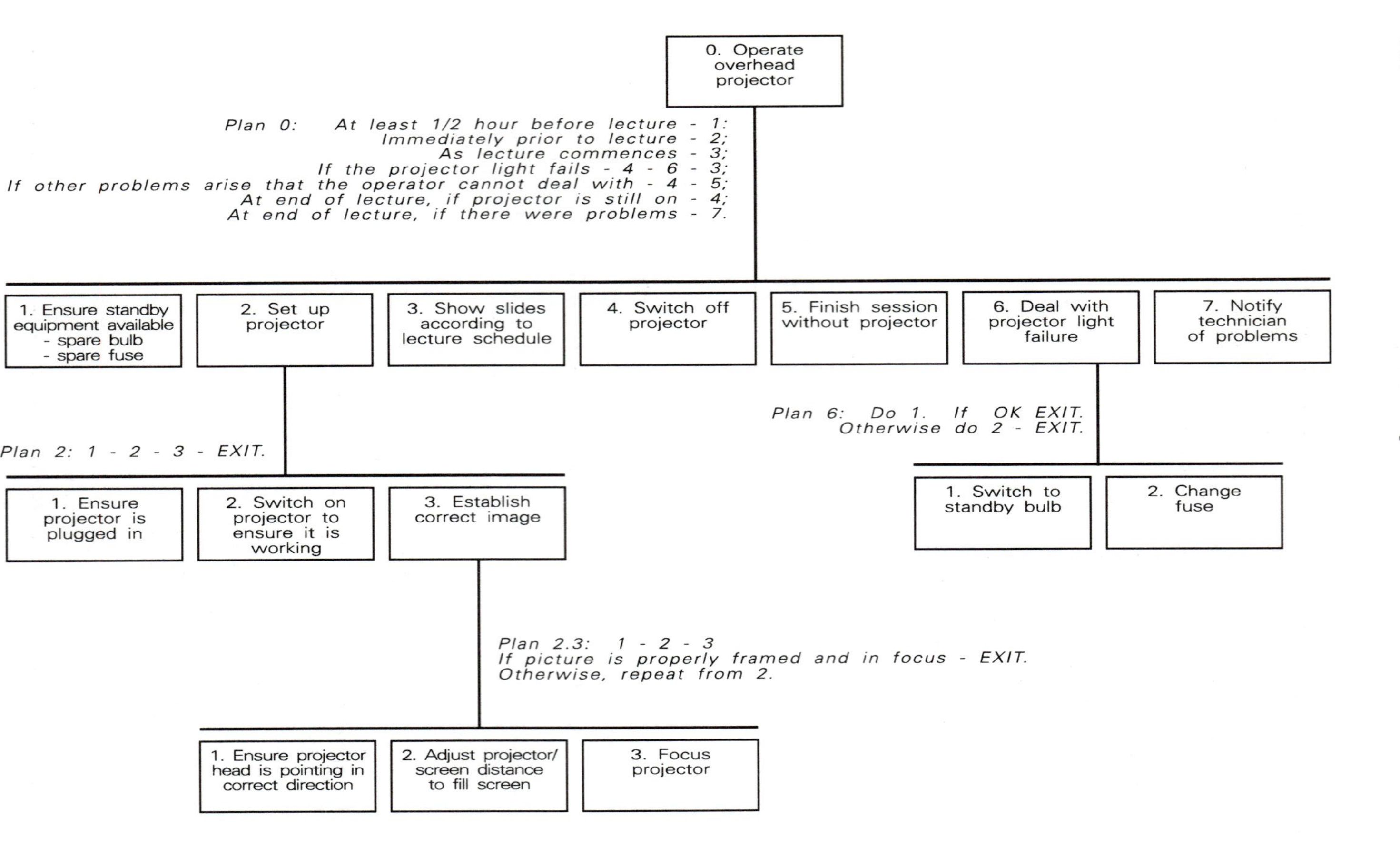

Figure 3.13 HTA for operating an overhead projector

Table 3.2 Hierarchical task analysis in tabular form. This represents the same analysis as Figure 3.13. Benefits of the tabular form are primarily that the analyst is not constrained by space in expressing plans, operations or notes concerning performance improvements.

Super-ordinate	Task Analysis Plans/Operations	Notes
0	OPERATE OVERHEAD PROJECTOR	
	Plan 0: *At least 30 mins before lecture: 1,* *Immediately prior to lecture: 2,* *As lecture commences: 3,* *If projector light fails: 4, 6, 5,* *If another problem occurs that cannot be dealt with: 4, 5,* *At end of lecture: 4 (if on:), 7 (if problem) EXIT*	
	1. Ensure standby equipment available (spare bulb, spare fuse)	Get replacement from technician
	2. Set up projector	
	3. Show slides according to lecture schedule	
	4. Switch off projector	
	5. Finish session without projector	This should never occur, but be prepared
	6. Deal with projector light failure	This is only fault lecturer should deal with personally
	7. Notify technician of problems	Failure to do this may cause problems for others later
2	SET UP PROJECTOR	
	Plan 2: 1 – 2 – 3 – EXIT	
	1. Ensure projector is plugged in.	
	2. Switch on projector to ensure that it is working.	
	3. Establish correct image.	
6.	DEAL WITH PROJECTOR LIGHT FAILURE	
	Plan 6: Try 1. If that does not work, try 2, then EXIT: *If 2 does not work, still exit, since this contingency is dealt with in plan 0*	
	1. Switch to standby bulb	
	2. Change fuse	

Table 3.2 (Continued)

Super-ordinate	Task Analysis Plans/Operations	Notes
2.3	ESTABLISH CORRECT IMAGE	
	Plan 2.3: 1 – 2 – 3. If OK, then EXIT If not OK, start cycle from 1 again	A demonstration and a bit of common sense is all that is really necessary here.
	1. Ensure projector head is pointing in correct direction 2. Focus projector 3. Adjust projector/ screen distance to fill screen	

Tasks

The method which is adopted to attain the goal, in any instance, is constrained by a number of contextual factors, including: availability and cost of suitable materials; equipment and facilities provided; availability and costs of services; constraints such as time and legal obligations; and managerial and staff preferences. A person's task may be seen as trying to attain a goal in a particular context, by utilizing the facilities and resources available while overcoming any constraints which may have been imposed.

Operations

Operations may be defined as *any unit of behaviour, no matter how long or short its duration, and no matter how simple or complex its structure, which can be defined in terms of its objective.* This means that an operation is something which people actually do to attain a goal. Implicit in the behaviour entailed in an operation is:

- The capability of the person to carry out actions appropriate to attain the specified goal
- The capability to locate and encode information necessary for selecting appropriate actions
- The capability to locate and encode feedback (i.e. more information) which can show the person whether the goal has yet been attained, or, what else needs to be done to attain it. This action–information–feedback structure can be used systematically within HTA in order to focus on aspects of performance difficulty. For example, it might become clear that people cannot perform effectively because there is no adequate feedback to enable them to judge the effectiveness or consequences of their actions.

Task analysis

At this point, the term *task analysis* should become clearer. It is the process of critically examining task factors – the operator's resources, constraints and preferences – in order to establish how these influence human operations in the attainment of system goals.

Hierarchies of goals and sub-goals

HTA exploits the fact that goals can be described at various levels of detail. For example, to 'inspect a vessel', we may need to 'obtain a permit to work', 'ensure the vessel is empty, 'ensure vessel has been evacuated', 'put on safety equipment', 'open a man-lid', and so on. Each of these activities may in turn be broken down – to ensure the vessel is empty, we may need to 'inspect instruments to determine zero levels' and 'open a drain valve', etc.

Plans and the organization of sub-goals

From the operations just described in inspecting a vessel, it becomes obvious that personnel would have to don safety clothing before opening the drain valve. Hence it is important to record when things are done – this is the function of plans. Plans state the conditions which specify when each of a set of sub-goals should be carried out. When to do things is often as important as what to do. Plans can include sequences of actions or sets of actions conditional upon time, process condition or instructions. Figure 3.14 shows a representative HTA of a continuous process control task, while Figure 3.15 shows a representative HTA of a batch process control task. These figures illustrate plans which are typically encountered in the analyses of process plants.

Stopping rules

Redescribing goals into plans and sub-operations should only be undertaken where necessary – otherwise a great deal of time and effort is wasted. A useful feature of a hierarchical description is that the analyst can stop where he or she feels that it is justified – sometimes operations need to be described in considerable detail; sometimes they can be described very briefly. The analyst may decide which stopping rule to utilize for a particular task analysis.

One important stopping rule, which is useful in training applications, is the $P \times C$ rule. P refers to the probability of inadequate performance of the operation concerned and C refers to the cost of inadequate performance. The stopping rule entails that redescription of an operation/task is unnecessary if the probability of inadequate performance multiplied by consequences or costs is acceptable. Therefore, for each operation/task, the following question may be asked: 'Is the operator capable of carrying out the operation to attain this goal to the criterion implied by the $P \times C$ rule, or, can we think of any acceptable measures to ensure this competence'? Cost in this context is not simply a direct financial consideration but should include anything the system values (e.g. personal accident, damage to equipment, degraded product). $P \times C$ can rarely be treated as a rigid formula suited to numerical analysis, since determining a complete value for C is rarely possible, while P suggests the rigour and assumptions of a human reliability analysis. However, it remains an effective rule-of-thumb for stopping, since it prompts the analyst to consider both of these factors in deciding whether analysis need continue. For example, however likely a person is to make a mistake, it hardly matters if the consequences of inadequate performance are trivial. Conversely, high estimates of C prompt attention even when P is low.

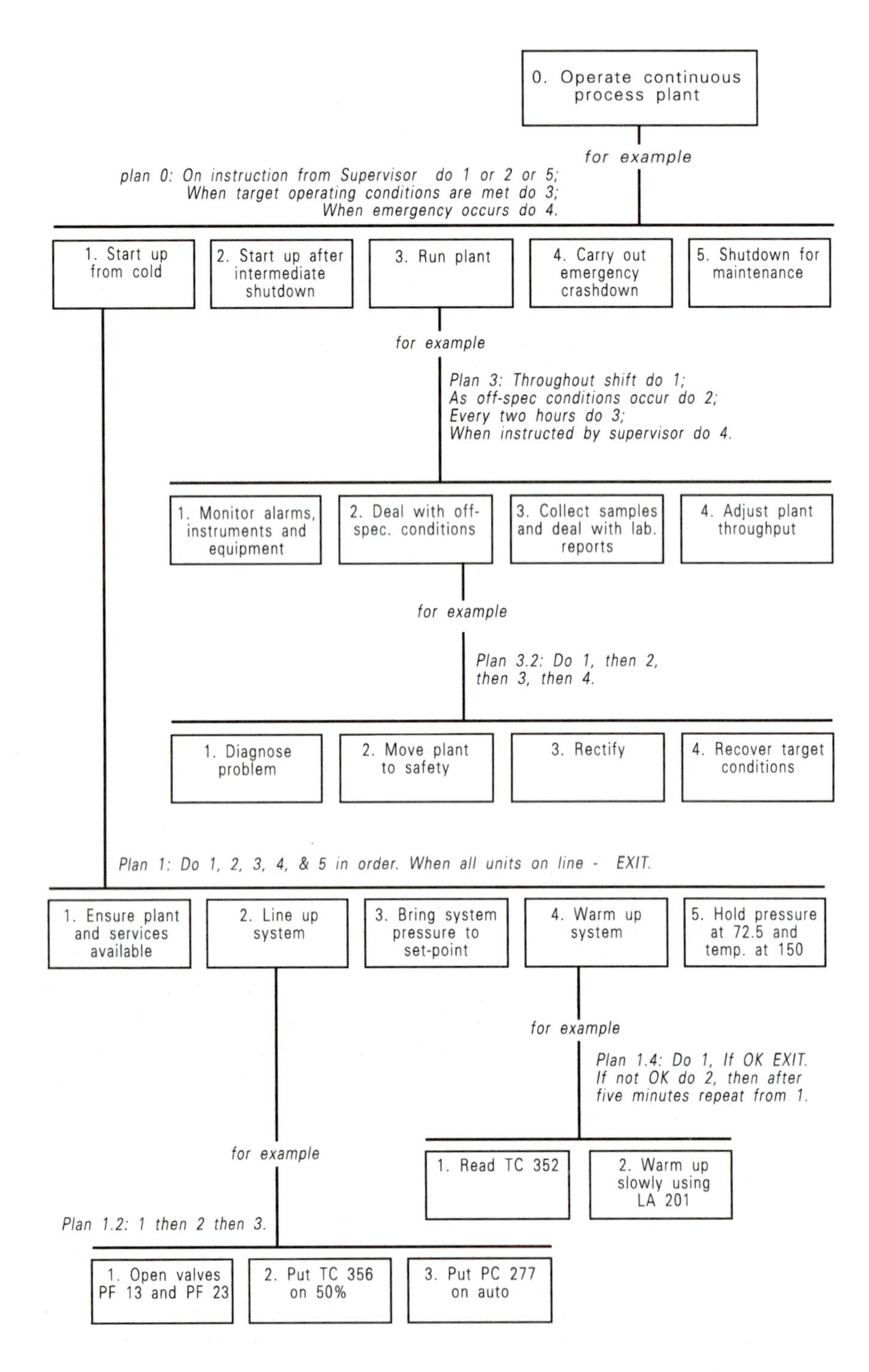

Figure 3.14 HTA for a continuous process control task

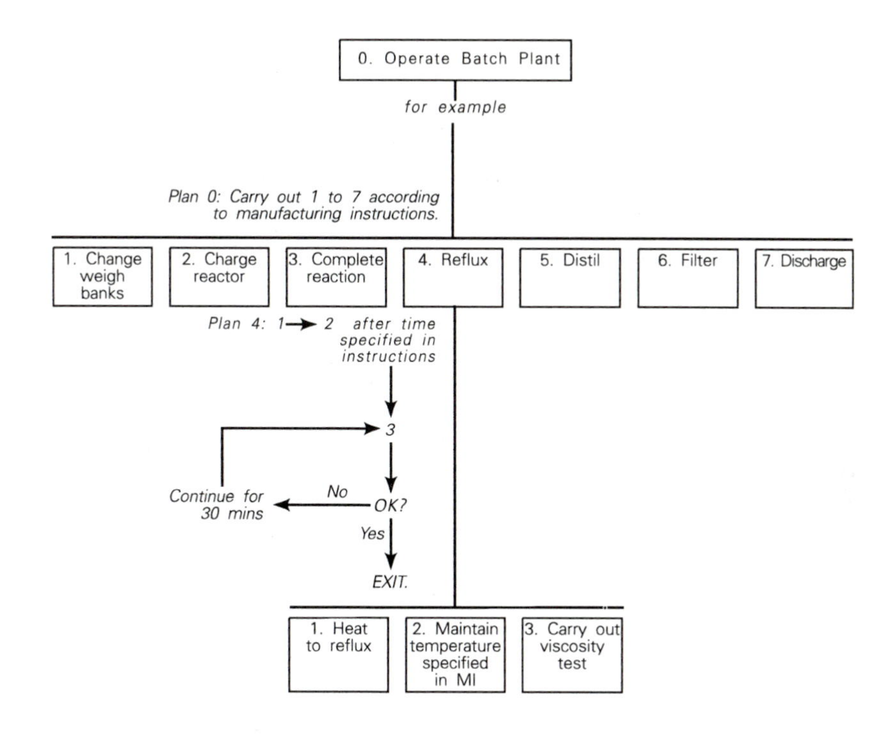

Figure 3.15 Hierarchical task analysis for a batch process

The P × C rule is not appropriate to all situations. In a training context the P × C rule is usually suitable because it is possible to stop anywhere that it is possible to ensure satisfactory performance. However, in investigations of how people interact with a system, such as in the design of displays, or the development of operator manuals, redescription must continue until one describes interfacing responses (i.e. operations which explicitly refer to the interface between the operator and the system), such as 'switching on pumps from a control panel', 'controlling valves on plant', or 'communicating with people via an intercom'. In this way, the precise detail of operator action is described.

Representation of hierarchical task analysis

HTA provides a set of rules for effective task description which also serve to guide the process of data collection. However, the HTA must be recorded in order that its adequacy can be checked and to provide a record of design suggestions, or concerns, made during the analysis. To do this tables and diagrams can be used. Both have their benefits, and it is often best to use both approaches to record a task. The main concern in recording tasks is to ensure clarity and consistency, and to enable clerical convenience.

Hierarchical diagrams

Figure 3.13 shows the hierarchical diagram of operating an overhead projector. Diagrams are very useful as a means of conveying the structure of the hierarchy. The double underlining of boxes is used to emphasize that no further redescription is offered. Diagrams are limited if the analyst wishes to make copious design notes and comments.

Numbering the analysis

It is important to adopt a rational numbering system to keep track of the work. Any numbering system that works may be adopted. The system used in Figure 3.13 has proven quite effective. The overall goal is numbered 0. Its sub-goals are numbered from 1 to whatever is necessary (7 in this case). The plan governing its sub-goals is given the same number as its super-ordinate goal (in this case *plan 0*) and can then refer to the sub-goals in terms of their numbers. Operation 2 is redescribed and its sub-goals are numbered from 1 to 3; its plan is labelled *plan 2*. Its operation 3 is further redescribed; its sub-goals are numbered 1 to 2; and its plan is given the label *plan 2.3*. Alternatively a unique number may be given to each operation. Some analysts may find it clearer to write down a full number for each operation, rather than just a single digit (e.g. they may prefer '*switch to standby bulb*' to be numbered 2.3.1). An objection to this is that numbers can become unwieldy if the analysis is taken to several levels – often the case in process control tasks. Another advantage of numbering operations with single digits is that it is easy to edit an analysis numbered in this way, because modifying the numbers of higher level operations will not affect the numbering of lower level operations. The numbering system described here has been very effective in a considerable number of applications but no matter what approach is taken to numbering, it is most important that the analyst clearly states how the adopted numbering system works for each HTA recorded.

Tabular formats

Table 3.2 shows one way of representing an HTA in a table. Diagrams are more easily assimilated by people, but tables are more thorough, because detailed design notes can be added. It is often best to use a tabular format both to record and communicate the analysis.

Functional analysis system technique (FAST)

FAST may be seen as a sub-technique belonging to HTA, and is useful in the early stages of design (e.g. when allocating functions between operator and machine). FAST was developed as a quick means of analysing and structuring problems (Creasy, 1980), and looks similar to HTA in appearance, but is solely directed at system *functions*. Each function is defined in terms of a statement containing an item (e.g. shaft drive), a verb (e.g. transmit), and a noun (e.g. torque), which together define the function of the items (it may sometimes be useful to give the full sentence as well). All functions for a given system are first identified and defined as described (item–verb–noun), and then the functions are arranged from left to right (or hierarchically). For example, starting with a basic function, the question 'how do I' (verb) (noun)?', (e.g. 'how do I transmit torque'?). The answer to this question will be in terms of other related system functions. The second question that is asked is 'why do I' (verb) (noun)?' (e.g. 'why do I transmit torque'?). The 'how' question, if modelling the problem hierarchically, denotes the functions that should appear underneath the current functions being analysed. The 'why' question denotes the functions that should appear above the current function being analysed. When all how and why questions can be answered by the hierarchical layout achieved, then the diagram is complete, and the analyst will have a coherent model of the system, including the inter–relationships of the functions and the rationale of the system's functional design. This is a sound basis upon which to consider the allocation of functions. The use of **AND** and **OR** gates (see pp. 188) can also be usefully added to the FAST methodology, and can be beneficial when examining complex systems with redundant or diverse sub-systems (e.g. a main feed water pump and a standby or reserve pump).

FAST appears to be a useful and quick variant of the HTA concept, and is probably most pertinent in the early stages of design.

Practical advice

The process of first collecting data and then organizing it is not recommended when carrying out HTA, because it will result in a mass of unstructured information. Although HTA is primarily a prescription for describing tasks in terms of hierarchies of goals, operations and plans, it is also immensely valuable in organizing the search for relevant information, examining documentation, discussing the tasks with experts, observing job holders and issuing questionnaires. Handling the process of conducting an HTA comes with experience, but some guidance can be offered.

The first stage in any analysis is to establish its purpose and the rules to be used for deciding where analysis should stop. The next step is to state the main goal to be analysed. This is not always easy and one may find oneself changing one's view regarding the precise goal to be analysed. The chosen goal may be too narrow. This can occur, for example, when examining a task that is carried out as part of a larger team effort – it is often necessary first to consider the broader role of the team in order to establish the precise boundaries of the main task of interest and how the operator should interact with other personnel in the system. Sometimes the chosen goal is too broad, or too imprecise. Stating an initial goal often entails examining the wider task context and its constraints.

Analysing complex tasks, especially those which entail considerable human skill, is usually best done in collaboration with a task expert. Anybody knowledgeable about a particular job, or task, might be used. This includes an experienced job incumbent, or a person who is responsible for the task, such as a manager, supervisor or engineer. Information can be collected from a variety of sources, including other forms of task analysis, such as talk-throughs (pp. 162), verbal protocol analysis (pp. 71), video recording (pp. 55), activity sampling (pp. 41) and critical incident technique (pp. 47). It is rarely a good idea to rely solely on observing performance as a prime source of task information, especially in tasks involving substantial decision making, since the operator's intentions and information gathering strategies are seldom apparent.

In order to ensure that the task analysis is reliable, it is useful to use different sources of information whilst developing and rechecking the task analysis. The analyst might start off talking to an operative or user, then check certain information from a document, then re-examine the analysis with a technical expert or a supervisor. Documents which may be available as sources of information could include job descriptions, operating manuals, safety and emergency procedures, maintenance records, manufacturers' manuals, etc. Reference to such documents may be useful at early stages in the analysis to inform the analyst about the overall nature and breadth of tasks which have to be carried out. Later, as the detail of the task becomes established, such documents serve to provide crucial information. The analyst will be directed to various parts of the documents in accordance with the emerging HTA.

There will often be many ways that a *super-ordinate* task element can be organized and broken down, but the analyst should be wary about using too many task elements to describe the super ordinate task, because this will tend to confuse, rather than clarify. If it is found that a plan is a straightforward fixed sequence plan, it may be easy to cope with a reasonable number of subordinate operations (e.g. as many as nine or more). If, however, the plan entails a number of choice points, it is useful to keep the number small (say three or four), so that options are clearer. It is often easier to state complex plans in terms of a hierarchy of simple plans (each governing only three or four operations). Together, these simple plans can encompass very complex

When the redescription of a particular task element has been completed, each subtask should be considered in turn to determine whether further redescription is justified. This is where the analyst may need to consider the stopping rule to decide whether existing performance can be carried out satisfactorily under the prevailing task conditions. Where further redescription is not justified, this should be clearly indicated, and reasons for stopping should be noted. In this way, the whole task analysis can be tackled systematically. Figure 3.16 shows the flow of the HTA process in an 'ideal' form, though this should not be followed rigidly. Often it is necessary to check and modify work already done. Sometimes it is necessary to work upwards from a given set of operations and tease out an effective hierarchy of plans to control them.

When the analysis has been completed, it should be checked for adequacy and consistency. It is best to take task experts through the analysis step by step, rather than leaving them to proofread a table, or hierarchical diagram, in their own time, because otherwise they may only skim over the analysis. Reviewers of task analyses may see familiar words and project their own meanings onto them which may not be the intended meanings. It is therefore necessary to trace through each plan in a conscientious manner.

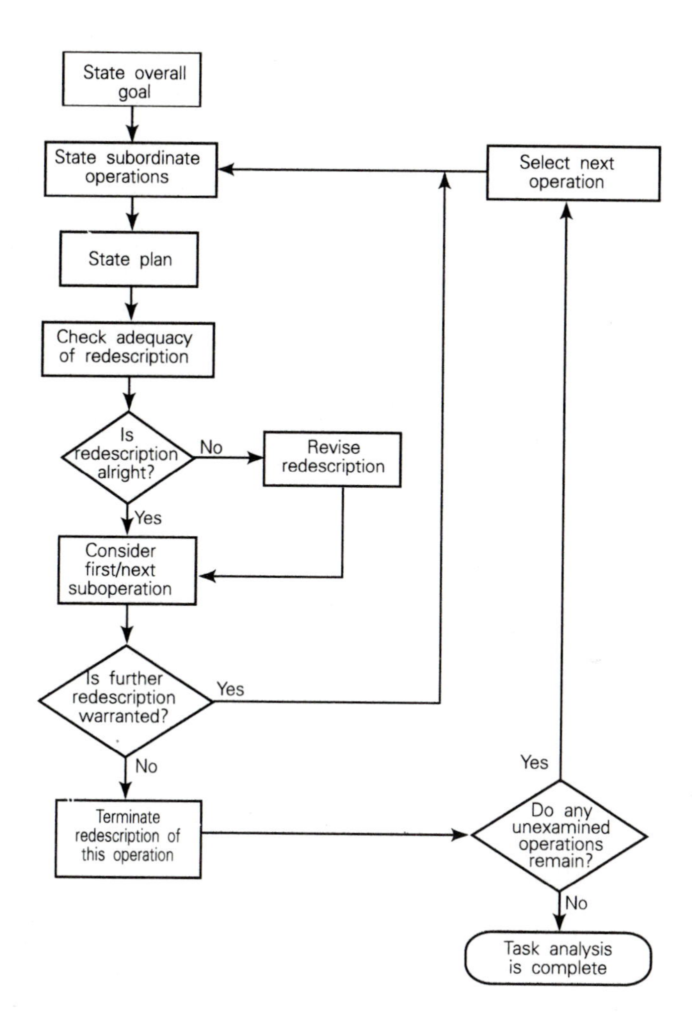

Figure 3.16 The flow of activities in redescription during HTA. In practice it may not be possible to follow this flow exactly (adapted from Shepherd, 1985).

In any diagrammatic presentation of the HTA, the analyst should try to avoid crowding the diagrams, by using several diagrams in preference to a single unwieldy diagram. However, if this is done, these diagrams must provide some overlap, and sets of diagrams should preferably all start at the same level and be clearly labelled. For instance, a first diagram could show the first levels of an HTA, and then a set of diagrams could be provided which each started at one of the third-level tasks.

Resources required

An HTA will typically be undertaken by one analyst, although more than one analyst might collaborate for larger tasks. In addition to the analyst(s), there is need for collaboration with operating personnel. It is useful to think of the following roles or functions that must be carried out: there should always be a *client* for whom the work is being done; the client should make available one or more *informants* or *experts* to provide information for the task and, eventually, to confirm the accuracy of the analysis; and the client must accept responsibility for providing the technical information and managerial preferences.

Because of the necessity of relying upon the co-operation of operating personnel, who have other demands on their time, it is useful to agree some form of contract with the client to ensure that such co-operation will be forthcoming.

HTA requires no special equipment other than items of stationery for recording analysis. However, various computer applications can be helpful, especially in preparing tables and hierarchical diagrams. For example, 'text outliners', which may be found in a number of word-processors or as stand-alone applications, are helpful in preparing tables. Object-oriented drawing packages are particularly useful in preparing hierarchical diagrams, since they are very easy to revise.

Courses can be arranged where guidance can be given by experienced HTA practitioners. Alternatively, it is easy to practice HTA using simple tasks, such as a small procedure (e.g. less than 20 task steps) or various domestic tasks (e.g. decorating, or maintenance of household equipment). It is also obviously necessary to practice on real applications, since the novice analyst should experience the manner in which information must be gathered and how practical decisions for progressing analysis are made in a real context.

When practicing HTA on a real application, it is best, first of all, to try to negotiate something reasonably straightforward (e.g. such as a moderately sized procedure which is fairly well defined), working with people who are likely to be sympathetic to the aims of HTA. Once mastered, the skills may be used in a wide range of situations to support a wide range of purposes.

Links with other techniques

HTA serves as a useful framework in which other task analysis techniques, or knowledge/expertise in human factors, can be applied effectively. Within the process of conducting HTA there are frequent requirements for collecting information to identify when operators must do things, how quickly they should do things, what sorts of difficulty they may have in carrying out the task(s), etc. This information may be established by discussion with an informant (e.g. pp. 66), or by simple observation (pp. 53). On the other hand, it may be beneficial to adopt any of the many approaches of data collection, several of which are included in this *Guide* as task analysis methods (see Chapter 2). For example, verbal protocols (pp. 71) can be used to examine a complex cognitive operation, prior to organization of operations and plans.

Another frequent activity within the context of HTA is making design decisions. Often, these are made on the basis of judgement by the analyst, but they can often be helped by applying appropriate checklists (e.g. see pp. 217), or decomposition methods of task analysis (pp. 95).

HTA is a general approach to examining tasks based on the redescription of operations. There are a number of other published hierarchical approaches which adopt similar structures to HTA, but which incorporate specific constraints to deal with a particular class of problem, (e.g. FAST). Several other approaches focus on commercial rather than industrial applications. These include approaches in the area of human computer interaction, such as Task Analysis for Knowledge Descriptions (TAKD – see Appendix) which focuses on the recording of task knowledge appropriate to human computer interaction tasks. Another method is the Goals, Operations, Methods and Systems approach (GOMS – see Appendix) which uses a stereotyped set of actions to describe how computer users interact with their machines.

Advantages

- HTA is an economical method of gathering and organizing information since the analyst needs only to develop the parts of the hierarchy where it is justified.
- The hierarchical structure of HTA enables the analyst to focus on crucial aspects of the task within the context of the overall task.
- HTA provides a context in which other specific approaches to task analysis (e.g. for data collection or for modelling design possibilities), may be applied to greater effect.
- HTA is best developed as a collaboration between the task analyst and people involved in operations. Thus, the analyst should operate in accordance with the perceived needs of line personnel who are responsible for effective operation of the system.
- HTA offers two distinct training benefits to people engaged in the analysis. First, analysts can use the technique rapidly to gain insight into processes and procedures entailed in plants and organizations generally – some companies have used HTA to help graduate trainees gain insight into company operations. Second, it has training benefits for people collaborating with the analyst, since they are required to express how they think tasks should be carried out, thereby articulating their understanding of systems.
- Another advantage is that HTA forms the basis of many other assessments e.g. communications analysis.
- Because each task element is only broken down into a limited number of sub-elements, the analyst is provided with a convenient check that no task elements have been omitted at each stage.

Disadvantages

- The major disadvantage of HTA is that the analyst needs to develop a measure of skill in order to analyse a task effectively – the technique is not a simple procedure that can be applied immediately. However, the necessary skills can be acquired reasonably quickly through practice.
- HTA has to be carried out with a measure of collaboration from managers, engineers and other operating staff. This is necessary in order to ensure adequacy of information and to confirm that the HTA complies with managerial requirements. While this collaboration is in most respects a strength, it entails commitment of time and effort from busy people.

References and further reading

Annett, J., Duncan, K.D., Stammers, R.B. and Gray, M.J. (1971). *Task Analysis.* Training Information Paper No. 6. London: H.M.S.O. *An early paper which sets out the ideas by the technique's originators.*

Creasy, R. (1980) Problem Solving, The FAST Way. *Proc. SAVE (Society of Added-Value Engineers) Conference*, 173-175. Irving, Texas, USA: SAVE. *Describes the FAST approach and cites an application.*

Duncan, K.D. (1974). Analytical Techniques in Training Design. In *The Human Operator in Process Control*, Edwards, E. and Lees, F.P. (eds.), pp. 283-320. London: Taylor and Francis. *A paper by one of the original HTA authors, describing its application to process control tasks.*

Shepherd, A. (1986) Issues in the Training of Process Operators. *Int. J. Ergonomics*, 1, 49-64.

Shepherd, A. (1989). Analysis and Training of Information Technology Tasks. In *Task Analysis for Human Computer Interaction*, Diaper, D. (ed.), pp. 15-54. Chichester: Ellis Harwood. *A recent discussion of HTA which includes more detailed advice about its practical application.*

See Chapters 7, 8, 11 and 12. See Chapter 7 for an example which utilized the FAST approach.

Link analysis

Overview

Link analysis is a technique used to identify the relationships (links) between an individual and some part of the system (Chapanis, 1959). In its broadest definition, a link between two parts of the system will occur when a person shifts his or her focus of attention, or physically moves, between two parts of the system. What constitutes a part of the system will depend upon the task itself, and could range from individual instruments to more broadly defined plant areas. It could also encompass communications, or other interactions between individuals.

Links can also be defined much more precisely, so that for instance, separate link analyses could be undertaken for the movements of each hand. Link analysis provides the basis of a systematic method of optimizing the arrangement of all links in a system.

The core of link analysis is a representation technique, providing the means to record and represent the nature, frequency and/or importance of links within a system. However, it allows some statistical analysis, usually using frequency data, of the relationships between system components. It also relies on the collection of certain key information, which is usually independent of application. Consequently, link analysis can be looked upon as one of the few self-contained task analysis techniques. It is also very simple to use and apply.

Application

Link analysis can be used to study communication links between individuals, in the assessment of operations where time is important and operators are expected to move between different plant areas, or as a hardware-oriented technique, which focuses on the task environment and the equipment used. It is particularly useful for system applications where the layout of instruments is important, and where transitions between locations, or individuals, play a significant role in efficiency or accuracy of system performance.

Typical applications of link analysis include equipment layout for offices and control rooms, and the layout of display and control systems. The method of analysis makes the results very context specific, so data cannot easily be generalized between systems, or tasks. It is easiest to obtain the necessary information from an existing system, or an accurate simulation, but it is also possible to use detailed layout drawings.

Description of technique

The general method of undertaking a link analysis, and data collection and representation issues, are described below.

General method

The first step in a link analysis is always to collect the background information that describes which parts of a system are used during particular task steps. The analyst can then develop a comprehensive list of the links between the individual elements of the equipment. The analyst needs to remember that a link only occurs when two system components are activated sequentially, one directly following from the other. It may be possible to establish these links from a procedural description of the task, in the case of a well-defined task with clear behavioural steps. However, it will often be necessary to observe task performance for some time, for example if a task involves reaction to system states which vary unpredictably over time.

Display design presents a particular problem for data collection, since the only reliable objective evidence of the operator interacting with the display – if real time observation is required – would be through the recording of eye movements.

Once the links are established, they are recorded by drawing lines between pairs of components on an equipment diagram or some other representation of the workplace. Once this has been done, it is necessary to compare the link information with the perceived importance of each link. A tabular method is appropriate for this. For the purposes of estimating the importance of each link, the frequency with which each link is made may be the only factor which needs to be considered. However, many other factors may influence the criticality of each link, and it may be necessary to solicit expert opinions if these factors are not fully understood by the analyst.

Definition of links

Before information collection commences, it is necessary to define the nature of the links which are to be studied. Usually these are defined as movements of attention, or position, between parts of the system, or they may be communications with other workers. However, for some purposes, it may be preferable to define links more precisely. For instance, they may include only control movements, or eye movements, or communications. It is also necessary to specify the level of detail required. For example, with a short task the analyst may wish to consider the links between individual items of equipment, while for more complex tasks, it may be sufficient to look at links between relatively large groups of equipment, such as complete control panels, or even links between different workstations altogether.

Information collection

Once the analyst has defined the links of interest, he or she should observe a task through a technique such as a walk-through (pp. 160), or else examine the task step by step from written procedures, or lists of task elements derived from other forms of task analysis. Whichever approach is taken, the objective is to record sequentially each link as it is made.

From this raw data, the analyst should count the number of times that each part of the system, or each person, is linked with another. These values must be recorded, and one way of doing this is shown in Table 3.3, which is known as a link association table.

As well as recording direct links which occur in a normal task sequence, it could for some purposes be useful to record conditional links, which would only occur under certain circumstances. For example, although an operator may normally move from one display to another which is close by, in particular situations (such as following an equipment malfunction which necessitates obtaining an alternative source of the same information, or if a control action is necessary in some circumstances), the operator may have to move to a different position. These conditional links should be recorded separately.

The following simple example illustrates the process: a control room consists of seven displays (labelled A to H), which are positioned from left to right along one wall, with some gamma radiation monitors to the right of panel G, and some panels of electrical supply indicators on the opposite wall. In the centre of the control room is an operator's desk. Observations were made of the usage of the different parts of this control room during a walk-through of a trip sequence, and these were then recorded. In addition, it was noted that at several points during the trip sequence the operator would have to undertake additional tasks if the automatic systems had failed, and so links between the original monitoring position and the location at which these recovery actions would be carried out, were recorded as conditional links. In Table 3.3 all the links which occurred between different panels, or parts of the control room, are presented.

From this table it is clear that the operator frequently has to move between panels B and F during the trip (9 links reported), and there is also a close link between panels A and B (6 links and 1 conditional link), though as these latter panels are adjacent to each other, this would not involve the operator in much movement, or delay.

Table 3.3 The link table for Figure 3.17. Conditional links are in parentheses.

	B	C	D	E	F	G	H	Gamma Monitors	Desk	Elec. supply
A	6 (1)	-	1	1 (1)	2	-	-	-	-	-
B		1	-	2	9	-	-	2	2	-
C			1	1	1	-	-	-	1	(1)
D				-	-	-	-	-	-	-
E					2(1)	-	-	-	(1)	1
F						3	1	1	-	1
G							-	-	-	-
H								-	1	1
Gamma radiation monitors									1	-
Central desk										1

Representation of information

For task analyses which involve only a small number of items, or in which there prove to be only a small number of dominant links, it is relatively easy to interpret the link analysis directly from the link association table. However for more complex analyses, or for analyses where temporal information is important, it will be preferable to use some form of graphical representation. There are two basic ways in which this can be done: schematic link diagrams and spatial link diagrams.

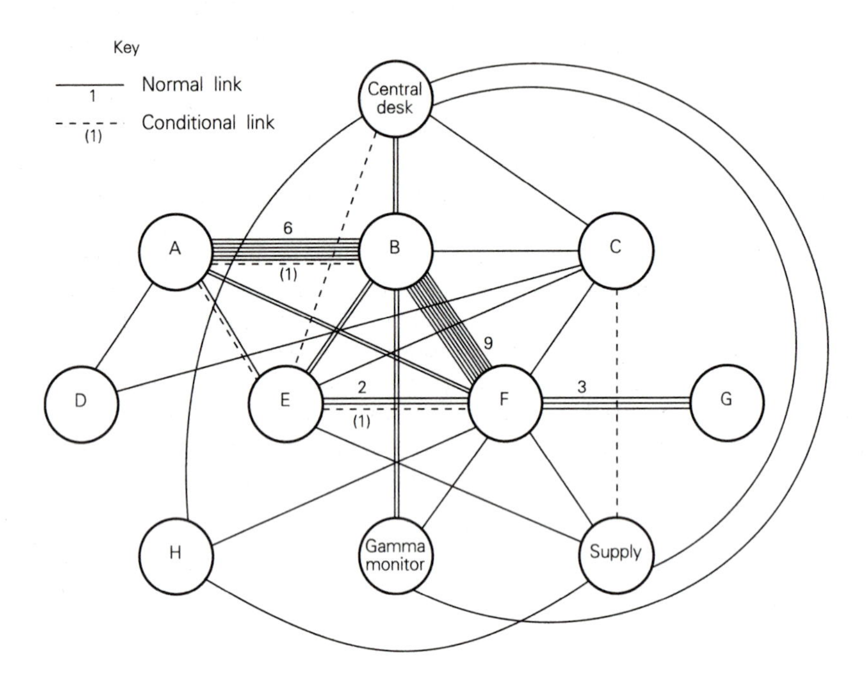

Figure 3.17 Link diagram. Adapted from Ainsworth, 1985.

Schematic link diagrams

In these diagrams, each part of the system (or each person) is represented by a box or a circle, but their relative positions need not correspond to the real situation. Every time that two items are linked, a line is drawn between them, so that the more frequently linked items will be joined by more lines. If it is necessary to show different kinds of links, for instance in multi-person tasks, or for conditional links, different line patterns, or colour coding, can be used.

Figure 3.17 shows a schematic link diagram for a task at a power station, using the data given in Table 3.3.

Although schematic link diagrams do not accurately represent the relative positions of the items, they are often sufficient to establish which are the most frequent links. They are also particularly useful for clarifying situations in which the links overlap considerably, or for representing links between several individuals.

Spatial link diagrams

If the analyst also requires information about the topographical relationships between the links, some form of spatial link diagram must be used. These are similar to schematic link diagrams, except that they are based upon an accurate spatial representation of the system, rather than a more symbolic representation. They can be produced either on a floor plan, or on a drawing of the equipment layout.

Although spatial link diagrams do add this extra dimension, they may prove much more difficult to draw when there are many links crossing a relatively restricted area. Therefore, the advantages of showing the topographical relationships must always be balanced against the possible lack of clarity if many links cross.

Spatial link diagrams are very similar to spatial operational sequence diagrams (see pp. 130), and the two are often confused. The important distinction is that spatial link diagrams aggregate all the information about particular combinations of equipment, and so they are useful for identifying situations where two equipment items are regularly used together regardless of what these sequences entail. On the other hand, spatial operational sequence diagrams permit the assessment and identification of individual task sequences which may be important, but in doing this the global view is often lost. As a general rule, spatial operational sequence diagrams should be used when the analyst wishes to look at manipulating the order of steps in a particular procedure, while spatial link diagrams will be most appropriate for looking at the gross effects of different equipment arrangements.

Practical advice

When collecting task data, it is important to ensure that a representative sample of data is collected. In order to avoid bias, this may involve the observation of several different operators performing the same task. Similarly, any system must be studied over its full range of functions, which may involve generating a representative matrix of tasks. This is to ensure that a system is not optimized for one task at the expense of another.

In order to draw a schematic link diagram, it is often helpful to place the item which is linked with the most other items at the centre of the diagram, and to arrange the other items in a fairly regular manner around this. This is likely to optimize the clarity of the diagram.

As the analysis is very context-specific, it is not normally possible to add information, or make inferences, if the system changes. It is therefore very important to ensure that the system being analysed is stable, because any changes will probably result in the analysis having to be totally restarted.

Most of the procedure of link analysis is very straightforward, and requires no specialist knowledge beyond some briefing as to what should be recorded and how. However, some of the variations which are demanded by specific applications may call for expert assistance (e.g. judgement of the relative importance of groups of links would require expert system knowledge). Later in the procedure, the application of the results of the analysis to design changes would in most cases require specialist design expertise in user interface design.

Resources required

The resources required for a link analysis itself are very limited. It is basically a pencil and paper exercise, and even the calculations involved are normally easy.

The main resources which will be required are those associated with the collection of the initial information, and its subsequent analysis. The former may be a large exercise, depending on the nature of the data required and the environment from which it must be extracted. However, for simple well-documented systems, very little time may be required for data collection. The analysis again depends heavily on the types of factors being considered. Obtaining expert judgements of the importance of various links, for example, can be a difficult and lengthy process. Also, for more complex systems, it may take several iterations before an optimal design can be achieved.

Links with other techniques

The usual first step in link analysis (construction of a pictorial representation of the environment or equipment) is effectively equivalent to starting the production of a spatial operational sequence diagram (pp. 130). However, the subsequent analysis is different, because link analysis aims to quantify some very specific types of link between the elements in the workspace.

Also, depending on the nature of the task, several data collection techniques (Chapter 2) can be used to provide the necessary quantitative information. In the case of well-defined, repetitive procedures, the rudimentary links between workspace elements can often be established from existing procedural documentation. If no such documentation exists, a description of the task can be derived from a talk-through (pp. 162).

If actual performance data are required, some other form of observational technique which allows unencumbered performance of the task must be used. In the example above, for instance, if the *time on target* information is critical for determining the importance of each display, the operators must be observed and recorded while they are performing a task. A suitable method of data acquisition in this case might be event logging (see pp. 41 and 53).

In the case of undefined tasks, for example, those which are performed in response to changes in system state, an observational technique (pp. 53) must be used to record the links between elements over a reasonable period of time.

Advantages

- Because it is hardware oriented, link analysis itself does not always depend upon the observation of people doing a task, and therefore may not require active participation by operators. Often, a description of the procedures involved provides sufficient basis for constructing link analysis charts and tables. However, some information required for particular analyses may give rise to the need for of observational data collection.

- The method relies upon observable and measurable data, and is therefore very objective.
- It is a straightforward technique, which can be practiced by analysts with little formal training.
- No expensive facilities or resources are required, beyond the analyst's time.
- It is a useful technique for looking at multi-person systems and communications.

Disadvantages

- The technique requires preliminary data collection by another means, to establish basic task procedures as a basis for deriving links. In the case of tasks which are essentially random and not system paced, this process itself may require extensive observation.
- Link analysis considers only the basic physical relationships within systems. For performance optimization, other relationships (such as sensory modality, conceptual compatibility, etc.) may be more important.
- Because of the complexity of the graphical and tabular representations, only fairly simple sub-systems can be represented using this technique.
- The technique equates the importance of links with their frequency, and if this is not a valid assumption the analysis could be misinterpreted.
- Link analysis only shows the frequency with which links are made, and provides no information about the amount of time available for making them.

References and further reading

Ainsworth, L.K. (1985) *A Study of Control Room Evaluation Techniques in a Nuclear Power Plant.* Applied Psychology Rept. 115. University of Aston.

Chapanis, A. (1959) *Research Techniques in Human Engineering*, Baltimore: Johns Hopkins.

Kantowitz, B.H. and Sorkin, R.D. (1983) *Human Factors: Understanding People-system Relationships*, chap 18. Chichester: John Wiley.

See Chapters 9 and 16, which made brief use of a form of link analysis.

Operational sequence diagrams

Overview

An operational sequence is any sequence of control movements and/or information collecting activities, which are executed in order to accomplish a task. Such sequences can be represented graphically in a variety of ways, known collectively as operational sequence diagrams. Together, these constitute a group of representational techniques by which data can be organized for presentation and subsequent analysis.

In its simplest form, an operational sequence diagram is a flow diagram linking the various task operations in the order in which they are normally carried out. For most simple applications this will be a predominantly linear flow, conventionally drawn from top to bottom. However, a variety of refinements can be introduced. For instance some limited branching, looping and other conditional activities can be shown.

Application

The simplest form of operational sequence diagram can be used to identify the operations associated with a task (grouped into categories if required) and the order in which they are performed. However, the versatility of the technique is such that several derivatives are in use, each aimed at clarifying and representing a particular characteristic of the operational sequence. Three such derivatives are temporal operational sequence diagrams (which illustrate the temporal relationships between operations), partitioned operational sequence diagrams (which focus on very specific criteria) and spatial operational sequence diagrams (which show the spatial relationships between items of equipment on which the operations are performed).

Operational sequence diagrams can be used to represent tasks at any stage in the cycle of design, development and operation. The constraint will invariably be the collection of appropriate information, and as long as there is an adequate means of studying the task and collecting information, operational sequence diagrams will usually be capable of representing that information.

However, some tasks, particularly highly cognitive ones, produce particular problems for the classification and identification of discrete operations. Such tasks will generally not permit the production of operational sequence diagrams.

Extremely complex tasks also cause problems, as they tend to do with most graphical methods of representation. Diagrams of complex sequences very soon become incomprehensible, particularly if they are not highly linear. Thus looping, and heavily repeated operations, can make any flow diagram virtually impossible to draw, and even more difficult to interpret.

Description of technique

The basic type of operational sequence diagram and its three major derivatives are described below.

The basic operational sequence diagram

In its original form, an operational sequence diagram is a simple flow chart based upon categorizing operations into a limited number of different *behavioural elements*, which are represented by specific symbols. This is achieved by splitting an activity into a series of *operations* which fully define the activity, and then categorizing these by applying output criteria (what is accomplished) and methodological criteria (how it is achieved). The five symbols which are commonly used to represent manual operations in operational sequence diagrams are illustrated in Figure 3.18, but some authors have introduced slight variations to this symbology.It can be noted that these are very similar to the symbols which are used for process charts (see Figure 3.3). However, operational sequence symbols define information flow and behaviour, rather than observable processes, and besides modifying the definitions, this has also necessitated the addition of specific symbols for *the receipt of information* and for *decisions*. Another variation from process charts is that operational sequence diagrams are not restricted to using a single symbol at each element in a sequence, and often two or more symbols can be grouped together.

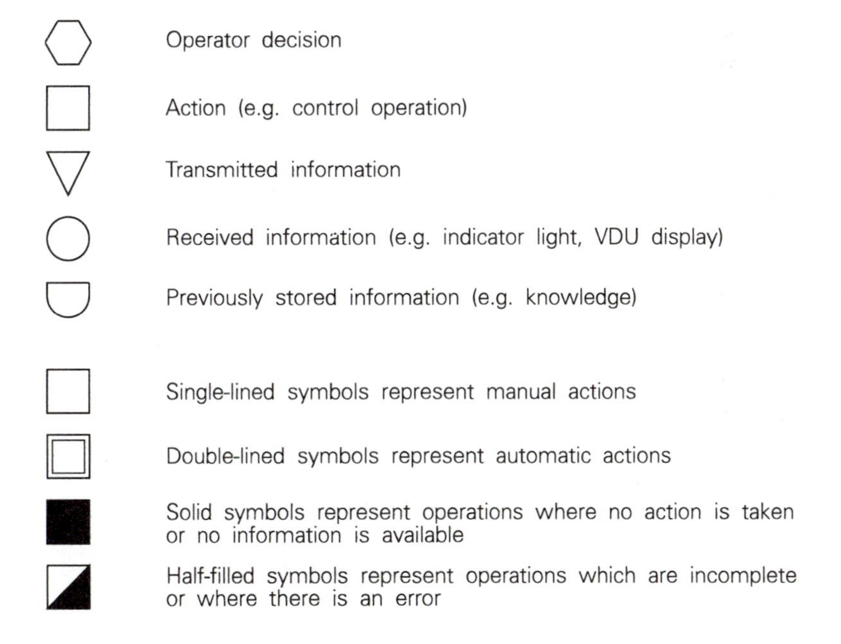

Figure 3.18 Symbols which are used for operational sequence diagrams

The original purpose of operational sequence diagrams was to represent complex, multi-person tasks (Kurke, 1961). Therefore, it is necessary to distinguish between manual and automatic operations, and to highlight manual operations which might be particularly vulnerable to error. This is achieved by modifying the basic set of symbols according to the following conventions:

- Automatic operations are represented by using double lines around the symbols.
- Where there is no action undertaken or no information is available, the symbol is filled in solid.

- Where the information is incomplete, or an operation is incorrect, due to error sources in the system, the symbol is partially-filled in, by filling it solid on one side of a diagonal line through that symbol.

When there are logical relationships between inputs and outputs, these are represented on the operational sequence diagram by the manner the lines between inputs and outputs are drawn. Thus, if the lines between elements are linked either before entering, or after leaving, an **AND** condition is represented. However, if there are separate lines entering an element, there is an **OR** condition.
An example of a portion of a basic operational sequence diagram is shown as Figure 3.19.Figure 3.19 Operational sequence diagram for an airport ground .

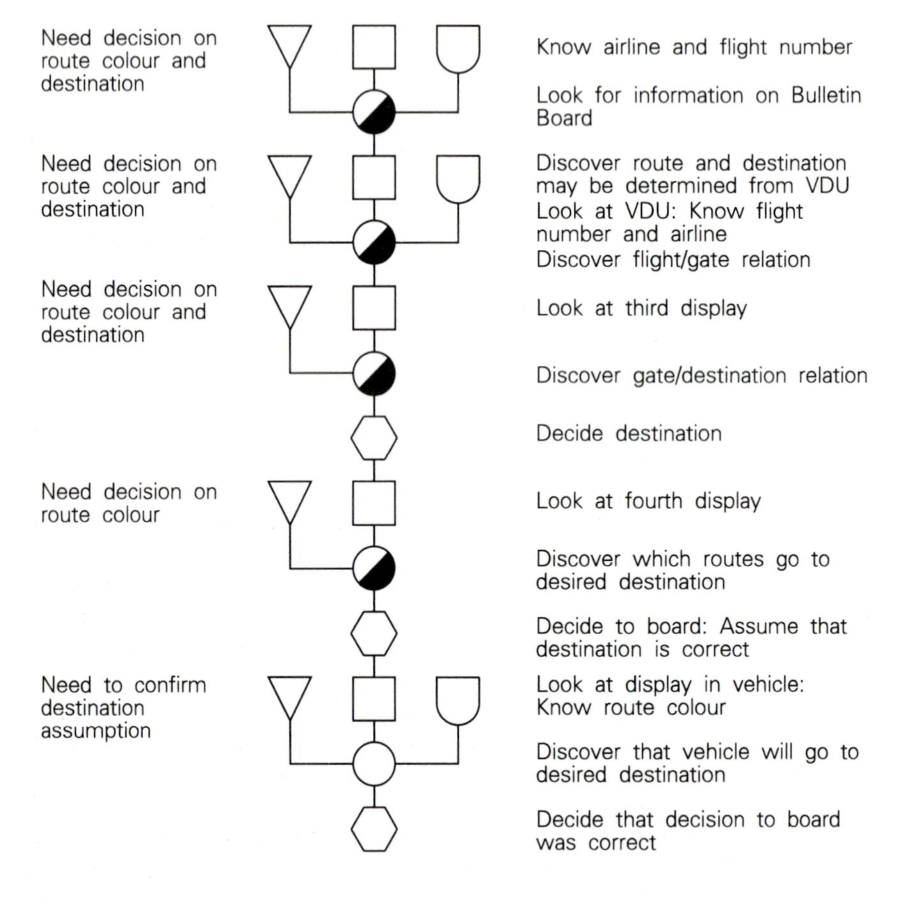

Figure 3.19 Operational sequence diagram for an airport ground transportation system, Bateman (1979). Adapted with permission from Proceedings of Human Factors Society 23rd Annual Meeting.

Temporal operational sequence diagrams

The simplest variation on the basic operational sequence diagram is to represent the time necessary for each action on the vertical axis, so that the vertical separation of operations relates directly to the amount of time between them, as opposed to merely indicating their relative order. This is known as a temporal operational sequence diagram, and it is really a specific type of vertical timeline (see pp. 137). By looking at the relationships between task elements on a temporal operational sequence diagram, it is possible to determine whether there is any potential for time stress (i.e. pressure on the operator(s) to perform tasks with minimal time available). It should be noted that on these diagrams, the repetition of operations cannot be represented by looping, as would be done in normal flowcharts, since any linking line moving upwards represents going back in time! Thus, if this variant is used for highly repetitive tasks, the analyst must accept an equivalent level of repetition in the diagram.

Partitioned operational sequence diagrams

These involve producing separate operational sequence diagrams for different dimensions of a task, and then presenting all of these together, with different columns for each dimension. For instance, in a multi-person task, operational sequence diagrams could be produced for each team member, and then these could each be presented as separate columns in a partitioned operational sequence diagram, from which it would be relatively easy to determine how all the team members were interacting with each other. Other dimensions which could be used to partition an operational sequence diagram could include the sensory modality which was used (e.g. aural, visual, tactile), or the extent to which each operation is performed by the operator, or the machine (which is particularly useful for looking at the man–machine allocations).

On these diagrams each dimension into which the operational sequence diagram has been partitioned is shown as a separate column, in such a manner that all operations which are being carried out together will appear in the same horizontal row of the diagram. If more precision is required, the left column can be used to provide the same timing information as there would be in a temporal operational sequence diagram, to form a partitioned, temporal operational sequence diagram.

Another feature of partitioned operational sequence diagrams is that information flows are represented on them by flowlines, and these are particularly useful for showing interactions between the different individuals or dimensions. Further information about these interactions can be provided by adding symbols or labels to the flowlines. For instance 'E' could be added to show that a link was electrical, while 'V' could be used for a visual link. However, in practice, the labelling of the individual links will not add very much useful information for a human factors analysis.

Job Process Charts (Tainsh, 1985) are a specific example of a partitioned operational sequence diagram, which are composed of the following columns:

- the left column is for machine operations
- the central column is used to show the functional requirements of the man–machine interface which links the two
- the right column is for tasks which are undertaken by the human operator

An example of a job process chart taken from Tainsh is presented as Figure 3.20.

Spatial operational sequence diagrams

These diagrams represent the spatial, or geographical, links which exist between operations. This is achieved by producing some form of map, or panel diagram, of all the items of equipment with which the operator interacts during the task, and then plotting the sequence of use of displays and/or controls onto this map. Figure 3.21 shows a spatial operational sequence diagram for a touch screen for a task in which speed was vitally important. A spatial operational sequence diagram of the original design showed for this key task it was necessary to make six page changes and to undertake thirty checks or actions which often necessitated seemingly random moves across the display pages. Figure 3.21 shows how the software was improved to enable the task to be undertaken on a single display page with only twenty checks or actions.

The usual operational sequence symbols can also be shown on these diagrams, but this can be confusing, and so it is usually preferable to omit them, or else to limit the number of symbols which are used (e.g. circles could be used for information sources and squares for control actions). The map itself does not have to be very accurate, provided that the general topographical relationships between items of equipment are shown, and indeed it may be helpful to use a larger scale for areas where many items are close together.

The sequence of use of controls and/or displays can be shown in many ways, but the simplest way is to plot a single path which represents the normal sequence of equipment usage. If there are important alternative sources of information, or independent feedback, links to the normal information source can be shown by using broken lines. Other variations could include plotting links between display and control equipment separately, or superimposing the operational sequences for several operators onto the same diagram. In either case, it must be very easy to differentiate the line designs which are used. If it can be achieved without cluttering the diagram, it is also desirable to use some form of sequential numbering system, and this is particularly important for diagrams of multi-person tasks. Another useful feature which can be presented on spatial operational sequence diagrams by using two different line styles is that it is possible to differentiate between instruments and controls which should always be used and those in conditional links which will only be used if the main equipment is malfunctioning, or if relatively infrequent conditions are met.

For VDU-based displays, the path through each display page can be shown as a box, with a new box for every page change. As the number of boxes for a

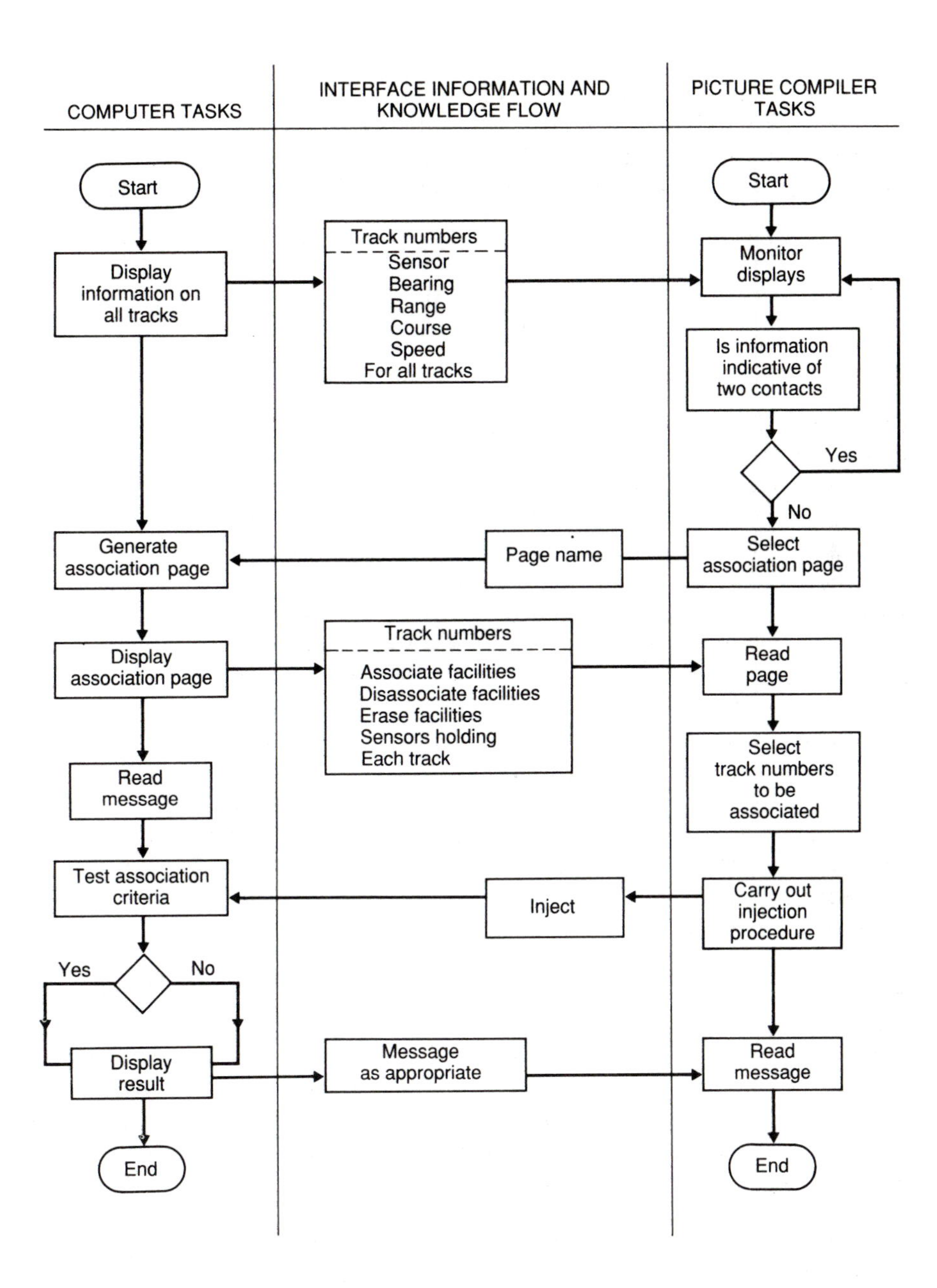

Figure 3.20 Job process chart for a man–computer dialogue concerned with the association of the estimated tracks of vessels.
Adapted from Tainsh, 1985. © Crown copyright/MoD, reproduced with the permission of the Controller of HMSO.

task proliferates, it will clearly be shown when a particular task demands an excessive number of page changes.

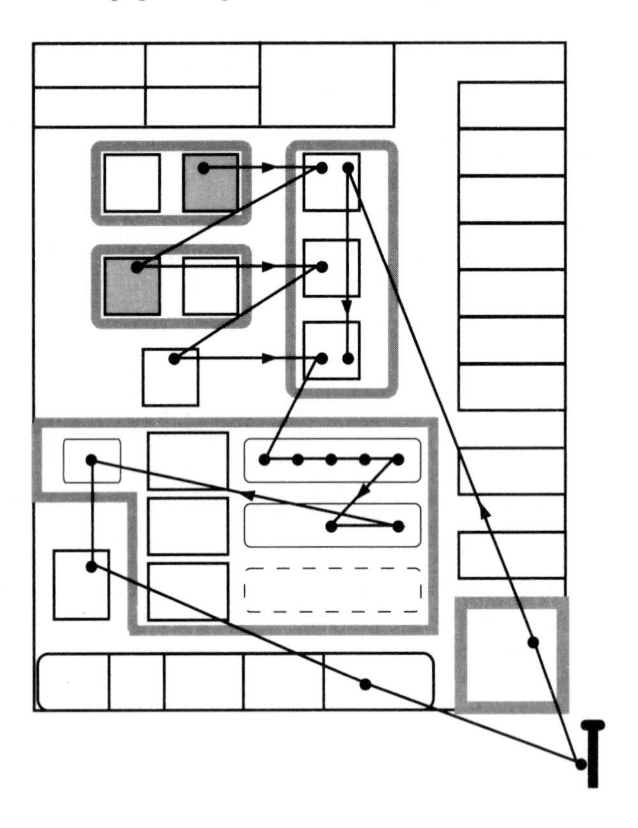

Figure 3.21 Spatial operational sequence diagram for an improved touch display for a time-critical task, Unpublished data from Ainsworth and Mullany, 1987.

Spatial operational sequence diagrams highlight situations which call for excessive movement between different parts of the system during a task. However the inclusion of specific temporal information is virtually impossible to do in any legible way. It has been suggested that the line style which is used on a spatial operational sequence diagram could be altered at predetermined times into a task sequence, but in practice this approach is very confusing. Thus, although a spatial operational sequence diagram should highlight parts of a task where excessive movement between locations was required, the same diagram cannot easily show how much time there was available for the associated actions. Therefore, the information from a spatial operational sequence diagram must be used in conjunction with information from an alternative method (e.g. timeline analysis, or, decomposition methods, pp. 135 and 95 respectively) before it is possible to make inferences about the existence of time stress.

There are many similarities, and hence some confusion, between spatial operational sequence diagrams and spatial link diagrams (see Link Analysis, pp. 122), and hence it is worth clarifying the differences between these two techniques. The similarity between them is that they both use a form of map to show sequential relationships. However, the spatial operational sequence diagram shows the full sequence of operations, so that it is possible to identify a single task sequence which may be particularly important. On the other hand, the spatial link diagram aggregates all the information pertaining to a particular combination of equipments and so it should identify situations where two items of equipment are commonly used sequentially, regardless of what these sequences each entail.

Practical advice

For the development of complex systems, the diagram will probably require frequent updating and alteration before it is complete. The simple expedient of using a soft pencil on tracing paper will allow many alterations without affecting the quality of the diagram. If colour coding is required, it can be marked in text and added later.

When constructing complex spatial diagrams, it is worth considering how easy it will be to relate the diagram to the equipment, if this is necessary later on. It may be necessary to give the location of the items of equipment in some detail, to minimize the time spent later trying to find them.

Resources required

When using operational sequence diagrams, it is useful to decide upon the suite of data collection and data representation techniques to be used before starting the operational sequence diagram. That way, information collection can be tailored to suit the representational method, and construction of the diagram can be relatively straightforward. As illustrated earlier, inadequate data collection can make subsequent representation and analysis difficult, or even impossible.

The implication is that resources should be invested early (i.e. in planning and information collection) to reduce the resources required to construct a diagram. However, very complex or large diagrams can be highly time-consuming to draw, particularly for spatial diagrams.

Links with other techniques

Care should be exercised when choosing a data collection technique to feed into an operational sequence diagram, to ensure that it can provide all the information required to construct the diagram. For example, workload analysis can be done effectively by using a temporal operational sequence diagram. However, if a simple talk-through (pp. 162) is used to elicit information, accurate temporal data will be missing. A real-time walk-through (pp. 163) would be far more effective.

Operational sequence diagrams are similar in principle to many other flow-charting techniques, but have particular advantages and shortcomings. Generally, they are capable of organizing information in more useful and powerful ways than simple process charts (pp. 84). However, there are other more formal techniques, such as Petri nets (see pp. 91), which are specifically adapted to representing highly conditional sequences, or to showing an accurate snapshot of the state of a system at a given time.

Operational sequence diagrams can however form a useful adjunct to some other techniques where these benefit from the organization of information in a graphical form. For instance, hierarchical task analysis typically requires the use of *plans*, which are effectively sequential statements representing how each individual goal is achieved. By presenting these plans using some form of partitioned operational sequence diagram, it is possible to obtain a more informative representation in certain circumstances.

Spatial operational sequence diagrams in particular are closely associated with link analysis (pp. 118). Although link analyses are typically used only to identify general problems with equipment location, the same *map* of workspace elements forms the basis of the much more specific optimization of relationships which spatial operational sequence diagrams allow.

Advantages

- Operational sequence diagrams are particularly useful for showing the relationships between operations, which are more difficult to show by textual or tabular representation. They can show various types of relationships between operations – such as temporal, spatial or even conceptual – with equal facility. To some extent, more than one type of relationship can be shown simultaneously, but this can give rise to excessive complexity.
- They can be used for multi-person tasks, as they can capture information derived from several operators on one diagram.
- The diagrams themselves can be prepared off-site, so they will not generally necessitate any use of the operational system, or the operators time, once the data have been collected.
- Operational sequence diagrams, especially spatial ones, have a high face validity.

Disadvantages

- Although operational sequence diagrams can illustrate many types of relationship between operations, no one format can show all of them together. This is not usually a serious shortcoming, but it can mean that considerable caution must be exercised when interpreting information. For example, a diagram may reveal that two successive operations are performed on items of equipment which are spatially distant from each other. However, that diagram cannot also easily show how much time is available to cover this distance, and therefore if this represents a real problem.

- Operational sequence diagrams are prone to becoming cluttered and confusing when used for complex or highly conditional tasks. Great care must be taken both during construction and during use, and it is particularly important to ensure that the analyst is working to the level of detail which is most appropriate. Spatial operational sequence diagrams also can become very difficult to read if individual pieces of equipment are used many times.

References and further reading

American Society of Mechanical Engineers (1972) *ASME Standard – Operation and Flow Process Charts*. ANSI Y15.3-1974. New York, USA: ASME.

British Standards Institute (1987) *Specification for Data Processing Flow Chart Symbols, Rules and Conventions*. BS 4058. Milton Keynes: British Standards Institute.

Gilbreth, F.B. and Gilbreth, L.M. (1921) Process Charts. *Trans. Amer. Soc. Mech. Engrs.*, 43/1818, 1029-1050. *The first reported use of process charts.*

Jordan, N. (1963) Allocation of Functions Between Man and Machine in Automated Systems. *J. Applied Psychol.*, 47, 151-165.

Kurke, M.I. (1961) Operational Sequence Diagrams in Systems Design. *Human Factors*, 3, 66-73 *This is one of the original references about operational sequence diagrams, and it provides a good overview of the ways in which they can be used.*

Tainsh, M.A. (1985) Job Process Charts and Man–Computer Interaction within Naval Command Systems. *Ergonomics*, 28/3, 555-565. *Describes Job Process Charts, their benefits and how to use them. Explains how they focus on the interactive component between the human operator and an automated control system, dealing with remote events and objects.*

Timeline analysis

Overview

There is a standard definition of timeline analysis from the American National Standards Institute, which describes it as 'an analytical technique for the derivation of human performance requirements which attends to both the functional and temporal loading for any given combination of tasks'. Basically this means that it can determine what a human operator needs to do to complete a task (the functional requirements), and how quickly it needs to be done (the temporal requirements).

Timeline analysis is really a set of principles, rather than a rigorously prescribed procedure. It provides some rules for the representation of information, in a graphical form which allows an analyst to determine the timing and functional requirements of systems.

Application

In practice, this can usually be done from observation and measurement, as long as it is possible to identify a representative set of tasks which will give an adequate profile of system performance. In many cases, this set of tasks may be very narrow and fixed. For example a two-person assembly and checking task, which involves sharing facilities and passing incomplete work from one person to another, would be very straightforward. All the operations in the task would be clearly allocated, performed in a fixed order, and the amount of time required would be measurable.

Systems which support a more flexible range of tasks, and in which the operators do not have clearly defined roles, present more difficulties. If the system permits different operators to undertake different roles depending on unpredictable factors, meaningful timeline analysis becomes extremely difficult. The analyst must be able to treat the relevant sub-system as having relatively predictable and repeatable performance.The primary value of timeline analysis lies in predicting, or rectifying, human, hardware and software resource allocation problems in systems. Timeline analysis can be used to aid in allocating tasks both between different operators, and between operators and machines. For maximum efficiency and quality, the goal in any system should be that resources are not under-utilized, nor are they overstretched. The former is wasteful of resources, and in the case of human systems can result in other performance problems due to low job satisfaction. The latter can result in errors, failure to meet targets and, in extreme cases, complete system breakdown.

In cases where resources are fixed, the object of using the technique would be to predict completion times for tasks. This would be true for instance, in the case of a computer system with a fixed 'architecture' and software. Alternatively, if resources are flexible but target times are critical, then the object will be to allocate adequate resources to ensure task completion within the specified system time limits.

Description of technique

The basic timeline chart and specific technical issues associated with its construction, are described below.

The basic timeline chart

A timeline is a line, or bar, whose length is proportional to the amount of time which is necessary for a particular task, or task element. In its simplest form, a timeline chart shows either a staggered line, or several timelines for each of a set of tasks, undertaken by a single person. On such charts, the time axis can be either horizontal, or vertical.

When a horizontal time axis is used, the task elements are listed on the left side of a graph, and then each timeline is shown individually as a horizontal bar to the right of its associated task identification, as shown in Figures 3.22 and

3.23. This format gives the advantage that the timelines can be drawn to a relatively fine time-scale, and several task elements can be shown on a single page. The horizontal format is also particularly useful when some task elements are repeated, because these only need to be identified once in the list of task elements, thereby saving space and emphasizing the importance of these repeated task elements. However, these advantages must be balanced against the fact that the amount of additional information which can be shown on a horizontal format is severely limited.

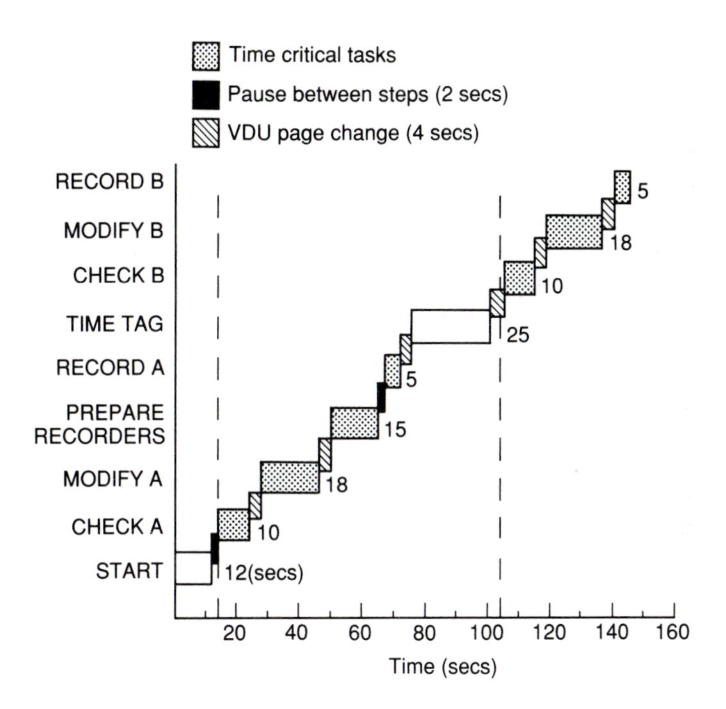

Figure 3.22 Timeline for initial method of undertaking a complex task. From Ainsworth and Mullany, 1986 (unpublished data).

In the vertical format, all the constituent timelines are joined together to form a continuous vertical bar, and some form of identification for each task element is given in the column alongside its associated section of timeline, as in Figure 3.24. This gives the advantage that there is room for a short task description, or other information, to be presented alongside the timeline. However, apart from very short task sequences, the vertical format can soon become very unwieldy, requiring several pages on which it is only possible to use a relatively coarse-grained time-scale. Therefore, vertical timelines should only be used for very short task sequences, or when the analyst wishes to present a lot of additional information alongside the timelines.

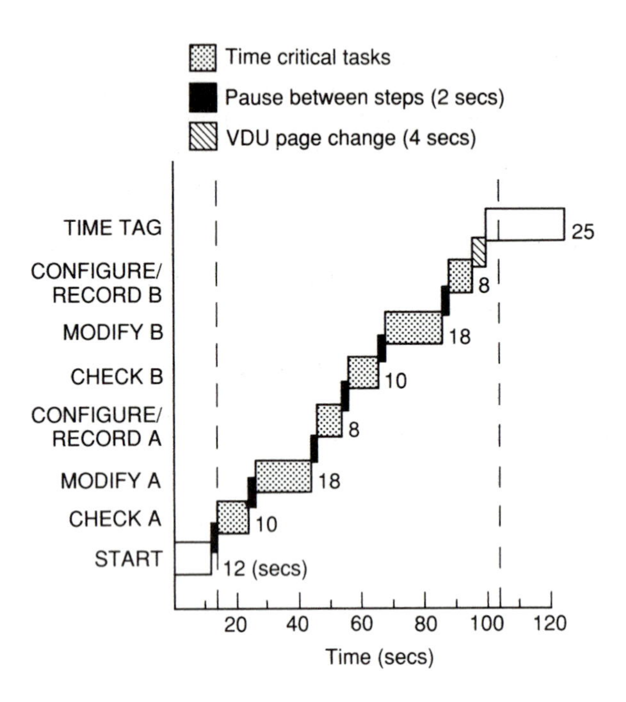

Figure 3.23 Timeline for improved design of the task shown in Figure 3.22. From Ainsworth and Mullany, 1986 (unpublished data).

On both formats, it is possible to differentiate between different types of tasks, or tasks undertaken at different locations, by using different shading for some of the bars.

Temporal information

The time which is associated with the timeline for each task element can be derived either from direct observation, or by estimation. Unfortunately, it may be difficult to obtain realistic time data, particularly for short task sequences, on which an individual is aware that his/her performance is being studied. Ideally therefore, the performance of several individuals should be measured, as unobtrusively as possible (see pp. 55). If this is not possible, specific time allowances should be estimated for decision-making, system response times, communications, etc., and such allowances must be clearly defined and consistently applied. In the examples shown in Figures 3.22 and 3.23, the time which elapses between requesting a VDU page change, and refocusing on the new page, was estimated rather than observed.

If the timelines are to be based upon a single time estimate for each task element, either because only a single individual has been timed, or because only one set of subjective timings has been made, the timelines can be drawn directly from these times. However, if more performance time data have been collected from a series of individuals, it will be necessary to decide what times to allocate to each task element. One possibility is to use the average time for each task element. However, if there is much variation in speed between different personnel, it is likely that timelines based solely upon average performance will present an optimistic (or at least non-conservative) view, and so if possible, some measure of variability should be shown on the timelines. One way to show this variability is to produce different timeline charts for separate individuals, such as the slowest and the fastest persons.

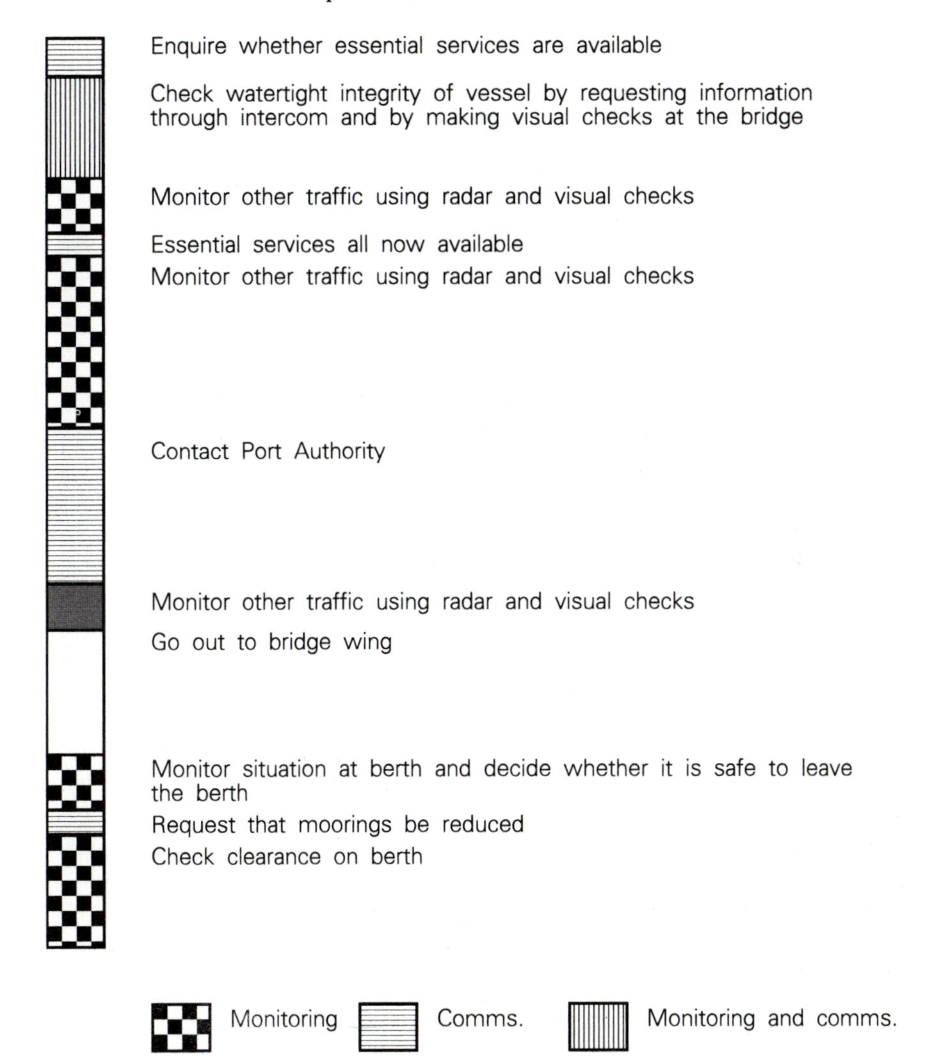

Figure 3.24 Vertical timeline. From Ainsworth, 1988 (unpublished data)

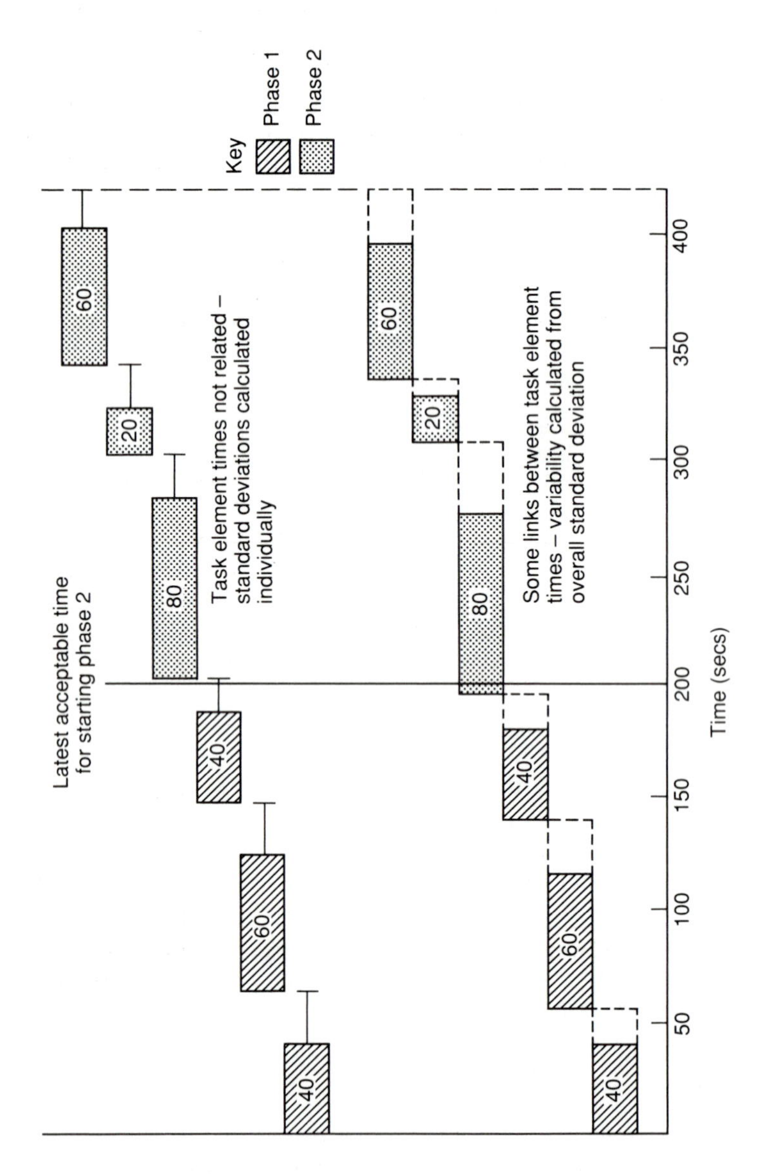

Figure 3.25 Hypothetical timelines to contrast two ways of showing variability

Another approach is to present both the average times and some statistical measure of variability, such as the standard deviation, on the same timeline chart (see Figure 3.25). However, how this is done depends upon how the task elements are associated with each other. If the analyst believes that the time taken for each task element is independent of the time taken for the other task elements, the standard deviation times can be calculated separately for each task element. The usual convention for showing such statistical variations of times on the timeline charts is to extend the timeline bar as a single line whose length is equivalent to the standard deviation time, and to end this line as a 'T'. It is usual for the next timeline to start from this T, but for certain purposes an analyst may prefer to align the mean time periods of adjacent task elements.

However, if the analyst has any reason to suspect that the time taken on one task element may influence the time taken on the subsequent task element, standard deviations cannot properly be estimated for the individual task elements. This situation is likely to occur when operators have to make decisions while undertaking other tasks. For instance, one operator may initiate the first action quickly, and defer decision-making until this task has been completed, whilst another operator may take longer on the first task because decision-making is done before starting the task. In such situations, relatively sophisticated methods are necessary to obtain any valid measures of variability for the individual task elements, and so it is preferable to base the timelines upon the standard deviation time of the overall task sequence. The analyst should then use the average times for each task element for drawing the timeline sections, but the time from the start of the task sequence to its conclusion, should be lengthened by leaving gaps between each timeline section, which aggregate to one standard deviation of the entire task duration. The size of each of these gaps should be carefully estimated by first calculating the proportion of the total average task time which is taken by the preceding task element. The following gap should then represent the same proportion of the standard deviation time. Thus, if the average time for a particular task element is twice as long as another, this should be followed by a gap which is twice as long.

In Figure 3.25 a hypothetical example has been used to illustrate how these two methods of showing variability can make a practical difference in certain circumstances. This example shows that the system requirement to start phase 2 of the task within 200 secs, could only be met in the case where the task element times were not interdependent.

External time constraints

In many situations, as noted above, the system imposes certain constraints upon when particular tasks can be undertaken. For instance, certain task elements may have to be completed within a limited *time window*. Such external time constraints can be shown very clearly on a timeline chart, by drawing a distinctive line, or lines, at right angles to the time axis, at the appropriate point. An example of this is shown in Figures 3.22 and 3.23, where a system requirement entailed that it had to be possible to be recording on machine B, within 100 secs of starting to check machine A. Figure 3.22 shows that with the original task design this was not possible, but by altering the nature of the tasks, and leaving the non-critical time-tagging task until later, as shown in Figure 3.23, the critical tasks could be completed within the required time.

Multi-person timelines

For some team tasks, it is helpful to present timelines for different members of a group on the same timeline chart. This can be done relatively easily on either a vertical, or a horizontal format, by using different bar shading for each person. Such charts can then be used to assess the effectiveness of task allocation between group members. They are also particularly useful for identifying communications requirements.

Showing additional information on a timeline

On a horizontal format, it is possible to add a little extra information, by additional coding, but this can soon become unwieldy. It is preferable to add such information onto a vertical format: several columns of extra information can be added in this way without unduly cluttering the diagram. Another possibility for adding information to a vertical format is to superimpose standard process control symbols on a timeline. This approach can then be slightly extended to produce temporal operational sequence diagrams (see pp. 129).

Another way to add more information to a basic timeline display is to convert the timelines into histograms, in which the heights of each bar are related to another quantitative variable which varies over time. A typical example of this approach are *workload analysis profiles*, which have been described by Parks (1979). This technique requires that an 'expert' (i.e. someone who is very familiar with these tasks or with timeline analysis of similar tasks) must make a subjective assessment of an operator's workload (i.e. the amount of a person's capacity which is currently being utilized) while this operator is undertaking particular tasks. These assessments are usually made on a five or six point scale, which ranges from a workload of zero to one of 100%. Several such assessments can then be presented as a histogram such as Figure 3.26 which shows a workload assessment for two operators undertaking a process control task. It was considered that the task represented in Figure 3.26 might be accomplished by a single operator, but although the workload assessments did cycle between the two operators, there were points at which the workload of both was relatively high, and so it was concluded that it would not be feasible for a single operator to undertake these tasks unless the task was markedly redesigned. The assessments which are used to make up a workload analysis profile are either made for each significant task element, or for longer duration tasks, assessments can be made at fixed sampling intervals (see pp. 42). In either case, if the workload profile is presented directly below a normal horizontal timeline chart which is drawn to the same scale, it is easy to identify the most demanding task elements. However, care must be taken in interpreting workload profiles, because long periods of low workload can be detrimental, though it is usually acceptable to have very short periods of high workload. As a rule of thumb, for sustained tasks, workloads of between 50 and 80 percent can be considered acceptable. This method of presenting temporal information could also be extended to represent quantitative variables other than workload.

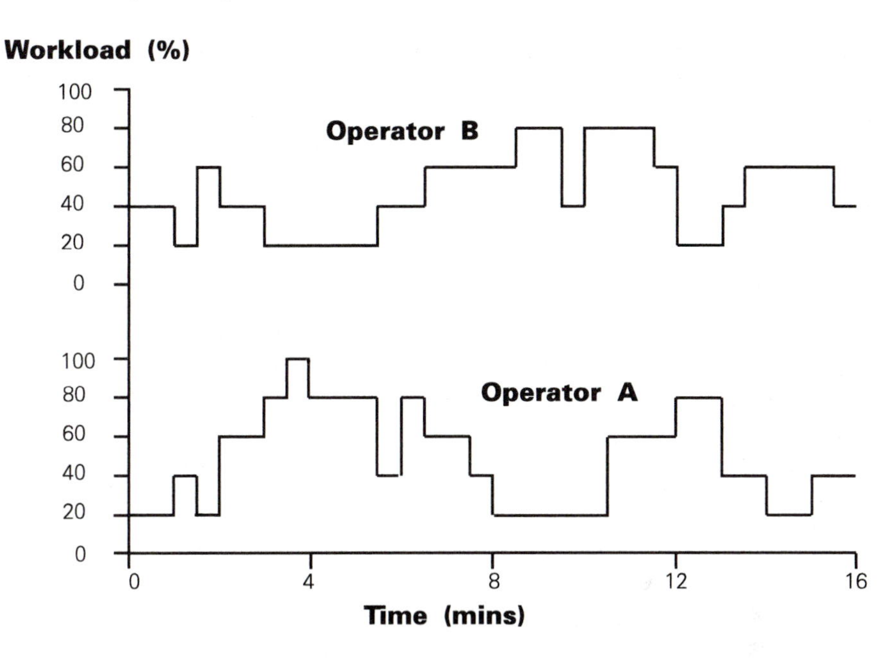

Figure 3.26 Workload analysis. Adapted from unpublished data by Ainsworth, 1989.

Practical advice

Any technique whose objective is to optimize system performance by saving time, is bound to be subject to scrutiny by the people whose performance is in question. The most important thing to remember with the timeline analysis of systems which include human operators, is that human operators may perceive (not necessarily with any justification) that they have something to lose from the study. If they have nothing to lose, it is up to the analyst to convince them of the benefits, and gain their trust.

If there is sufficient space, it can be helpful to write the actual durations on each timeline section, as has been done on Figures 3.22 and 3.23.

Because the amount of information which can be shown on a timeline chart is often limited, it is useful to relate the timelines directly to other information, by using the same reference numbers.

Resources required

For existing tasks, the times taken for the individual task elements can be obtained very easily, provided that the analyst has the forethought to time the task elements whilst gathering other information. If not, it will be necessary to carry out some additional timing exercises. For tasks which are still being planned, the times for individual task elements will have to be estimated. This can be done relatively easily by using a set of general timings for different types of activity, or, if greater accuracy is required, subjective time ratings for each task element could be compared with similar ratings of standard tasks to establish more precise timings. It is difficult to provide guidance as to how long it will take to make relatively accurate subjective timings, but once the judges are practiced in making such assessments, between 5 and 10 min should be allowed for each task element.

For the visual production of timeline analysis, no special equipment or experience is required. In comparison, with some other graphical techniques, drawing timeline charts is neither an esoteric nor a complex procedure, and hence it is not particularly time-consuming.

The point at which resources become unpredictable is often when an attempt is made to optimize a process on the basis of a timeline analysis. Then, the number and effects of various options can increase exponentially as other factors (e.g. the number of operators) increase. In such cases, considerable gains in efficiency can be made by making use of a proprietary computer program written for the purpose.

Links with other techniques

As timelines are a representational technique, they rely heavily upon other techniques for gathering the basic input information. Therefore, if timelines are to be used, it is necessary to ensure that the techniques which are used to gather the basic task information, such as verbal protocols (pp. 71), or observational techniques (pp. 53), make some sufficiently accurate record of the times taken for various task elements. It will then be highly useful if an HTA (pp. 104) is carried out, prior to commencing the timeline analysis. The HTA usefully breaks down the tasks into subtasks and operations, for which task durations can then be sought or estimated.

Temporal information can be added to several flow charting techniques (pp. 82) to achieve the same, or even more informative, results than timeline analysis. An example of this approach would be timed Petri nets, which have the advantage that they can represent conditional events and the dependencies between one event and another. Neither of these are possible with conventional timelines.

Another technique which is similar to timeline analysis is the temporal operational sequence diagram (pp. 125). The particular feature of timelines that makes them appropriate for analysing workload issues, is simply the ability to represent a number of sequences of activity side by side on the same time-scale.

Advantages

- Timeline analysis is very simple conceptually, and can be understood and administered by practitioners with very little training. Moreover, it is a form of representation which is used by people in a wide range of disciplines, and therefore, it can be readily understood in the context of human factors.
- It is a pencil and paper technique (usually) requiring very few resources in application.
- As it uses a graphical form of representation, it has high face validity when presented to non-specialist clients.
- The potentially complex process of optimizing temporal relationships between tasks can be accomplished by computer, and several proprietary software packages are available.

Disadvantages

- Timeline analysis looks only at temporal relationships, and does not indicate how problems can be resolved. Further analysis, or expertise, is often required to reveal the underlying causes of problems and their solution.
- It can be difficult to obtain objective data on temporal factors in task performance because of the problems in gaining the trust of the individuals being studied.
- Timeline analysis shows when things happen, but is not well adapted to showing interdependencies between them. More complex flow charting techniques are needed to represent such data.
- If timelines are being used to optimize the allocation of tasks between a team, there is a danger that other important influences on task allocation might be ignored. For instance, a person who is currently not actively undertaking any task, may be unable to undertake another one because he, or she, does not possess the necessary skills, or because of currently being in a different location.

References and further reading

Laughery, K.R. and Laughery, K.R. (1987) Analytic Techniques for Function Analysis. In *Handbook of Human Factors*, Salvendy, G. (ed.), pp. 329-354. New York: Wiley and Sons. *This paper provides a description of timeline techniques in general, and their application mainly in computing. Contains useful worked examples.*

Parks, D.L. (1979) Current Workload Methods and Emerging Challenges. In *Mental Workload. Its Theory and Measurement*, Moray, N. (ed.), pp. 387-416. New York: Plenum Press. *Gives a description of workload analysis profiles.*

See Chapters 10 and 16, which used timeline analysis, and Chapter 11, which used a computerized workload assessment system.

Chapter 4
Task simulation methods

Computer modelling and simulation

Overview

Computer modelling and simulation involves the use of computer programs to represent operator and/or system activities or features (e.g. of the environment – Siegel and Wolf, 1969). Human performance data that have been previously collected, or estimates of task components, error probabilities, etc, are entered into the computer program. The program can then either simulate graphically the environment and workspace, or can dynamically 'run' the task in real or fast time as a way of estimating complete cycle times and error likelihoods, etc. This has proven to be a convenient way of assessing the potential contribution of alternative configurations of tasks, equipment and team organizations.

Application

Simulation techniques have been used in a number of different contexts, such as task design assessment, workload assessment, equipment layout assessment and human reliability assessment. They can be used in the early stages of system design. Here different task configurations, team structures and complements, operating procedures etc, could be evaluated. This approach could also be used at a later stage where particular operational bottle-necks have been identified. Task activities can be tried out in order to estimate the likely effects of altering procedures, etc. The simulations would typically be used where it would be very expensive to run real trials under the different conditions. In particular it might be used as a way of generating information for design decisions, for example, in assessments of different task and/or equipment configurations.

Description of technique

There are two main types of computer simulations in human factors engineering; those which simulate the dynamic aspects of the operators task (simulation models), and those which are used in the process of ergonomically laying out workplaces (design packages). These are discussed below.

Simulation models

The methods used in this context differ greatly in their procedures and structures. It is obviously possible for a computer program to be written from scratch to carry out a particular function. Alternatively there are a number of commercial packages available for use in this area.

Some packages provide a flexible programming facility whereby a particular task can be modelled. This means that task components and their interrelationships are represented within the computer. Information is then entered into the model, which can then be run, and the result produced as output. A good example of this approach is Micro-SAINT (Laughery, 1984), a development of the SAINT (system analysis of integrated networks of tasks) approach. Task structures are represented in the software, and task data, (real or hypothetical), can be used to execute a model of task performance. For example, different procedures for carrying out a task can be compared in terms of cycle times, error rates etc.. Other techniques are available which simulate performance in maintenance tasks (maintenance personnel performance system, Siegel *et al.*, 1984), and there is an expert-system-driven model of the human operator carrying out diagnosis in a nuclear power plant (CES: cognitive environment simulation, Woods *et al.*, 1990). These two latter developments are in the field of human reliability.

Design packages

It is also possible to use a computer package as a design tool. For instance, an interface could be produced into which stored anthropometric data (on human physical dimensional ranges) could be introduced. The feasibility of the design could then be determined, (see Porter *et al.*, 1990 for a more detailed review of computer packages for workspace design). Probably the most well known example is SAMMIE (system for aiding man–machine interface evaluation). This allows three-dimensional representation of manikin human forms (of appropriate anthropometric dimensions) to be seen in interaction with accurately sized workplaces and pieces of equipment. Designs can thus be evaluated without the need to build equipment or use human subjects.

Practical advice

Problems will exist in determining the accuracy of any model of a complex task. As suggested above, the quality of the output is dependent not just upon the human performance data, but on the nature of the model building and the effectiveness of the model building package.

Unless potential users are expert computer programmers, or have access to such individuals, they are well-advised to first determine how usable they would find off-the-shelf packages before specifying unique software for a particular project. This in turn requires an informed access to information about currently available software. This area, in common with many other areas of software, is a rapidly changing one, and up-to-date information can save a great deal of time and effort. Assessing the ‘usability’ of the software by the potential customer should have a high priority.

Resources required

Although simulation software is becoming increasingly available, it can be expensive. Therefore, the potential user needs to budget not only for the hardware on which to run the programs, but also for the software itself and the time required to learn to use such programs effectively. An alternative would be to use consultancy groups that have a wide experience of using such facilities.

Links with other techniques

The most obvious links would be with any technique which is used to generate, or estimate, data about individual performance (e.g. observation or activity sampling, pp. 53 and 41). Such data can then be run in computer simulations firstly to determine its accuracy, and secondly to use it in a predictive manner. Data obtained from simulations may then be input to other forms of data representation techniques (e.g. timeline analyses, pp. 135).

Advantages

- Advantages come in the form of relatively inexpensive ways of collecting large amounts of data about predicted human performance.
- Given an adequate database, a great deal of exploratory work can be carried out with different human–machine configurations, without the expense of physically producing them.

Disadvantages

- The main disadvantage is that the quality of the data produced by the process will be dependent both on the quality of the original estimates put into the model and also on the proficiency of the model itself. The software packages that are used in this area may in turn be limited, but it is more likely that limitations will lie in the effectiveness of the model building procedures embodied within the program.

References and further reading

Buck, J.R. and Maltas, K.L . (1978) Simulation of Industrial Man–machine Systems. *Ergonomics*, 22, 785-797. *Reports on the application of a modelling program.*

Chubb, G.P., Laughery, K.R. and Pritsker, A.A.B. (1987) Simulating Manned Systems. In *Handbook of Human Factors*, Salvendy, G. (ed.), pp. 1298-1327. New York: Wiley and Sons. *An overview of modelling techniques.*

Laughery, K.R. (1984) Computer Modeling of Human Performance on Microcomputers. *Proc. Annu. Meet. Human Factors Soc.*, 884-888. *An example of the use of one technique (Micro-SAINT).*

Moraal, J. and Kraiss, K.F. (eds.) (1981) *Manned Systems Design.* New York: Plenum. *Includes chapters by Lane et al. on the Human Operator Simulation (HOS4) (pp. 121-152), and by Chubb on SAINT (pp. 153-200).*

Porter, J.M., Case, K. and Bonney, M.C. (1990) Computer Workspace Modelling. In *Evaluation of Human Workplaces: A Practical Ergonomics Methodology,* Wilson, J.R. and Corlett, E.N. (eds.), pp. 472-499. London: Taylor and Francis. *A comprehensive review of the use of computer packages for developing workplace designs.*

Siegel, A.I., Bartter, W.S., Wolf, J.J., Knee, H.E. and Haas, P.M. (1984)*Maintenance Personnel Performance Simulation (MAPPS) Model: Summary Description.* Rept. No. NUREG/CR 3626, Vol. 1. Washington D.C.: US Nuclear Regulatory Commission.

Siegel, A.I. and Wolf, J.J. (1969) *Man–Machine Simulation Models.* New York: Wiley and Sons. *The classic book on simulation techniques.*

Woods, D.D., Pople, H.E. and Roth, E.M. (1990)*The Cognitive Environment Simulation as a Tool for Modelling Human Performance and Reliability.* Rept. No. NUREG/CR 5213. Washington D.C.: US Nuclear Regulatory Commission,.

See Chapter 11 which documents a workload assessment using Micro-SAINT.

Simulators/mock-ups

Overview

Under this heading a range of techniques can be considered which involve the development and use of some form of simulation of systems (Stammers, 1986). The simulation in question might range from a full-scale high fidelity or full-scope simulator through to some simple mock-up of a single piece of equipment. The simulation may vary in the extent to which it represents the true demands of the task and the environmental conditions under which it is performed. The main objective is to observe operator activity and to record performance in a representation of the task environment.

Application

The technique will usually be used when the real equipment and/or system is not available for analysis work. This might be because of the inaccessibility of the task due to the nature of the working conditions. Alternatively, the nature of the real-life process may mean that the tasks are too hazardous to allow for intervention on the part of the analyst. Simulators or mock-ups also enable an analyst to study systems in advance of the real system being designed, or being operational.

The simulation would typically be used to establish appropriate working methods, ergonomics of control layout and design and identification of potential sources of error, or to derive training recommendations.

A simulator may therefore take many forms: flight simulator; a central control room simulation which responds in 'real time'; a simulator which shows a range of system functions and variables on a console which are representative of those found in the real situation, but not identical in format; a simulator used to demonstrate principles important in training; or a mock-up of the situation which may or may not have the dynamic indications or features that occur in the real task.

The technique will be used when the real situation is unavailable for study or cannot be used cost-effectively for training, or does not yet exist. It therefore has a particularly important part to play in the early stages of system design and development. As aspects of the system emerge in the design process, their evaluation in mock-up situations can be invaluable. This should occur with various iterations of the design.

Behaviour observed in the simulation may depart from that seen in reality for several reasons. The simulation may not contain all the features of the real task quantitatively or qualitatively, and therefore behaviour may be somewhat distorted. In addition, behaviour may not be typical because of the operator's attitudes and motivation towards the simulator. The final problem may be the lack of trained personnel with the relevant experience. Against all this must be weighed the possibility that a simulation is the only the place from which any data can be collected.

Description of technique

It is difficult to be specific about how the technique would be used in practice. It would require the application of a range of other analysis and observational techniques to a particular embodiment of a task in the simulation. In some cases, for example to assess workplace layout, a physical model of desks, etc., may be used. This is usually termed a 'mock-up'. For looking at more task-based activity static simulations, for example a control panel representation, may be used. To assess the dynamic outputs of tasks, and their psychological demands on individuals and teams, a dynamic simulation may be needed. It is possible to represent a complete task environment (full-scope simulation), alternatively only a part of the environment may be required (part-task simulation).

The level of detail required (or the fidelity of simulation) is notoriously difficult to specify accurately. It should be possible to collect useful data with low levels of fidelity, and then to progress to more realism with critical tasks.

The situation involves an appropriate analysis of the task in order to determine the nature of the simulation to be used. In particular there is the need to consider which aspects of fidelity (or simulation realism) require simulation to yield the appropriate performance. This has to be weighed against the cost of the simulation, which will rise quite dramatically as further fidelity dimensions are built in. It is also important to consider which aspects of fidelity are relevant at different stages of system design. In the early stages it may be possible to go for low fidelity, but as the system design becomes increasingly refined, more realistic and complete simulations may be required.

There are a number of different dimensions along which the fidelity of a simulator can vary, and Stammers (1981) suggested that these could be classified as follows:

Stimuli
Included here would be the dynamic characteristics of displays as well as their 'face validity'.

Responses
This would concern the degree of realism of the control and its operation.

Stimulus/response interactions/intrinsic feedback
These would be difficult to separate from the above, but would concern features arising from relationships between displays and controls.

Task complexity
This would refer to the degree of representation of the rule structure, or number of characteristics of the real task which were simulated. For example, are all the faults that can occur in a piece of equipment simulated? Is it possible to take the system into all possible states? Is only part of the real system represented, albeit faithfully?

Temporal aspects
For most task analysis purposes it will be necessary for simulator trials to be undertaken at a very close approximation to 'real time'. However, there are two notable exceptions to this. First, if a task involves long periods of passive monitoring while waiting for the system to reach a particular condition, the simulator trials can be considerably shortened by speeding up the simulation for such periods. Second, for tasks where errors are possible but infrequent, it is sometimes possible to obtain more information about errors by speeding up the simulator and thereby increasing work stress.

Environmental stressors
It has been pointed out that there are a number of sources of stress in real situations. Attempts are made to control these in simulations. There is the stress that comes from the task situation itself, which may be hazardous to life, or novel in some way. There is the stress that comes from carrying out a complex task at an early stage of learning. In addition there is the possibility that the job context has environmental features such as heat, noise or poor illumination, which are not conducive to good performance. These factors will often be given as reasons for using simulators in a controlled way, to cope with them.

Situational pay-offs

Closely related to the above is that simulators are often built to protect the individuals from the results of inadequate decisions and actions that will occur during learning. These negative task pay-offs might concern life and limb, the environment, lost production or wasted materials. The need to control such variables is evident, but it is equally important to adequately prepare the individual to handle these same stressors before being transferred to the job environment.

Social environment

Some trainees have difficulties in coping with the social environment of the workplace following transfer from the training situation. This is a variable that is often ignored, but it becomes important with a broad view of simulation. Some social skill situations and team training situations would be also relevant here.

Analyst control

For a simulator or mock-up to be particularly useful during task analysis, the analyst should have control over the conditions presented to the operator under observation. This control might be the ability to adjust events with which the operator has to cope. This would include enabling a representative set of events to be presented and enable task loading. In this way, the analyst could study those aspects of the task that are of particular importance. The control might also permit adjustment of the rate of signals or 'noise' in the system. Analyst control would also cover the ways in which information can be recorded. There are a number of ways that the analyst might wish to control or record events to facilitate, or focus, data collection for task analysis. It is important to try to anticipate these before a simulator is designed, so that they can easily be included.

Full-scope simulators have been used to collect information on the nature of human errors in the operation of nuclear power plant control rooms (Beare and Dorris, 1983). For this activity, a high degree of realism is important. In other situations, a reduced scope or part-task simulator may be adequate.

It should be clear from the above that some task analysis should be undertaken before the simulation is devised, in order to establish which aspects of the task need to be simulated. In turn, task analysis techniques can be used to study behaviour in the simulator.

Strictly speaking, simulators/mock-ups are not task analysis techniques, but are safe or controllable environments in which operators can be observed, so that other task analysis techniques can be applied. Indeed, the use of simulators can be very beneficial in circumstances where behaviours are complex, where task demands are such that critical behaviours will only occur very rarely, or where a system has yet to be built. It should be clear that simulators are tools to be used in various activities. Their main use will be in situations where access to the real task environment is impossible, inconvenient, costly, or dangerous.

Practical advice

The main practical problem facing an analyst upon using a simulator is the sheer quantity of data. In order to ensure capture of the pertinent information, thorough planning of a session is essential. Before any simulator trial, the analyst should identify important target actions and try to anticipate any complex task events. Use several observers, including experienced observers as well as ergonomists. A simple pro forma for recording observations is preferable to an elaborately coded checklist. Video can provide valuable backup and may prove essential for the investigation of fast, complex activities. Exploit facilities provided by the simulator, such as freeze, snapshots, acceleration of real time, and event logs. Consider simple aids, for instance zoom lenses may be invaluable in a large control room simulator.

Resources required

These will range from a very simple paper and pencil simulation, or some computer-based interface prototyping tool at a relatively low cost, through to the high cost of a full-scale simulator, such as that required for a full control room. Obviously in the latter case, any simulator may subsequently be used for training purposes and, therefore, the costs may be already planned for in the provision of training for a new system. Similarly, an existing training simulator may provide an excellent facility for the collection of data about human performance.

Links with other techniques

As mentioned above, this technique would be used to provide the basic task context in which a range of other techniques, observational in particular, will be utilized. It can provide the source of data for information collected from structured interviews (pp. 66) and questionnaires (pp. 58) and other such methods. They can also be used for walk-throughs (pp. 160) etc. It is possible to look at performance under load by speeding up the event rate, or adding a secondary task and thus feeding into workload assessments via timeline analysis (pp. 142). It is also possible to examine performance on infrequently encountered tasks or potentially dangerous ones as part of task behavioural assessment (Chapter 5).

Advantages

- Advantages for this technique lie in its ability to be used to assess operator activities in a situation which is not normally easily observable.
- It may be relatively cheap compared with the possible mistakes that could be made in developing a large-scale system without any form of evaluation.

Disadvantages

- The behaviour observed may not be fully realistic. An example of this would be that a static simulation may not reveal the true nature of the operators' dynamic interactions with a system. Also, behaviour in a simulator may not fully replicate that found in the real situation because of the absence of real stressors found in the actual task (e.g. risk to life and limb, criticality of the process and presence of other operators and/or supervisors).
- If very high fidelity is provided, simulations can be extremely costly.

References and further reading

Adams, J.A. (1989) Human Factors Engineering (Chapter on Training and Training Devices). MacMillan: New York. *The chapter on training contains simulation and computer-based training, with some evidence of the former's effectiveness in the field of aviation.*

Beare, A.N. and Dorris, R.E. (1983) A Simulation-based Study of Human Errors in Nuclear Power Plant Control Room Tasks. *Proc. Annu. Meet. Human Factors Soc.*, 170-174. *A report on studies where simulation was used for collecting human factors information.*

Life, M.A., Narborough-Hall, C.S. and Hamilton, W.L. (1990) *Simulation and the User Interface.* London: Taylor and Francis. *Conference on the use of simulation in a range of application areas.*

Stammers, R.B. (1981) Theory and Practise in the Design of Training Simulators. *Programmed Learning and Educational Technology*, 18, 67-71. *A review of the problems associated with using simulators for training purposes.*

Stammers, R.B. (1986) Psychological Aspects of Simulator Design and Use. *Advances in Nuclear Science and Technology*, 17, 117-132. *An overview of the use of simulation in human factors.*

Table-top analysis

Overview

The essence of a table-top analysis is that a group of experts who have an understanding of a specific aspect of a system, meet together as a discussion group to define or assess particular aspects of a task. However, the discussions of such a group must be directed around some basic framework, which is often provided by working through a set of procedures, with a specific set of questions.

Application

In theory, a table-top analysis could be undertaken at any stage during the life cycle of a system. However, the prime use of the method is to expand upon task description information, and so it is most likely to be used during the design stage, at any point following the derivation of skeletal task descriptions. The technique can be used to deepen task knowledge of a system (e.g. to generate training or procedural knowledge) and also to resolve an identified problem (e.g. if the level of instrumentation in a plant is being changed, and the analyst wishes to explore the impact of this on task performance and/or safety). Table-top analysis can therefore create (on-line) detailed task information and/or can analyse that information in a problem-solving and explanatory way. In this way it is to some extent an on-line simulation of the task, if a somewhat ethereal and shifting one. The technique's usefulness lies in its use of expertise, so that detailed systems analysis and other knowledge can be quickly and powerfully focused on particular aspects temporarily being discussed,

Description of technique

Table-top discussion groups should be composed of experts who have a knowledge of the specific technical or operational aspects of the system in which the analyst is particularly interested. Therefore, the first decision which the analyst must make is what technical expertise is needed within the group. The analyst should then assemble a group of individuals who have the necessary expertise, but in doing so, should try to ensure that there is no more than one representative for each specialist area, otherwise there is a danger that one sub-group might tend to dominate the discussions.

Ideally, there should be between three and seven members in a table-top discussion group. However, if many different systems are involved, it may be necessary to exceed this number. In such situations, it may be preferable to arrange a series of meetings, so that no individual has to wait too long before being called upon to give opinions/information on the matter being addressed.

Once the analyst has decided upon the composition of the group, each member should be told what is required, and asked if they are willing to participate. Then the discussion session should be arranged for a date and location which is convenient to all group members.

The actual meeting should be chaired by the analyst, who should adopt a reasonably rigid structure to the discussions. One of the most convenient ways to do this is to base the discussions upon a limited set of operating procedures or problem descriptions, and then to ask for specific information about each step or aspect in these. Such procedures or problem definitions need not be very detailed, and indeed, the table-top discussion process can be very effectively used to develop more detailed procedures from skeletal ones. Another approach is to work through a problem scenario which has been prepared to elicit information of interest to the analyst. However, whichever approach is taken, in order to optimize the use of the experts time and to avoid gathering information which could have been gathered more economically elsewhere, any initial information should be as complete as possible. The table-top discussions can then concentrate upon gathering specialist information, or resolving particularly complex issues.

During the discussions, key information should be recorded as it is elicited, and providing that all the group members are agreeable, a tape recording can be made or the discussion transcribed on-line (i.e. by a stenographer). At a later date, this information can be transcribed onto a task decomposition data sheet (see pp. 95).

Group methods

There are various means of running group expert sessions. The simplest is to have the group meet and discuss all aspects of the problems being addressed, and to agree by unpressured consensus on whatever agreements are required of them. This single group approach has the primary advantage that experts can share each other's expertise and come to a rational unified decision. However, in practice there are some major limitations which tend to occur with expert groups. The first is that experts often do not agree, and, either because of communication failures or personality conflicts, rationalization and resolution of these disagreements does not occur, the argument simply ending in stalemate.

The second major problem is that of bias, which can occur in several ways. One member of the group may appear highly confident and persuasive, and even if incorrect, may sway the opinions of the other group members. Additionally, 'group dynamics' may occur so that one person out-talks others, and is assumed to be correct simply by having 'held the floor' for so long. Alternatively, an idea may be discussed at length and even though incorrect, it appears worthwhile simply because of its duration of discussion. Also, some groups may go to extremes, ending up with a decision or estimate which appears right at the time and in the group, but which would actually far exceed any member's own original decision or estimate.

The above and other biasing factors such as memorability of information, conservatism, and imaginability of scenarios (e.g. see Tversky and Kahneman, 1974) can to an extent be reduced by using a good chairperson to run the group, called a *facilitator*, who is experienced with teams and how to counteract biases or compensate for their effects.

Two alternatives to the above group consensus approach are the Delphi method and the nominal group technique (Gustafson *et al.*, 1973). The Delphi procedure has three main characteristics:

- Anonymity of groups
- Interaction with controlled feedback
- Statistical group response

The Delphi method first elicits judgements from experts individually, and then feedback of all judgements from all experts is (anonymously) fed back to each expert. The experts then reassess their judgements, individually and these are then aggregated. If they are quantified judgements (e.g. for the purposes of human error quantification) this is done statistically.

The nominal group technique (NGT) is similar, except that limited group discussion is allowed for clarification after the first feedback session. While this is potentially an advantage over the Delphi method, the latter has the advantages that it does not require geographical co-location of the experts and entirely avoids any personality conflicts or influences.

The degree to which the Delphi method and NGT are useful for table-top analysis depends upon the extent to which each participant has equal expertise, or enough expertise individually to make an assessment. If the participants do have such expertise then the Delphi method and NGT may be seen as useful. If, however the group is a problem-solving group for which all members are required to participate and share information (such that the 'whole' group is more than the sum of its 'parts'), then significant group discussion is essential, at least in the early stages. Furthermore, a group that goes through a table-top exercise and arrives at a consensus will tend to be bound more to the decisions from the exercise, and hence table-top discussions can be seen as consolidating a set of beliefs and understandings amongst the participants, which in itself may be useful.

Overall, the Delphi method and NGT are useful considerations when utilizing table-top discussions for two reasons. Firstly, they raise the problems of bias and personality conflicts which otherwise may be overlooked. Secondly, if table-top discussions are primarily between experts who each have sufficient expertise to make assessments, or between personnel who cannot easily be co-located, or if discussions break down through conflicts, etc., they can be used instead of the table-top method or as an adjunct to it.

Practical advice

In order to encourage full and frank discussions, the group members should be assured that their contributions will remain confidential to the analyst. If any tape recording is made of the discussion session, the analyst should also assure the group members that this will remain confidential.

Strong chairperson skills are essential to avoid individuals dominating the group, and to keep to the points which are of interest to the analyst.

Especially in complex systems, it is often useful for the chairperson to give a verbal précis of his or her understanding of particular points, in order to ensure that this is an accurate summary of the experts' comments and perceptions.

The analyst must not try to change opinions. In particular, the chairperson/analyst should not be rigid in insisting on reaching a consensus view on each point. If there is no consensus, this should be recorded, and the decision will have to be made by the analyst and 'supervisor' or project manager afterwards.

Resources required

Any table-top discussion will require a suitable meeting room with facilities for recording the sessions (e.g. tape-recorder, writing materials, secretary) and for displaying information (e.g. flip charts, blackboard). However, the main resources used by this technique will be the experts' and the analyst's time during the meeting and for later transcription and clarification. Although the actual table-top session will involve the relatively intensive use of human resources, the very nature of the discussions should mean that this is a very effective way to gather information from experts, as opposed to individual discussions with the experts, because they will be given a clear framework, and can respond directly to comments from other experts, or to complex issues which might require some compromises.

It is feasible in the future that tele-conferencing will expand the role of table-top analysis, as this would not require co-location of experts. Such a development awaits suitably widespread technology, and will doubtlessly raise new psychological problems of its own concerning how experts interact through such media. Therefore, while a tele-conference can facilitate table-top discussions if co-location is problematical, it is recommended that wherever possible, experts for table-top discussions are physically brought together.

Links with Other Techniques

In order to provide a framework for table-top discussions, it is necessary to collect some basic task information, and to present this in a form which is usable by the group. This can be done by conducting an initial hierarchical task analysis (pp. 104), or by using existing operating procedures.

The output from a table-top discussion can feed directly into task decompositions (pp. 95), and a task decomposition data sheet can provide a useful way for recording many of the comments obtained during a table-top discussion session.

In many respects a table-top analysis will be similar to a talk-through (pp. 162). However, in a talk-through only those who are actually undertaking the task make a contribution, and there is no discussion element. Hence, a talk-through should provide mainly operational information about a task, while a table-top analysis should be used to ensure that the task takes due account of various technical and operational requirements and limitations.

The table-top analysis is also very similar in its approach to the HAZOP technique (pp. 194), and can be used to consider performance adequacy, particularly if some change in task environment is about to occur or has already occurred.

Advantages

- Table-top discussions enable many different points of view to be considered together, so that potential conflicts can be identified early.

- Table-top discussions provide an economical way to obtain information from experts, because the discussions themselves form a good background context for these experts.

Disadvantages

- As in all interpersonal discussions, the great enemy is bias. The dangers of this can be reduced to a large extent by ensuring that the chairman of the group steers the discussions carefully, and obtains the views from all the members of the group, without permitting any individual to dominate the discussions.
- Unless there is strong chairing, there is always the risk that the group could get side-tracked into discussing peripheral issues.
- If the analyst has not undertaken enough preparation before the table-top discussion session, much time might be wasted collecting information which could have been obtained from readily available documentation or other sources, without involving the experts.

References and further reading

Gustafson, D., Shulka, R., Delbecq, A. and Walister, G. (1975) A Comparative Study of the Differences in Subjective Likelihood Estimates Made by Individuals, Interacting Groups, Delphi Groups, and Nominal Groups *Organisational Behaviour and Human Performance*, 9, 280-291. *A comparative study involving Delphi and Nominal Group Techniques.*

Tversky, A. and Kahneman, D. (1974) Judgement Under Uncertainty: Heuristics and Biases. *Science*, 185, 1124-1131. *Reviews biases in expert opinion.*

Walk-throughs and talk-throughs

Overview

Walk-throughs require personnel to perform some sort of demonstration of a task in realistic surroundings, without necessarily undertaking that task. Thus in a large system walk-through, subjects would probably walk round the plant pointing in turn to the controls and displays which they would use for successive steps, and explaining how and why each of these would be used. A talk-through is very similar, but it is undertaken more remotely from the normal task location, so that the tasks are verbalized rather than demonstrated (Meister, 1986).

Application

Walk-throughs and talk-throughs are generally used to describe and verify the principal observable components of tasks. Walk-throughs can take place in *real time* with a minimum of comment from the person who is performing the walk-through, or they can be paced by the respondent to permit more extensive comments and also, if necessary, to allow the analyst to request additional information or clarification. The analyst must use his or her discretion to determine the nature of the information which should be reported. Real-time walk-throughs are only required if temporal factors are of interest, such as task times or workload.

If the walk-through is not centred on an actual task performance on the real system, target tasks do not need to be real, and can be invented by an investigator. In such cases the techniques can reveal what operators would do in novel circumstances, or can allow them to role-play potentially stressful or dangerous situations.

Beyond straightforward gathering of descriptive data, these techniques typically feed forward into other evaluative functions, for example, whether a particular working practice is capable of achieving the desired goal, or whether procedures or training need to be enhanced.

Walk-through and talk-through exercises can conveniently be considered as lying on a continuum which represents the degree of fidelity with which a task and its interfaces are represented. At the highest end of this fidelity continuum would be real time walk-throughs on an actual system without the walk-through exerting any influence on either the nature, or the timing, of task performance. Talk-throughs sit at the opposite end of this continuum, because they use a low fidelity representation of both the task and the interfaces. Between these two extremes there are more intrusive walk-through techniques, which can use mock-ups, simulators or real plant.

The degree of fidelity which is necessary depends upon the issues which are being investigated. For instance, a low-fidelity talk-through exercise should be sufficient to enable an analyst to verify that all the controls and displays which are necessary for a task have been provided. However, in order to assess whether an operator is likely to be able to cope with the workload demands of a task, it will be necessary to use greater fidelity and avoid interruption of the respondent by the analyst.

Description of technique

A walk-through, as the name implies, involves a respondent *walking through* a task by demonstrating it on a real system, or an adequate simulation. If the system is large and distributed, the task may involve a significant amount of movement between locations. Walk-throughs of tasks confined to a single or limited control location still require the physical actions associated with the task to be demonstrated, though this can usually be accomplished merely by pointing, rather than by actually manipulating controls. Talk-throughs, on the other hand, involve only a verbal description of the respondent's actions. For both techniques the four stages outlined below are necessary.

Select method

There are many different ways in which these techniques can be applied, and it is essential for the analyst to develop a methodology which will ensure that all the required information is obtained. To assist in selecting the most appropriate method for any particular application, it is useful to categorize walk-through/talk-through methods using the following four classifications. For limited applications it should only be necessary to use one of these approaches, but for more comprehensive applications the analyst may have to develop an integrated walk-through/talk-through programme which utilizes more than one of the methods.

Talk-throughs
The main advantage of talk-throughs is that they do not require any special task environment, though it is helpful to have some technical documentation available, such as procedures or control/display panel drawings. This gives the advantage that talk-throughs can be undertaken early during the design stage before any mock-ups, simulators or real plant have been produced. However, for very complex plants some practical difficulties are likely to be encountered if talk-throughs are attempted too early in the design cycle, because no single person may have a sufficient understanding of the system to provide a useful talk-through. In such situations, it is likely to be more profitable to utilize the knowledge of several individuals and to undertake a table-top discussion (see pp. 155).

Another important, but often neglected, use of talk-throughs is as a means of obtaining an orientation prior to undertaking a more extensive walk-through. This will be particularly useful when the walk-through requires expensive resources.

Detailed walk-throughs
These are walk-throughs in which the emphasis is placed upon gathering information about particular aspects of a task, so that the task is paced by the rate at which responses are given, which will not necessarily correspond to the normal rate of task performance. Prior to such a walk-through the analyst should explain to the respondent what sort of information as required, and during the actual walk-through the analyst may wish to prompt the respondent for particular information or for clarification.

Detailed walk-throughs can be undertaken in any task environment from a very basic mock-up to a fully finished system, and they can be used to gather a wide variety of task information. Thus they are a useful means of assessing the interfaces which are to be provided for particular tasks, at any stage during the design cycle.

Real-time walk-throughs

In real-time walk-throughs the main aim is for the respondent to closely emulate the way in which a task would be undertaken. An essential element of this is that the walk-through should be undertaken at a realistic pace, and so for most tasks will be limited the depth of the comments which the respondent is able to provide.

In order to enable a person to undertake a real time walk-through at a realistic pace, it will be necessary to undertake such walk-throughs in a relatively high fidelity task environment. Although it will be helpful if the system response times are simulated accurately in the task environment, this is not essential, provided that the analyst can obtain an accurate assessment of the most significant system response times. Similarly, it will generally be beneficial to have some form of dynamic displays, but for many real time walk-throughs, static displays should be adequate.

The main reasons for undertaking real time walk-throughs will be to determine whether the workload which is imposed by a particular task will be feasible for an operator, and to assess the effects of dynamic factors upon task performance.

Team walk-throughs

These are real time walk-throughs which are undertaken by a team of workers who are either undertaking the same task, or a group of highly related tasks. Such walk-throughs are useful for investigating the interactions and communications between individuals in complex multi-person tasks, and assessing how effectively the constituent tasks can be linked together. In order to be effective, team walk-throughs require a relatively high fidelity simulation, and all team members must be experienced with the system and its interfaces.

Preparing for a walk-through/talk-through session

For existing systems, the most appropriate personnel to perform the walk-through/talk-through will generally be individuals who have some operational experience of it. However, there are other criteria which could be used to select the respondents, especially when this technique is being used at an early stage in the design cycle. For instance, in some circumstances designers may have sufficient knowledge of a system to provide a useful commentary, while individuals who have absolutely no prior knowledge may be essential to evaluate systems which are intended for usage by the general population (such as an assessment of emergency evacuation procedures).

The analyst must also develop a scenario which will address the particular aspects of the task, or its interfaces, which are of interest. For some purposes this could be limited to a very brief statement of the starting conditions or the final goals which are to be achieved. However, for other purposes it will be necessary to prepare a much more detailed description of the scenario, which will list the values, or status, of key parameters at various points in the walk-through/talk-through.

Undertaking a walk-through/talk-through session

Walk-throughs are not an exact science, and so it will not be possible to predict in advance whether all the issues of interest will be adequately covered. Therefore, it will be particularly useful to run a short pilot trial before embarking upon an extensive programme. Any shortcomings can then be rectified and tested out if necessary, before proceeding. Similarly, for particularly complex tasks it may be helpful to undertake a brief preliminary talk-through to orientate the analyst before a more comprehensive walk-through.

At the start of each session the analyst should explain to the respondent what is expected of him or her, and should describe the roles of the analyst and any assistants. The analyst should also explain how any data which are collected will be used, and should ensure that the respondent fully understands what is required and that the exercise is an assessment of the system NOT a test of the respondent. If the analyst intends to prompt the respondent for any additional information it should be explained that the exercise may be interrupted, and similarly if no prompting is to be given, the respondent should be forewarned that the analyst will not directly acknowledge any questions.

There are four general ways of recording data from walk-through/talk-through sessions, and the choice of method depends very much upon the nature of the study.

Data sheets
Specially prepared data sheets can be useful for recording a limited number of specific points about all the steps in a task. However, unless these data sheets are designed very carefully, it can be difficult to fill them in adequately during a session. Therefore, prepared data sheets should not be used without first making some test of their efficacy.

Written comments
Although unstructured written comments offer perhaps the least elegant way of gathering information, they can be particularly effective, provided that the walk-through/talk-through can be interrupted when necessary to permit the analyst to make adequate notes. This means that this approach will generally not be tenable for real-time or team walk-throughs (unless the analyst has previously studied these tasks and has good reason to suppose that there will only be a limited number of problems encountered during the walk-through). To assist with the collection and analysis of data by written comments it is useful to use a very basic data sheet which has space for a step number and a short comment.

Audio/video recording
This provides the best way of ensuring that all the verbal information from the session is recorded. However, the analysis of such recordings can be somewhat time-consuming, and their use is probably best limited to situations where it may be difficult to make a written record, or where the recordings are only intended to supplement written records when the latter are unclear or insufficient.

Data logging
If facilities are available for using the system to automatically log the values of key parameters during a walk-through, it is possible for the analyst to obtain a very detailed impression of certain aspects of a task. However, such analysis will be very time-consuming, and is best limited to the examination of a limited number of key elements in a task, rather than being used to analyse task performance in its entirety.

Debriefing

Subsequent to any walk-through session it will generally be useful to hold a short debriefing. During this debriefing the analyst should raise general issues which have been noted, or should pursue any aspects of the task which appear to cause particular difficulty. The respondent should also be encouraged to raise any points which gave him or her concern. The motivational effects of doing this upon the respondent will generally in themselves provide justification for undertaking a debriefing.

Practical Advice

Before using either of these techniques for an investigation, it is advisable to spend some time becoming familiar with the basic characteristics of the system under investigation, and relating these to the objective(s) of the study. It is essential that the analyst has a clear impression of the type of information and the amount of detail which is required. It will also be helpful, if not essential, for the analyst to gain some orientation about the task prior to the walk-through.

If a full walk-through with comprehensive data recording is required, then all problems must be anticipated and solved before sessions begin. Unnecessary delays, or worse still, apparent unpreparedness on the part of the analyst, can jeopardize the success of any study which relies on goodwill and co-operation.

In any walk-through/talk-through it is possible that the respondent might omit an important point, particularly in talk-throughs, or limited fidelity walk-throughs, where the respondents may not be prompted by the familiar cues provided by a real system. Therefore, the analyst must be on guard against omissions, errors and inconsistencies in descriptions. This problem requires the investigator to acquire some knowledge of the system in order to assess the quality of the description which is being given. Hence it is also useful to have more than one respondent carry out the walk-through/talk-through independently.

In order to obtain a realistic picture from a walk-through or a talk-through, it is essential to gain the respondent's confidence by assurances that comments will be treated in complete confidence, or else there is a risk that the information will reflect the official approach to the task, rather than the manner in which it is actually performed.

Resources Required

For a walk-through, it is obviously necessary to have some form of system available on which the task can be demonstrated. For some purposes it will be necessary to use a fully operational system, but in other cases a relatively low-fidelity simulation, or a mock-up, can be sufficient. For instance, it is possible to perform an effective walk-through to study the dialogue structure of a computer-based system, by using a set of flip-charts, which an assistant moves to represent different screens in response to user requests. However, in order to look at time stress on the same system, fully operational software would be required.

In principle, no simulation whatsoever is required for a talk-through. However, in practice, some representation of the system is invariably needed to provide a means of describing the boundaries of the task to the respondent, as a prompt for the respondent, and to assist the analyst in following the respondent's explanations. Usually, a drawing will be adequate, but photographs may help, as may more technical representations such as Piping and Instrumentation Diagrams.

At least one respondent will be required for each walk-through/talk-through, and generally this person will require some familiarity with the system, especially for real-time studies. Also at least one analyst must be present throughout the session, but more may be required depending on how much supplementary manual data collection needs to be made. For real-time walk-throughs, each session should take only slightly longer than it would normally take to complete the task itself. However, if the respondent is being asked for a large amount of information, even a short task sequence could take a long time to describe adequately in a detailed walk-through/talk-through.

Some resources will be expended prior to the walk-through, or talk-through sessions, developing or identifying suitable sub-tasks or scenarios to focus on. In some cases, this may be a clearly defined task within a larger process, with a beginning and an end and clear procedures in between. However, if the aim of the study is to document task performance in unusual situations, the investigator will often need to spend time identifying and understanding some suitable system states and behaviours. In such cases, it may well be necessary to enlist the help of an experienced system engineer or designer.

Data recording resources must be provided according to the type of data required and the situation. With simple talk-throughs at a relaxed pace, a pencil and paper may be all that is required. However, for more demanding situations it can be useful to make an audio, or video recording, if only as a back-up to other data collection methods.

Links with Other Techniques

Walk-throughs and talk-throughs bear a close relationship to verbal protocols. However, whereas in a verbal protocol session (pp. 71) the respondent must actually perform the task and describe the unobservable mental processes, a walk-through requires a rationalized and selective description of the task steps without actually performing a task. In some cases, cognitive information (e.g. interpretations or problem-solving strategies) will be elicited from walk-throughs, but because such comments will be consciously selected and retrieved from memory, they will be much more limited and open to bias than verbal protocol information. A talk-through is one step further removed from actual task performance, in that the demonstrations and indications of task elements are supported only by a drawing (for example) rather than by a real plant or simulation. Walk-throughs may require some form of simulator/mock-up (pp. 150) if the system does not yet exist, or if it cannot be used for this purpose.

Walk-throughs and talk-throughs are useful data collection techniques, and the results can be fed into one of several analytical techniques (e.g. fault or event tree analysis, pp. 188 or 178; or hierarchical task analysis, pp. 104), either to describe performance or assess its adequacy. Walk-throughs and talk-throughs can be used in their own right to assess whether a task is feasible and can be carried out in time. These techniques can also be useful as training media, especially for scenarios which are infrequent and/or dangerous or difficult to simulate.

Virtually any of the graphical representation techniques described in this book can be used to redescribe the data collected in a more useful way. Petri nets (pp. 91), link analyses (pp. 118) and operational sequence diagrams (pp. 125) can all be constructed from sufficiently comprehensive walk-through, or talk-through investigations. If the task is performed in real-time, walk-through data can also be used in temporal representations such as timelines (pp. 135). However, it is possible that normal task times will become distorted during walk-throughs, so such information must be used cautiously.

Advantages

- Although it is not necessary actually to perform a task – only to demonstrate it – a walk-through is effectively an actual performance of a real task, and as such will yield a very accurate and straightforward description of the task. This is also generally true of talk-throughs, since the respondent is describing observable events and behaviours rather than inferred mental processes.
- In the case of talk-throughs, or if the walk-through is not undertaken in real-time, it is possible to 'interrupt' the process with periods of questioning or role playing as described above. 'What if' questions can be interposed, thus enabling a kind of informal one-to-one HAZOP study to be undertaken (see pp. 194).

- Like many observational techniques, walk-throughs or talk-throughs can be set up and executed quite rapidly.
- For straightforward walk-throughs, or talk-throughs, the investigator needs very little specialist training or knowledge of the system being studied.

Disadvantages

- In common with observational techniques, time saved in data collection is typically lost in data reduction and analysis, which can be extremely time-consuming.
- The technique is not normally used unless a skilled operator is available. The definition of 'skilled operator' can clearly be stretched to include anyone with a thorough knowledge of the system. However, the investigator must be satisfied that the quality of information will be as good as that which would be obtained from an experienced operator of the working system. The exception to this is if the system being studied is intended either for casual and untrained users, or for users trained on a different system. In such cases, the user's expectations of the way in which tasks should be performed will be a focus of interest. Thus, it will be possible to assess the discrepancy between the operator's assumptions and the design requirements, and use this assessment as a measure of system usability.
- In the case of 'interrupted' walk-throughs, the investigator needs to have considerable knowledge of the system, in order to pose sensible questions or scenarios.

References

Meister, D. (1986) *Human Factors Testing and Evaluation.* New York: Elsevier.

See Chapters 8 and 15, which used forms of talk-through and walk-through respectively.

Chapter 5
Task behaviour assessment methods

Barrier and work safety analysis

Overview

Two techniques of safety analysis are outlined. These are barrier analysis and work safety analysis. Barrier analysis (Trost and Nertney, 1985) focuses on the transfer of harmful energy to vulnerable objects (e.g. people), establishing what barriers should have been in place to prevent the accident, or could be installed to increase safety.

Work safety analysis is 'a systematic investigation of working methods, machines and working environments in order to find out direct accident potentials' (Suokas and Rouhiainen, 1984). Its primary goal is the identification of potential hazards, and appropriate corresponding protection measures.

Both of these techniques focus on the analysis of protective measures in systems. The difference between them lies in their perspective: barrier analysis looks qualitatively and functionally at the barriers that should be present to prevent unwanted energy flows reaching targets (people); work safety analysis looks in detail at each step of the task to see what hazards could occur, and to provide a rough quantitative calculation of their relative risks, and hence what barriers are needed.

Application

Barrier analysis

This technique can be used on its own or as part of the management oversight risk tree (MORT: pp. 208) approach for accident investigation or safety analysis. In an accident investigation it is used firstly to establish the nature of the accident in terms of what energy directly caused the consequences in terms of death, injury or loss. Barrier analysis then determines which barriers failed to prevent the undesired energy flow. In a MORT accident investigation, barrier analysis is used early on to determine what actually happened (i.e. the physical mechanisms which allowed the accident to happen). MORT itself would then try to determine why the barriers failed.

In a safety analysis, barrier analysis can be used prospectively to analyse qualitatively whether sufficient barriers exist to ensure adequate safety, or whether extra barriers should be in place. As with accident investigation, this analysis can be used on its own or as part of a MORT assessment.

Work safety analysis

The types of tasks most suited to Work Safety Analysis (WSA) are those with immediate risks, for instance industrial safety risks in systems involving:

- Production tasks
- Operating with moving machinery (including robotics systems)
- Repetitive maintenance tasks, etc.

WSA is also particularly appropriate if systems of work are being changed (e.g. due to modernizing plant equipment, changing plant sites, upgrading procedures or working methods, etc.). WSA is aimed at analysing either existing tasks, or tasks being designed. WSA only covers hazards which are more or less directly connected with the task step. Hence, an unwanted energy flow which is not related to the task steps under investigation (but which could none the less cause a fatality) would not necessarily be considered by WSA. Also, the emphasis is on physical and human subsystems: the information system is only partly reviewed, and the management system not at all (Suokas, 1988).

Description of techniques

Barrier analysis

Barrier analysis defines an accident as an unwanted flow of harmful energy or exposure to an environmental condition that results in adverse consequences. It further defines the four necessary ingredients for an accident to be able to occur:

- A harmful energy flow or environmental condition
- Vulnerable people or objects
- A failure or loss of protective barriers that keep the two above elements separated
- The events and energy flows that lead to the final accident phase (i.e. in which the consequences occur)

If one of the above is missing, then an accident cannot occur. The types of energies that may be harmful can take many forms, but usually fit into one of the following categories:

- Electrical
- Environmental condition
- Chemical
- Thermal
- Biological
- Radiation
- Kinetic

Haddon (1973) outlined a number of strategies or barrier types to keep unwanted energy flows from potential targets:

- Alternative energy source: e.g. in the case of Bhopal, utilization of a different chemical than MIC (i.e. with similar process utility, but without the harmful effects);
- Reduction of amount of energy and prevention of its build up: e.g. reduction of voltage; use of gas detectors; avoidance of large fuel repositories, etc.
- Prevention of sudden release: e.g. use of safety valves or bursting discs; use of containments such as pressure vessels; secondary containment, etc.
- Modification of the rate of release: e.g. slow the burning rate; or use a lower speed of rotation of machinery, etc.
- Separation of targets from energy via space or time: e.g. restricted areas; traffic lights; power cables out of reach, etc.
- Use of barriers between source and target: e.g. protective guarding; shielding, etc.
- Modification of shock concentration surfaces, or use of energy attenuation devices: e.g. acclimatization to heat/cold; or use of earplugs, etc.
- Strengthening of the target: e.g. fortified building structures, etc.
- Damage limitation measures: e.g. evacuation; alarm signals; sprinklers; etc.

The barriers themselves may be on the energy source, or between the source and the target, on the target itself, or through separation of source and target in time and space. Barriers may be physical or administrative (e.g. procedures, warnings or training, etc.). Barrier analysis can then be used to do the following:

- Define the energy sources involved
- Define all the barriers that should have been present
- Identify which barriers failed and how they failed
- Establish the sequence of events and barrier failures leading up to the accident

Barriers tend to fail (or be missing) for one of the following reasons:

- Necessity of barrier not realized
- Barrier not possible
- Barrier too expensive
- Physical barrier failure
- Operator error

Having defined energy sources, barriers and barrier failure mechanisms, the final phase is to reconstruct the series of events which led to the accident or, if carrying out prospective safety analysis, which could lead to an accident. In the former case, this will involve the use of accident investigation procedures such as MORT (pp. 208). In the latter case other methods such as fault/event trees (pp.

188 and 178) may be used to consider the nature and adequacy of barrier systems in preventing accidents.

Human error analysis of barrier systems

Barrier analysis may be used to focus on human errors which may overcome barriers (e.g. if carrying out a human reliability analysis). In this case, after barriers have been identified, ways in which human errors, intentional or unintentional, could jeopardize the barrier's effectiveness, can be considered. Such errors may be simple unintentional 'lapses' (e.g. forgetting to re-energize a protective system on leaving the protected area), or could be an intentional 'rule violation' (e.g. going into a restricted zone around rotating machinery to remove an obstructing piece of debris) which is carried out to get the job done quicker.

Table 5.1 Example of the use of a barrier approach to error identification (Bellamy et al., 1986)

Barrier		**Barrier failure**	
Function	**Type**	**Design features and assumptions**	**Human errors**
1.1 Dropped object protection (DOP)	Physical	Protective decks Drop out area *Assumptions:* Securing of heavy equipment No design errors	Drop equipment in unprotected areas Deck not constructed or installed as designed Deck not inspected and maintained as designed Inspection error (miss/false alarm) Maintenance error Failure to secure heavy equipment Leave DOP hatches open
1.2 Containment and isolation of flammables	Physical	Vessels Pipework Gas/oil tight decks etc. Hydrostatic barriers Cement Valves Blowout prevention (BOP) Interlocks *Assumptions*: No design errors	Valve operation errors Hose connection errors Failure to use hot work tent Failure to maintain hydrostatic well barrier Inadequate cementing of casing Equipment not connected or installed according to design Inspection error (miss/false alarm) Maintenance error Failure to operate BOP correctly BOP removed at wrong time Breaking into live vessel or pipework Leaving gas tight doors open Disabled interlocks

An example of this type of approach applied to an offshore system safety investigation is shown in Table 5.1 (Bellamy *et al.*, 1986). Tabular presentation is generally useful for barrier analysis.

Following such analysis of errors leading to barrier failure, other barriers may be designed into the system, in consultation with the design/safety/operations personnel responsible. Such 'barrier recovery systems' may mean extra physical devices (e.g. interlocks), or more robust administrative controls, including extra training to increase the level of compliance with such barriers.

Work safety analysis

WSA firstly needs a list of the work steps involved in the job. This can be obtained by carrying out a HTA. For each work (task) step, potential hazards are considered (Suokas and Rouhiainen, 1984 provide a list of hazards and examples of their causes). Each hazard is described in the WSA tabular representation (see Table 5.2) in a way which defines the consequences (e.g. worker crushed between two equipment items). Causative factors are then noted (i.e. factors which contribute to the hazard's occurrence). The analyst(s) must then judge the severity of the consequences and the likelihood of the hazard. Preventive and corrective measures are then developed by the analysts, according to the event's probability and severity. These basic ingredients of a WSA are defined further below, based on Suokas and Rouhiainen (1984).

Work step
In this part of the table the phase of work is defined. Additionally, the component/machine under investigation or observation is defined, as well as any other auxiliary devices so that such systems and subsystems can be precisely identified and documented.

Hazards and causative factors
The potential hazards and causative factors associated with the work step, its machinery and auxiliary devices, are noted. The aim is to find all hazards, whether they are caused by any of the following: the machine, the working method, working conditions, the operator, other operators working with equipment nearby, or environmental variations (e.g. temperature variations) or disturbances (e.g. strong winds affecting hazardous materials handling). Additionally, variations in working methods often contribute to accidents (i.e. when the task is being carried out under abnormal conditions). While such conditions will be infrequent, the risk of accident is likely to be significantly higher, making their overall contribution to risk relatively high. In particular, if equipment is less protected (i.e. usual protective systems are disabled) in a 'maintenance mode', then this may call for a significant WSA investigation.

Examples of hazards and causative factors are shown in Table 5.2, adapted from Suokas and Rouhiainen (1984). Typical operator-related questions which can be used in an analysis to generate potential hazards include the following:

- If the equipment can be used improperly at some time, it probably will be: what hazards will this cause?
- What short-cuts can be taken to overcome awkward procedures?

Table 5.2 Work safety analysis of a roll change (from Suokas and Rouhiainen, 1984)

Work step Machine component Auxiliary device	Hazard	Causative factors	Classification						Corrective actions
			Before			After			
			P	C	R	P	C	R	
Lifting the roll	A worker may get squeezed between the roll and the machine	The crane is not exactly vertical with the roll	3	5	15	1	5	5	A marking on the machine surface enabling the identification of the right position of the crane
Moving the roll with crane	A worker may fall down	The workers have to climb on the machine to protect the roll with planks Platform removed	3	3	9	0	3	0	Stationary pads are set on the roll, in which case the plan k control and falling between rolls are avoided
Cleaning the lower surface of the roll with compressed air	Litter may get into worker's eye	Compressed air makes litter fly around	3	2	6	1	2	2	Safety glasses are used
Change of the lifting ropes	The roll may fall down	The roll tilts and the rope slides	3	5	15	1	5	5	A lifting plank is used. Instructions: The belts must be placed carefully so that the roll cannot tilt
Setting down the roll on trestles on the floor	The floor may give way	The roll is set down in a wrong place	2	3	6	0	3	0	The proper place for the roll to set down is marked on the floor
Change of doctor blade	The blade may cut	The blade is touched	2	2	4	1	2	2	Safety gloves are used

- If maintenance is difficult, it will suffer and errors/omissions will occur: what hazards can arise from this?

Risk classification
Each identified hazard must then be classified to derive an appreciation of its relative risk. The relative probability is rated on a 5 point scale as follows:

0 – Hazard eliminated
1 – Very improbable (once in 10 years)
2 – Improbable (once in 10 years)
3 – Slightly probable (once a year)
4 – Rather probable (once a year)
5 – Very probable (several/many times a year)

The consequences are then categorized as follows:

1 – Insignificant (only first aid required)
2 – Little (1–2 days of disability)
3 – Considerably (3–21 days of disability)
4 – Serious (22–300 days of disability)
5 – Very serious (over 300 days of disability)

The relative risk (R) is then calculated simply by multiplying the probability by the consequences, as shown in the example in Table 5.2.

Corrective actions
Ways of reducing/eliminating risk are identified mostly during the investigation itself (e.g. simply by asking the operators how the system could be made more safe – *note this has the advantage that operators are unlikely to suggest awkward or over-constraining procedures which are then unlikely to be complied with*). The types of corrective actions feasible will generally fall into the same categories as for barrier analysis as discussed earlier.

Once corrective actions have been identified, the effectiveness must be checked, and if they are at all complex, it will be necessary to review work steps to see if new hazards have been introduced by the corrective actions themselves. Lastly, if new working methods have been introduced, often their initial introductory period must be checked to see if workers still use them (i.e. if the novelty has 'worn off'). Maintaining safety is a continual process.

Practical advice

Barrier Analysis

The identification of physical protective barriers is usually far easier than identification of administrative barriers. The former can be achieved by inspection of the workplace, design specifications, procedures and safety documentation. The latter can to an extent rely on procedural and safety technical information. However, particularly for training, supervisory and restricted area-type barriers, it is important to discuss with personnel their attitudes to, and knowledge of, such barriers. In practice, as incident and accident experience shows, administrative barriers are often overcome, particularly when they interfere with the actual efficiency of the task, or make the operator's task more difficult. Also, errors which can inadvertently overcome barriers must be considered by the analyst.

Designing barriers for protection is a difficult process and requires input from design, safety and operational personnel. The adequacy of the barriers will ultimately be judged in the context of the consequences of the potential accident (e.g. on a submarine it may be sufficient to provide a perspex cover over a pushbutton to prevent accidental raising of a periscope; but more substantial barriers, such as two independent keys and passwords, would be more appropriate for nuclear missile launch controls).

Work safety analysis

WSA in total is best carried out as a team approach, with members from management, workers and safety involved in the study. The analyst may need a fairly large 'reference' group to refer potential hazards and risk reduction measures to consider their appropriateness and realism, etc.

Suokas (1981) noted that slides and videotapes in particular proved effective tools in giving information on identified hazards to designers, and further aided worker training.

In WSA, the following task situations can be usefully investigated to carry out comprehensive work-related hazard identification:

- Work preparation
- Start-up
- Disturbance-related actions
- Stopping and shut-down
- Maintenance and repair

Corrective actions which are purely motivational in nature tend not to work, or else only work for a short time (e.g. asking operators to 'be careful', while laudable, will not necessarily be sufficient to prevent an accident from occurring). In general, the effects of risk reduction measures should not be overestimated by the analyst: measures rarely fully eliminate risk, they reduce it, and often introduce other potential hazards, albeit hopefully less risky ones. An analyst should preferably carry out a follow-up study (e.g. 6 or 12 months later) to determine how effective the measures have been.

Resources required

If barrier analysis is being used for accident investigation, then experienced and trained accident investigators are recommended. If it is being used for safety analysis, expertise in safety and human reliability assessment is necessary to define potential hazards and barrier failure mechanisms. Usually it is best in such situations to have a hybrid team available (e.g. safety assessors, human reliability analysts, personnel with operational experience, etc.).

WSA requires a fair degree of resources if a full evaluation (including video recording if appropriate) is being carried out, as it requires access (albeit on an intermittent basis) to a large number of technical personnel. However, WSA can be utilized to carry out a quick audit of a work method to gain an impression of the more observable hazards and risks inherent in the systems. This quick approach may be utilized, for example, if a 'near miss' incident has occurred and has raised concern about a particular part of machinery, and if a quick response is required to deal with the matter.

Links with other techniques

Barrier analysis has strong links with MORT (pp. 208) and can be used with fault or event tree analysis (pp. 188 and 178) and in particular with HAZOP (pp. 194) and failure modes and effects analysis (pp. 184). It also ideally needs techniques such as observation (pp. 53), structured interviews (pp. 66), and the critical incident technique (pp. 47) to gather information, and hierarchical task analysis (pp. 104) and/or task decomposition (pp. 95) to structure the data collected.

Work Safety Analysis has strong similarities with human HAZOP (pp. 194) and failure modes and effects analysis (pp. 184) and similarly to barrier analysis, utilizes observation (pp. 53), structured interviews (pp. 66) and critical incident techniques (pp. 47) to gather information, and hierarchical task analysis (pp. 104) and task decomposition (pp. 95) to develop the work steps (WSA's need for the latter two techniques is stronger than that of barrier analysis).

Advantages

- The advantages of barrier analysis in accident investigation are that it can be used to provide an unbiased description of what happened, including not only physical barrier failures, but also failures of administrative controls, and to identify the absence of additional barriers that, with hindsight, should have been in place.
- In safety analysis work, barrier analysis offers a useful meeting point for safety analysts, designers and human reliability practitioners, and focuses on all protective systems, with a realistic appreciation of their effectiveness and potential failure in predicted accident sequences.
- WSA offers a systematic method of providing a detailed view of immediate accident risks related to the work under surveillance, focusing on individual risk in the workplace based on all relevant operator tasks, frequent or infrequent.

Disadvantages

- A disadvantage with barrier analysis is that in predictive analysis for future systems, there may be a tendency to assume more compliance with administrative barriers than actually tends to occur.
- Both techniques, but particularly WSA, are limited in the extent to which they can consider complicated combinations of events which can lead to accidents. For such combinations, fault and event trees are more appropriate.

References and further reading

Bellamy, L, Kirwan, B. and Cox, R.A. (1973) *Incorporating Human Reliability into Probabilistic Risk Assessment.* 5th International Symposium 'Loss Prevention and Safety Promotion in the Process Industries'. pp. 6.1-6.20. Paris: Société de Chimie Industrielle, 28 Rue St. Dominique, F75000 Paris, France.

Haddon, W. (1973) Energy Damage and the Ten Counter-measure Strategies. *Human Factors*, 15/4, 355-366.

Suokas, J. (1981) *Experiences of Work Safety Analysis.* Paper presented at the SCRATCH seminar at Barnholm, Finland, May, Technical Research Centre of Finland.

Suokas, J. and Rouhiainen, V. (1984) *Work Safety Analysis – Method Description and User's Guide* VTT Rept. No. 314. Helsinki: Technical Research Centre of Finland.

Trost, W.A. and Nertney, R.J. (1985) *Barrier Analysis.* Rept. No. US DOE-76-45/29, SSDC-29. Idaho Falls, Idaho, USA: EG & G.

Event trees

Overview

Although event tree techniques were originally developed for the analysis of system reliability (see Henley and Kumamoto, 1981), they can be effectively used to study human reliability. As a technique for task analysis, event trees can be used to:

- Investigate the ways in which sequences of operator actions can develop
- Identify the possible end points and consequences of these sequences

The event tree is a tree-like diagram consisting of nodes and connecting lines, as illustrated in Figure 5.1. The nodes correspond to the different stages in a sequence, and the lines which lead out of the nodes correspond to the operator's possible modes of behaviour. There are two basic ways that this behaviour can be depicted on the event tree. The traditional and simplest approach is to depict the action of the operator as either *success* or *failure*, in which case two paths would lead out of each node. The alternative approach is to represent the alternative modes of behaviour explicitly, in which case several lines could lead out of a node. (as in Figure 5.1) The latter approach is more complicated, but it can provide a more explicit and detailed representation of more complex human behaviours, and may be necessary to model, for example, the different potential diagnoses that an operator could make during an emergency. Generally however, the 'binary' mode of event tree will suffice.

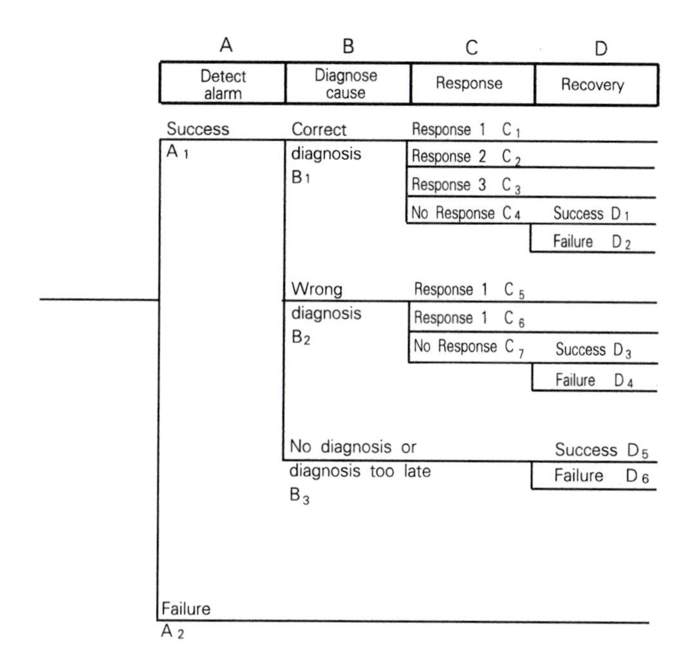

Figure 5.1 Event tree

Event trees show the relative importance of different tasks and errors and they indicate how these impact upon system safety and reliability. Event trees also provide the framework for probabilistic analysis, whereby the probabilities of the various actions are assessed. These can then be combined mathematically to calculate the probabilities of the different action/event sequences, and ultimately the overall risk of the system.

Application

Event trees are used to depict graphically the different permutations of operator behaviour that may occur during a procedure and to identify the various outcomes that are possible.

Event trees can also be used to identify the tasks and errors which have the greatest effect on the safety or reliability of the overall system. This can be done through qualitative analysis by visual inspection of the event tree, or through quantitative analysis by calculating the probabilities of the different human errors and sequences.

Description of technique

The technique employs a tree-like structure to represent graphically the various paths of behaviour which are possible in an activity. Each task in a behavioural or event-driven sequence is represented by a node in this tree structure. The possible outcomes of the task are depicted by paths which lead out of the node. Traditionally, two paths corresponding to *success* and *failure* are drawn from each node. An alternative approach is to show the modes of behaviour explicitly, in which case many paths could lead out of a node. This latter approach would be used for example if looking at diagnosis, in which case diagnosis could be correct, or fail, or could lead to misdiagnosis (a third option) and hence a new event sequence. In such cases this third branch or path would have to 'transfer' to another event tree (see Hannaman *et al*., 1985).

Many variations of event trees have been developed including operator action trees (OATs) (Hall *et al*., 1982), enhanced operator action trees (EOATs) (Hannaman *et al*., 1985), Human reliability analysis event trees (HRAET) (Bell and Swain, 1983) and generic accident sequence event trees (Kirwan and James, 1989). However, while they differ in their mode of presentation their underlying philosophies are the same. Figure 5.2 shows the generic accident sequence event tree, which shows in generic fashion the human contributions (positive and negative) towards accident progression. The HRAET is a simpler form of event tree, usually with a diagonal line representing success, and individual branches leading diagonally off the success diagonal representing failure at each point in the task step. It is used mostly in conjunction with a specialized technique for predicting error probability called THERP (technique for human error rate prediction; Swain and Guttmann, 1983).

The steps of the sequence, and the possible behaviour at each step, are identified by human factors reliability methods of error identification (see *Links with other techniques*). The level of detail of the task elements can vary as it would do within a hierarchical task analysis. At a high level, a typical step definition could be *initiate emergency shutdown*. At a lower level the component substeps such as detection and diagnosis, would be defined explicitly.

Recovery paths can be drawn where opportunities exist for the operator to correct errors. These are positioned in the event tree wherever they provide the most meaningful logic structure. In Figure 5.1, the final recovery actions are positioned at the end of the sequence, whereas in other situations it may be appropriate to model the recovery step elsewhere. For example, in a task including two operators, recovery of an error may occur by the second operator acting as a checker, and in this case it may be appropriate to model the recovery step immediately after the first operator's error.

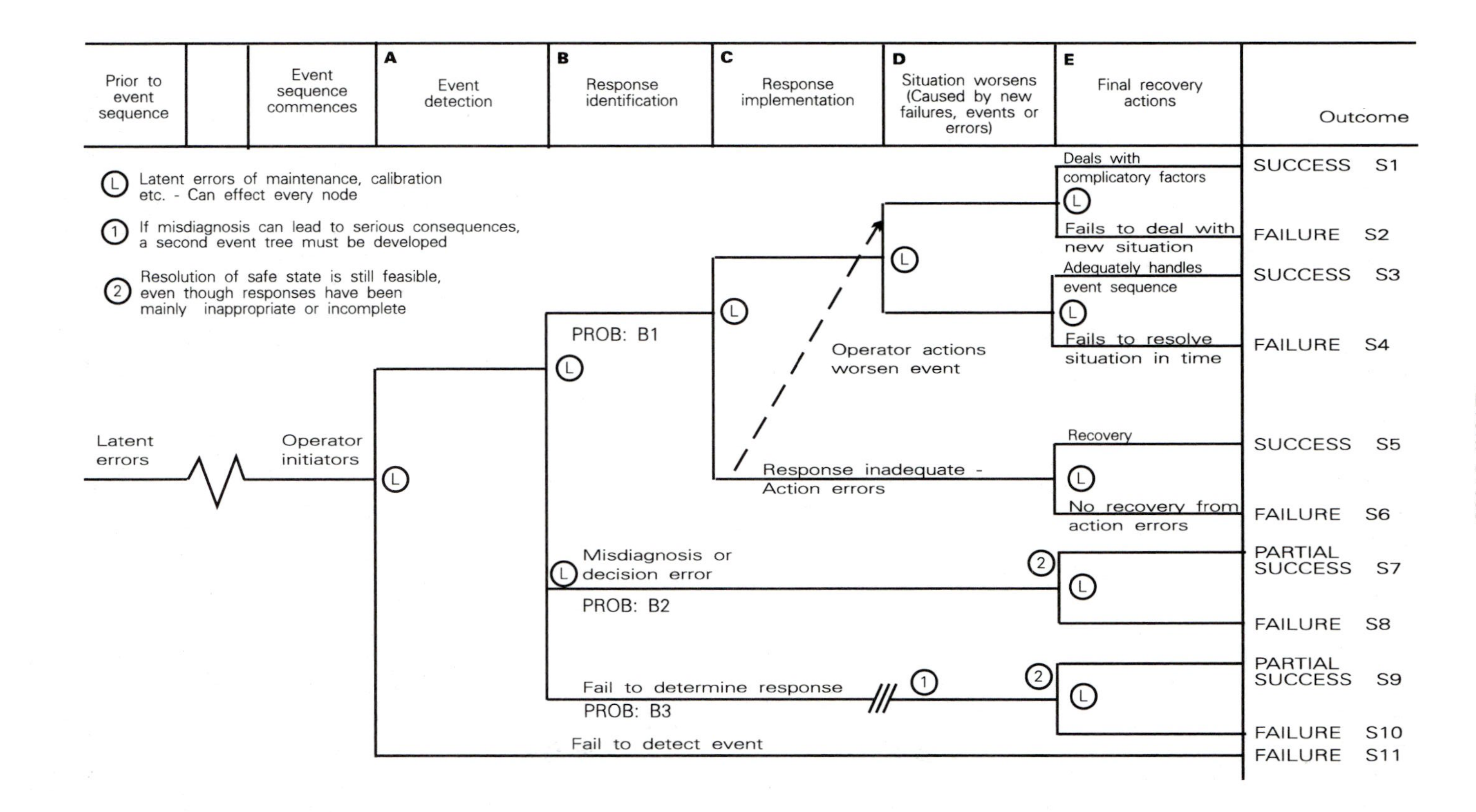

Figure 5.2 Generic accident sequence event tree

The errors that have severe consequences can often be identified by visual inspection of the event tree. This can often be supplemented by quantitative assessment in which the probability of each sequence is assessed. The overall probability of a particular sequence occurring, can be estimated by multiplying the probabilities of the constituent steps along an event 'path' through the tree. Simple checks on accuracy can be made by ensuring that these probabilities always add up to unity.

Therefore in Figure 5.1:

$$Probability\ (B1) + Probability\ (B2) + Probability\ (B3) = 1.0$$

The relative likelihood of the different sequences can be investigated and the most significant errors identified. The effect of various changes, such as improved design, can then be measured by re-assessing the probabilities of the errors and recalculating the sequence probabilities.

Practical advice

It can be difficult to achieve the appropriate level of task decomposition in an event tree. The event tree can become unwieldy if very small subtasks are used, making it difficult to follow and more difficult to gain insights from it.

Tasks are often decomposed to the level of:

- Detection (e.g. of alarms, or onset of an event)
- Response identification (interpretation and diagnosis of the events and/or the required response, etc.)
- Response implementation (responding to the events, attempting to regain control and a safe system state, through various actions)

However, for particularly important tasks, it may be useful to decompose the tasks further (e.g. identify the state of the system, decide upon the desired state of the system, select procedure, etc.).

While event trees are best used for the qualitative insights that are gained, they can also be used as a basis for the quantitative assessment of human error probability. The results of quantitative assessment can be particularly useful on a comparative basis (i.e. comparing the risk levels associated with various event sequences, or with various systems or procedures design).

Care should be taken in the use of recovery factors, because these can exert a very significant effect. Generally, recovery paths are appropriate where there is a specific mechanism intended to aid error recovery, for example an alarm, a supervisor check or even a routine walk-round inspection.

Resources Required

Although event trees are conceptually straightforward, skill is required to achieve the most appropriate level of detail and a logic structure which is meaningful and easy to quantify.

Additional resources will be necessary if the event tree is to be quantified. A range of techniques that can be used for this purpose can be found in Humphreys (1988).

Some training in probability theory is required, if quantitative assessment is to be performed.

Links with other techniques

Event trees are methods for the representation of failure, and are useful within an overall systems analysis for identifying where the system is vulnerable, or where critical errors are likely. Hierarchical task analysis (pp. 104) provides a useful framework for the identification of requisite tasks, and can also help the analyst clarify the appropriate level of task decomposition.

The initial identification of errors, or failures, can be assisted by techniques such as the critical incident technique (pp. 47), HAZOP (pp. 194), decomposition methods (pp. 95), questionnaires (pp. 58), computer modelling and simulation (pp. 147) and failure modes and effects analysis (pp. 184). In addition, confusion matrices (Potash *et al.* 1981) can be used for the prediction of diagnosis errors.

Event trees have connections with other techniques for human reliability analysis, such as fault trees (pp. 188). Whereas event trees work forwards to identify the consequences of errors, fault trees work backwards to identify the causes of errors. It can be useful to 'hang' fault trees from the various nodes of an event tree. For example, Figure 5.3, a fault tree, shows the causes of an operator's failure to detect an alarm. This could be used to determine the cause of failure in task A (fail to detect alarm) in Figure 5.1.

Advantages

- The logic which underpins event trees is usually based on sequences of events as they proceed through time, and event trees lend themselves to modelling sequences of actions.
- Event trees provide a ready interface with systems analysis techniques and are commonly used by engineers.
- Event trees help to identify those tasks, or errors, which are most critical, and which have the greatest impact upon system failure or success.

Disadvantages

- Event trees can quickly become unwieldy for long sequences of tasks.
- Event trees rarely model the complexities and subtleties of human behaviour and often provide only a gross simplification of reality. For example, it would be unusual to exhaustively represent all the checks and adjustments that could take place in a prolonged monitoring and control task. Instead, the behaviour would typically be simplified as a small number of steps.

References and further reading

Bell, B.J. and Swain, A.D. (1983) *A Procedure for Conducting a Human Reliability Analysis for Nuclear Power Plants.* Rept. NUREG/CR-2254. Washington D.C.: US Nuclear Regulatory Commission.

Hall, R.E., Fragola, J. and Wreathall, J. (1982) *Post Event Decision Errors: Operator Action Tree/Time Reliability Correlation.* Rept. NUREG/CR-3010. Washington D.C.: US Nuclear Regulatory Commission.

Hannaman, G.W., Spurgin, A.J. and Lukic, Y.D. (1985) *A Model for Assessing Human Cognitive Reliability in PRA Studies.* Paper at Third IEEE Conference on Human Reliability. Monterey, California. USA.

Henley, J. and Kumamoto, H. (1981) *Reliability Engineering and Risk Assessment.* New York: Prentice-Hall.

Humphreys, P. (ed.) (1988) *Human Reliability Assessors Guide.* Rept. No. RTS 88/95Q. Warrington: UK, Atomic Energy Authority.

Kirwan, B. and James, N.J. (1989) The Development of a Human Reliability Assessment System for the Management of Human Error in Complex Systems. *Proc. Reliability '89*, Inst. Quality Assurance and National Centre for Systems Reliability.

Parry, S.T. (1986) *A Review of Hazard Identification Techniques and their Application to Major Accident Hazards.* Warrington: UK Atomic Energy Authority.

Potash, L., Stewart, M., Dietz, P.E., Lewis, C.M., and Dougherty, E.M.Jr. (1981) Experience in Integrating The Operator Contribution in the PRA of Actual Operating Plants. *Proc. ANS/IENS Topical Meeting on Probabilistic Risk Assessment.*

Swain, A.D. and Guttmann, H.E. (1983) *Handbook of Human Reliability Analysis with Emphasis on Nuclear Power Plant Applications.* Rept. No. NUREG/CR-1278. Washington D.C.: US Nuclear Regulatory Commission, Washington.

See Chapters 14 and 16 which utilize fault and event trees.

Failure modes and effects analysis

Overview

Failure modes and effects analysis (FMEA) is a straightforward method from the field of hardware reliability analysis, which can modified and used for the analysis of human reliability. It was primarily developed to investigate the effects of component failures on systems (Henley and Kumamoto, 1981), but can also be used to study human errors. It can be applied at different levels in a system, from operator functions down to specific operator tasks. It can be used in conjunction with a checklist of human errors and is presented in a tabular format.

Application

FMEA is a relatively simple technique which can help the analyst to consider the effects of human error on systems. It is very flexible and can be used to consider the failures of either individual operators, or teams, and it can be based upon either tasks or functions. As such it can be applied throughout the design process.

Description of technique

FMEA is a technique which enables the analyst to consider the reliability of the operator and the consequences of errors upon the system. The technique uses a 'bottom-up' approach (as opposed to the top-down approach of fault trees – pp. 188). This means that the technique starts at the lowest level (the task), postulates failure mechanisms and then investigates the consequences for the overall system. If the consequences are serious, then further investigation may be performed to prevent the error, facilitate error recovery or mitigate the effects of error on the system.

For example, a possible application may be to investigate the errors that are possible during the maintenance of a car braking system. Errors that would be investigated at certain steps in this procedure might include:

- Omit action
- Reverse action
- Perform action on wrong object
- Apply too much torque
- Apply too little torque

In this example, the effects of the errors may include;

- Reduced braking efficiency
- Immediate or delayed catastrophic brake failure
- Increased component wear
- Excessive component temperature, noise, vibration, etc.

A tabular format is used, and tends in practice to be based on the use of four columns:

- The task, or task step
- Possible errors
- The effects of the error on the system
- Comments and notes

However, this can be expanded according to the requirements of the situation and extra columns could include:

- The causes of error
- The criticality of error
- Compensating safeguards

- Recovery possibilities
- Means to reduce the error probability, or the effects of error

An example of the method is provided in Table 5.3. This table shows an FMEA approach being used to access a tanker-filling procedure. In this example the task step column is followed by the error 'mode', and the third column notes whether or not error recovery is likely (either it will state '*no recovery*', or else give the task step number at which recovery could take place). The psychological mechanism column denotes the underlying behavioural reason for the error. The remaining columns deal with the consequences of the errors and ways of reducing error likelihood or its impact.

Practical advice

It is recommended that a human error taxonomy should be used as a checklist to help the analyst to identify potential error modes (e.g. Rasmussen *et al.*, 1981). Descriptions of tasks, errors and their effects should be clear and comprehensive, because simplification or abbreviation may result in the loss of information, and make it more difficult to scrutinize the analysis.

Before the FMEA is constructed, it is often useful to organize the information in the same manner as in a hierarchical task analysis, although this is not essential.

Resources required

This technique is conceptually simple and little training is required (although experience in human reliability analysis is desirable). However, it can be rather time-consuming to apply for lengthy tasks, in which case the resources required will be high.

A human error checklist can be useful as an *aide-mémoire*.

Links with other techniques

FMEA has a number of similarities with the task decomposition approach (pp. 95) in terms of its tabular presentation and the stages in the design process at which it can be applied. The main difference between the methods lie in their scope. FMEA concentrates purely on reliability while a task decomposition tends to be much broader. FMEA is a parallel technique to HAZOP (pp. 194) but does not require a group approach and is more structured than HAZOP.

The tasks which are considered in an FMEA can be identified by standard methods such as observation (pp. 53), structured interviews (pp. 66), hierarchical task analysis (pp. 104), and barrier analysis (pp. 169).

Advantages

- It is conceptually easy to understand and apply.
- It is flexible and can be used at different levels of system detail.

Table 5.3 Example of a typical FMEA for operator error (Kirwan and Rea, 1987)

Task step	Error type	Recovery step	Psychological mechanism	Causes, consequences and comments	Recommendations		
					Procedures	Training	Equipment
51.1 Close supply valve	Action too late	No recovery	Place-losing error	Overfill of tanker, resulting in dangerous circumstances	Operator estimates time/records amount loaded	Explain consequences of overfilling	Fit alarm – timing/vol./ tanker level
51.2.1 Open N_2 valve	Action omitted	51.2.4	Slip of memory	Feedback when attempting to close closed valve. Otherwise alarm when liquid vented to vent line.			Mimic of valve configuration
51.2.2 Close tanker supply valve	Action too early	51.2.2	Place losing error	Alarm when liquid drains to vent lines	Specify time for motions	Operator to count to determine time	
	Action omitted	51.2.2	Slip of memory	As above and possible Over-pressure of tanker (see step 51.2.3).			Mimic of valve configuration
51.2.3 Close tanker vent vallve	Action too early	No recovery	Place losing error	If valve closed before tanker supply valve overpressure of tanker will occur.		Stress importance of sequence explain consequences	Interlock on tanker vent valve
	Action omitted	51.2.6	Slip of memory	Automatic closure on loss of instrument air.			Mimic of valve configuration.
51.2.4 C;lose N_2 valve	Action omitted	51.2.2	Slip of memory	Audio feedback when vent line opened.		Explain audio feedback	Mimic of valve configuration
51.3 Close pressure vent valve	Action omitted	No recovery	Slip of memory	Latent error.	Add check on final valve posns. before proceeding.		Mimic of valve configuration

(The task step numbers refer to steps in a hierarchical task analysis which is not shown)

- It helps the analyst to identify which errors are most important by showing their consequences.
- It uses relatively few resources and can be applied relatively quickly.
- It can accommodate descriptive information on the nature of the error, and its causes, which some other methods of human reliability assessment cannot.

Disadvantages

- On its own, it relies heavily upon the skill of the analyst, and therefore it is best used in conjunction with a checklist of human errors.
- The description of some errors in complex behaviour can become lengthy and verbose.

References

Henley, J. and Kumamoto, H. (1981) *Reliability Engineering and Risk Assessment.* New York: Prentice-Hall.

Kirwan, B. and Rea, K. (1986) *Assessing the Human Contribution to Risk in Hazardous Materials Handling Operations,* Paper at The First International Conference on Risk Assessment of Chemicals and Nuclear Materials, Robens Institute, University of Surrey, September 22-26.

Parry, S.T. (1986) *A Review of Hazard Identification Techniques and their Application to Major Accident Hazards.* Warrington: UK Atomic Energy Authority.

Rasmussen, J., Pedersen, O.M., Mancini, G., Carnino, A., Griffon, M. and Gagnolet, P. (1981) *Classification System for Reporting Events Involving Human Malfunction.* Rept. No. Risø–M–2240. Roskilde, Denmark: Risø National Laboratory.

See Chapter 9 in particular, which describes a human error analysis tabular format, similar to FMEA, and Chapter 16 which includes tabular scenario analysis, another variant of an FMEA-type approach.

Fault trees

Overview

The fault tree is a tree-like diagram showing how hardware faults and human errors combine using **AND/OR** logic to cause system failures (see Henley and Kumamoto, 1981; Green, 1983). This provides insights about the relative importance of various causes of failure and indicates potential weak links in system reliability. When quantified, fault trees allow system risk to be calculated.

Fault trees are used in human error analysis to analyse the causes of human error, and in systems analysis to assess the impact of operator error on system reliability.

The technique was originally developed in the aerospace and defence industries for the analysis of the reliability of complex systems (e.g. nuclear power plant), and is often used in conjunction with event trees (pp. 178).

Application

Fault trees are a major method of analysing risk in systems, and most probabilistic risk assessments use a fault tree approach, often together with the use of event trees. Fault trees can be used to assess the likelihood of an undesirable event or accident scenario, and typically they consider hardware faults, environmental stressors or events, and human errors as potential accident causes. Fault trees are powerful media for presenting graphically the relationships between the potential causes of accidents, and when quantified, yield the frequency or probability of the accident. Sensitivity analysis of the tree can then determine the relative importance of each contributor to the accident.

Fault trees can hence be used to investigate the causes of failures, involving the analysis of human errors and error recovery.

They have different applications depending upon the objective of the analysis. In a systems analysis, they can be used to assess the impact of operator error upon the safety, reliability and availability of systems. In a human error analysis, they can be used to analyse the conditions, factors and psychological mechanisms which combine to result in an operator error.

The combinations of fault events can be analysed using Boolean algebra to identify the individual errors and combinations of errors which have the greatest importance. This requires that the probabilities of human error can be estimated, so that quantitative analysis can be performed.

Fault trees can, in principle, be applied at all stages of system development, although at the early stages, the human failures will tend to be described in functional terms. Fault tree analysis generally occurs at, or later than, the detailed design stage.

Description of technique

Fault tree analysis is a powerful analytical approach, and only the basics can be outlined in this section. The reader requiring more information is referred to Henley and Kumamoto (1981) and Green (1983).

Essentially fault tree analysis involves defining one undesirable event at the top of the tree, the 'top' event, and deciding what can cause it, either alone, or in combination with other events, errors, etc. For each of these causes or intermediate events underneath the top event, the question of what could cause them is again asked. This reduction process frequently can be quantitatively assessed.

Basic events related to human errors will be quantified as a probability, called the human error probability.

The logic of fault trees centres around the top-down nature of the approach and its use of **AND** and **OR** gates which denote the relationship between an event and those events immediately below it and joined to it via the gate. An **OR** gate means that any of the events underneath that gate can, on their own, cause the event above the gate. An **AND** gate, in contrast, means that the event above the gate will only occur if *all* the events below the gate occur. As most events in a fault tree are usually independent events, it can be seen that an event above an **OR** gate is much more likely than an event above an **AND** gate if the basic events underneath each gate have the same likelihoods. In practice, the probability of failures of events underneath an **OR** gate are effectively added to denote the probability of the event above that gate (a correction to this procedure must be implemented if the failure probabilities are not small, e.g. particularly if they are larger than 0.1: see Henley and Kumamoto, 1980. This will yield a probability of failure larger than any single contributory basic event. However, event probabilities under an **AND** gate are effectively multiplied, yielding a probability of failure for the top events which is smaller than any of its contributors.

An example of a simple fault tree is shown in Figure 5.3. In this example the top event is 'Operator fails to detect alarm', which is one of the events contained in the example event tree used on pp. 179 (event detection failure in Figure 5.1). There are 12 standard symbols which are used in the construction of fault trees, but for simple event trees it will only be necessary differentiate between two types of gate and three types of event. Drawing can often be facilitated by the use of labelled circles for the type of gate, with boxes representing intermediate, basic and undeveloped events, as was done in Figure 5.3. Examples and explanations of the other, more specialized symbols can be found in Veseley *et al.*, (1981).

To analyse a fault tree, a mathematical/logical expression must be derived for the top event, based on those events lower in the line which contribute to the top event. In this expression, two simple rules are followed, namely:

- The output of an **AND** gate is equal to the product of its input probabilities
- The output of an **OR** gate is approximately equal to the sum of its input probabilities (NB this approximation only holds if the probabilities are small – i.e. significantly less that 0.1)

This expression is then reduced, using the standard rules of Boolean algebra (see Henley and Kumamoto, 1981). The basic Boolean expression for the top event in Figure 5.3 is:

$$B + C + (G + H + F) \,.\, (A + E + D) + D + (A + E + F + G) \,.\, (H + I + J)$$

This identifies base events B, C and D as being potentially important, because the occurrence of any one of these alone, will lead directly to the top

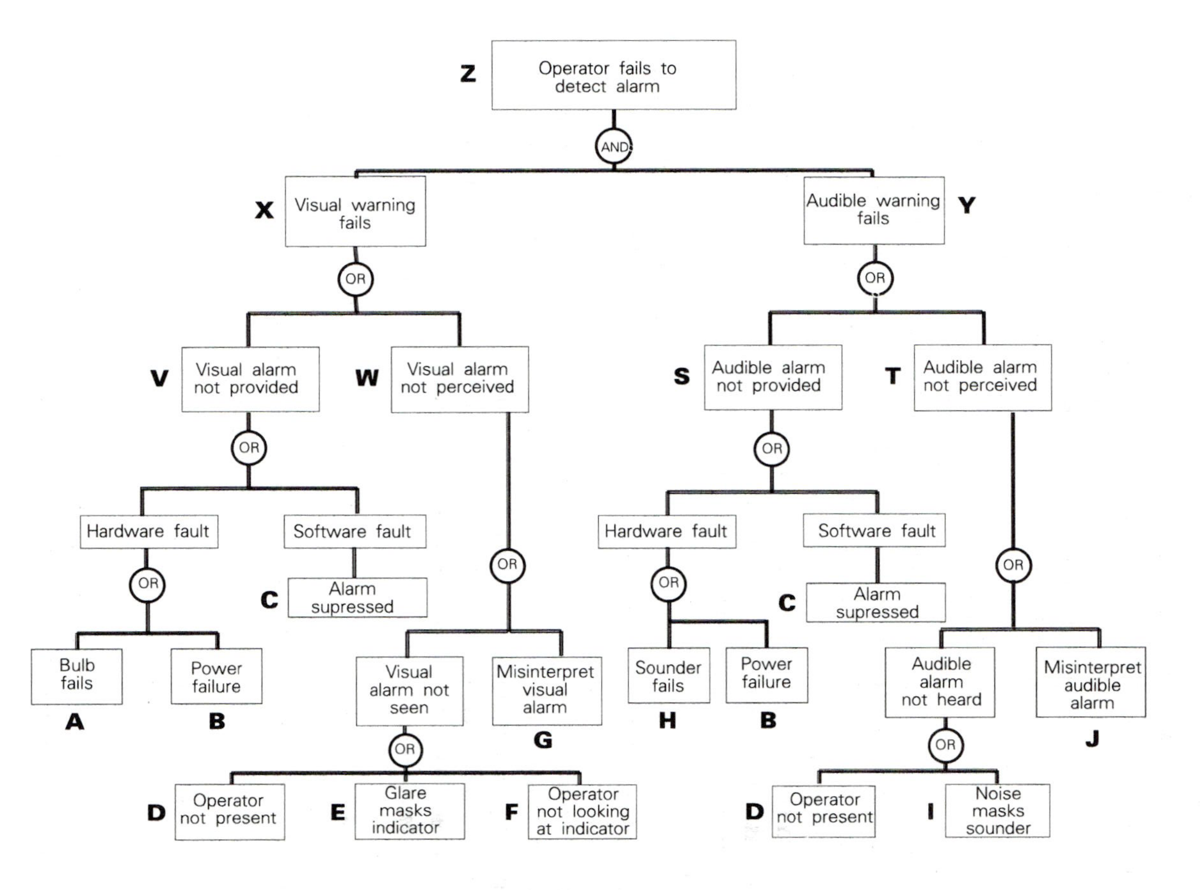

Figure 5.3 Fault tree

event. Other contributors (e.g. A, E, F, G, H, I and J) cannot cause the top event alone, instead they must occur in unison with another basic event.

Practical advice

Fault trees can be difficult to construct, and care should be exercised to avoid omitting the faults and intermediate mechanisms which occur between the top event and the various base events. **AND** and **OR** gates should always be separated by an intermediate event which is accompanied by a description. Therefore, the base of the fault tree should consist of basic events (i.e. the initiating faults or errors) or undeveloped events.

A methodical and meticulous approach will generally enhance the accuracy of the logic, and the overall utility of the method. In this respect, the statements of failure should be clear and unambiguous. The error, or failure description, should determine the size of the box, and not the other way round.

Gate to gate situations should not exist, because this either indicates sloppy analysis, or an inadequate understanding of the system.

It is important that the fault/error identification stage is thorough in order to achieve a 'complete' fault tree. An incomplete fault tree may lead to false conclusions. For example, if the fault tree is used in a quantitative assessment then an incomplete tree will result in the overall probability of failure being underestimated.

The limitations of quantitative assessment should be recognized, because its validity depends upon the accuracy of the basic human error estimates. While the results of quantitative assessment can be particularly useful on a comparative basis, the absolute accuracy of results should be treated with caution. A range of techniques that can be used to assess human error probabilities can be found in Humphreys (1988).

Resources required

For complex situations, considerable time may be required to organize the logic of a fault tree correctly, and this is likely to be the area where most of the time is spent. Drawing the fault tree can also be time-consuming, if many redrafts are necessary, although commercial software packages are available which can reduce the time considerably.

For most applications involving human reliability assessment, the Boolean reduction will be relatively simple and can be undertaken effectively without the need to use a software package.

Links with other techniques

Fault tree analysis is one of a number of techniques which can be used for the analysis of reliability, and it is often used in conjunction with event tree analysis (pp. 178), or with failure modes and effects analysis (pp. 184). Whereas fault trees use backwards reasoning to identify causes of failures, both failure modes and effects analysis and event trees use forward reasoning to identify the consequences of failures.

The causes of errors which are shown in fault trees can be identified by techniques such as questionnaires (pp. 58), structured interviews (pp. 66), critical incident analysis (pp. 47), decomposition methods (pp. 95) and hazard and operability analysis (pp. 194).

Advantages

- Fault trees help to identify aspects of tasks and errors that are critical to the system, and to differentiate these from the errors which are of less consequence. This can be determined quantitatively via sensitivity/importance analysis of all basic events in the fault tree.
- They provide a ready interface with hardware reliability, and are readily understood by engineers.

Disadvantages

- Fault trees can be difficult to construct, and considerable care is required to ensure that the logic is correct.
- Due to the complexity of human behaviour, and the multiplicity of potential error causes, fault trees can become complicated and unwieldy.
- Some training in probability theory and in the use of Boolean algebra is required in order to perform quantitative analyses.

References and further reading

Green, A.E. (1983) *Safety Systems Reliability.* Chichester: John Wiley.

Henley, J. and Kumamoto, H. (1981) *Reliability Engineering and Risk Assessment.* New York: Prentice-Hall.

Humphreys, P. (ed.) (1988) *Human Reliability Assessors Guide.* Rept. No. RTS 88/95Q. Warrington: UK Atomic Energy Authority.

Parry, S.T. (1986) *A Review of Hazard Identification Techniques and their Application to Major Accident Hazards.* Warrington: UK Atomic Energy Authority.

Veseley, W.E., Goldberg. FG.F., Roberts, N.H. and Haasl, D.F. (1981) *Fault Tree Handbook.* Rept. No. NUREG-0492. Washington D.C.: US Nuclear Regulatory Commission.

See Chapters 14 and 16, which both utilize fault trees.

Hazard and operability analysis

Overview

The hazard and operability (HAZOP) study technique (Kletz, 1986) uses a group of experienced personnel to systematically identify potential problems in a system's design and operational intent. This group consists of a chairperson to direct the discussions, and usually a small group (e.g. 4–6 persons) of safety, design and operations personnel, with a secretary to record the proceedings in a tabular manner. HAZOP involves the detailed and structured consideration of engineering diagrams (e.g. Piping and instrumentation diagrams: P & IDs), line by line, and uses a keyword system to identify problems which could occur (e.g. the keyword 'reverse' could be used to consider whether reverse flow would be possible in a pipeline). The causes of these problems, their consequences in terms of safety and operability, and potential solutions or means of deriving solutions, are determined by the group. All of this information is recorded by the secretary.

HAZOP was developed in the seventies as an intensive and powerful medium of design/operational intent assessment. Usually HAZOP studies are undertaken early in the design of a plant, but they can also be used later (e.g. to identify potential problems in existing systems). Recently HAZOP has been extended by some human factors practitioners to focus more effectively upon human error problems in systems design. Both conventional HAZOP and the more human factors-oriented Human HAZOP are described in this section.

Application

HAZOP can be defined as 'a disciplined procedure which generates questions systematically for consideration in an ordered but creative manner by a team of design and operations personnel carefully selected to consider all aspects of the system under review' (Andow, 1990). The basic philosophy of the technique is that safety and operational problems usually stem from the same sources, namely that of deviation from the design or operational intent of a system.

The HAZOP analysis method was developed in the 1970s primarily at ICI (UK), and has been used extensively in the chemical, petrochemical, oil and associated industries. The HAZOP technique and its 'sister' technique HAZAN (hazard analysis) are usually used before a system has been built, but can be undertaken at any stage within a plant's life. For successful completion, HAZOP requires the co-operation of a group of people with safety, technical and operating experience, usually over a period of several days (dependent upon the size of the system). Any venue can be used, provided that the team can be left undisturbed. It is preferable to have a large table in order to lay out the P & IDs or task analysis.

Description of technique

Input needs

The basic input needs for a HAZOP analysis are as follows:

- P & IDs, or equivalent, plus the following as required: flow-sheets (functional specifications), operating instructions, plant layouts, isometric and fabrication drawings, hierarchical task analysis
- Operating instructions (personnel or computer)
- Hierarchical task analysis (if considering potential human failures in depth)

In addition, the group must be composed of experienced personnel. If human errors are the particular concern of the study, then a human factors practitioner should also be present.

Procedure

HAZOP studies are a strict and systematic form of analysis. The team leader (chairperson) must prepare the strategy for the study after gaining an overall understanding of the process. The strategy must determine the sequence of the study and a provisional timetable. The study itself is most productive if spread over several sessions with each lasting no more than half a day.

Considering a typical HAZOP group reviewing a P & ID, the group would start with the first line of the first vessel, the first relevant *property word* and the first appropriate *guide word*. Property words refer to physical characteristics or functions associated with the vessel/pipeline/instrument being considered (e.g. pressure – see listing below). Guide words refer to deviations which can occur that lead to operability/safety problems (e.g. Less – too little pressure: see listing below). For this guide word, the group examines the possible causes of the deviation, and if, applicable the consequences and actions required. The group then moves on to the next guide word. Once all the appropriate guide words for a specific property have been covered, the group will consider the next property word in the same way. Similarly, when all the property words have been covered for a line, the group will start assessing the next line. Once all the lines have been checked, the HAZOP team will attend to any auxiliaries and finally the vessel itself.

Typical property words:
Flow, temperature, pressure, level, concentration, amount, absorb, dissolve, purge, inert, heat transfer, separate, react, reduce, trip, drain, viscosity, density, pH

Standard guide words:
No, less, more, part, also, reverse, other, early, late

These considerations are generally qualitative throughout the HAZOP process (i.e. no quantification of likelihood of occurrence of the deviation occurs), although the group may implicitly consider likely frequencies/probabilities of events. If quantification is deemed necessary (e.g. for major risk areas), this should take place as part of the HAZAN study. HAZAN involves fault and event trees and probabilistic safety assessment or reliability analysis, and full HAZAN is outside of the scope of this *Guide*.

Output of the analysis

Contribution to design decisions
HAZOPs enable the system designer and safety manager to check that the plant is safe and operable. If any areas of concern are discovered, preventative measures can be considered, costed and undertaken while still at the drawing board. If an existing plant is studied, recommendations can be passed to the design engineers for the design of modifications.

Information produced
Usually the output is tabular covering each line, each relevant property of the system (e.g. flow, temperature, pressure) for each line, each relevant deviation guide word for each property, followed by how the deviation could occur, the consequences, operator indications, how hazardous it may be and any preventative methods with their cost. The latter may be noted as an 'Action' for someone to determine, outside of the meeting. Human failures may feed forward into human reliability assessments, human error cause analysis and fault tree analysis.

Use of end results

The highlighted actions will be passed on to the relevant personnel so that design corrections can be made to the P & IDs and to the engineering drawings. If the plant is already built the suggestions have to be viewed in terms of design modifications, new drawings and structural alterations to the plant.

Human HAZOP

The HAZOP approach which has been described so far is the conventional format, but a variant has emerged which deals more specifically with human factors/human error issues. Firstly, although HAZOP will typically identify much human error potential, it has been felt by some human reliability assessment practitioners that the technique could benefit from modifications to direct the technique more closely to the identification of human performance problems. Typically this involves the addition of a human factors/human reliability assessment practitioner to the group, and the use of supplementary material on the operator's role in the system (e.g. in the form of a hierarchical task analysis and/or task decomposition).

Secondly, the HAZOP group, even if it can identify all human error problems, is not always able to determine adequate design solutions to those problems that would satisfy the human factors practitioner. In particular, sometimes HAZOPs yield a large number of recommendations to overcome errors by training and procedural solutions. While such solutions may be expected to reduce the error, sometimes to a large extent, they may not be as powerful as design solutions. The human factors practitioner is more often in a better position to judge what level of 'solution' is most appropriate, and the nature and appropriateness of a design solution. For example, recommendations relating to alarm systems or specialized displays in particular should involve a human factors practitioner.

Therefore, the human HAZOP approach is aimed at more comprehensive error identification, including the understanding of the causes of error, in order to achieve more robust error reduction.

Whalley (1988) utilized a HAZOP-type approach as part of a human error identification tool. This had a set of additional keywords in it, as shown in Table 5.4. In addition to such HAZOP-based keywords, other keywords which are more conventional to the human error identification field can also be utilized (e.g. those from Rasmussen *et al.* (1981) as shown in Table 5.5.

Table 5.4 Error types related to HAZOP key words

	Error Types		
HAZOP	Error	HAZOP	Error
1. No	1. Not Done	—	6. Repeated
2. Less	2. Less Than		7. Sooner Than
3. More	3. More Than	6. Reverse	8. Later Than
4. As Well As	4. As Well As		9. Mis-ordered
5. Other Than	5. Other Than	7. Part Of	10. Part Of

An example of a human HAZOP approach is given by Comer *et al.* (1986). This approach looks at problems in offshore drilling, and utilizes several additions to the conventional HAZOP approach:

- Human factors expertise in the development of the approach
- A form of task analysis showing the sequence of operations and personnel involved
- More information on the interface;
- An extended keyword set

In a study of an 'informal' HAZOP approach (i.e. group identification of human errors compared to the use of other methods), the group method was shown to be more powerful in identifying incident errors and causes (Whalley and Kirwan, 1989). While this study did not involve the use of explicit human HAZOP, it demonstrated the relative power of the use of group experts.

Table 5.5 SRK external error modes

Action omitted
Action too early
Action too late
Action too little
Action too much
Action too short
Action too long
Action in wrong direction
Right action on wrong object
Wrong action on right object
Misalignment error
Information not obtained/transmitted
Wrong information obtained/transmitted
Check omitted
Check on wrong object
Wrong check
Check mistimed

A third approach utilized a group human error identification HAZOP approach, using P & IDs of the system (a lifeboat evacuation system), with human factors, safety assessment and operational expertise. This approach was found by the participants to be highly effective in identifying significant errors and failure modes for a proposed offshore lifeboat design. Only two human HAZOP sessions were held with five people in each, supplemented by visits to a lifeboat training centre and interviews with personnel who had been involved in lifeboat evacuations (see Kirwan *et al.*, 1988).

An example of the tabular format from the drillers' HAZOP is shown as Table 5.6.

Human HAZOP therefore offers significant potential for identifying and reducing human errors, but it is currently still in its formative stages of development.

Practical advice

It is important to consider all consequences and recommended actions in relation to each other, otherwise what might solve one problem may lead to another.

HAZOP is very time-consuming and is dependent upon the technical experience of the team. The team leader needs experience in HAZOP participation in order to effectively guide the team, stimulating discussion rather than dominating – an inexperienced or ineffective team leader can seriously jeopardize the analysis. Unless the secretary has adequate recording skills as well as technical experience, information may be lost or team progress hindered.

Table 5.6 Extract from driller's HAZOP study

Ref. No.	Deviation	Causes	Consequences	Action notes/ recommendations
		5A/5D Lift and set aside kelly. Ensure hook is free to rotate		
5.1	No movement	Kelly valve not closed	Mud on drill-floor	A new (mud-saver) valve was discussed. (a pre-set pressure valve which prevents flow for pressure below 200 psi – i.e. a mud column equal to the length of the kelly assembly
		Forget to open hook to ensure free rotation	With stabilizer in the hole, the drill string will rotate when pulled out, which may cause damage to the hook	
5.2	Reverse movement	Lifting kelly too high	Damaged hoses, piping etc. Injuries to operators	Q.7 Should sensor alarms be provided on guide dolly? Q.8 How do we ensure that kelly is not lifted too high?
		Failure of lifting equipment e.g. drill line breaks, or operator error	Falling objects causing damaged equipment and injuries to operators	R.2 Consider safe location of driller's cabin in designing new rigs

In terms of human errors, unless the experience of a human factors expert is available the full implication of human error and its causes can be missed resulting in inappropriate or vague actions being requested (e.g. ensure procedures and training take this error into account).

Resources required

A mixed team covering design, engineering, production (preferably an individual with extensive operational experience), safety and human factors experience, are ideally required for HAZOP, as well as a study leader experienced with HAZOP.

A secretary to record the study is also required, with sufficient technical background to understand what is discussed in order to condense the information for entry in the tables.

It is usual to record the study on pre-printed sheets with five columns, property word, guide word, cause, consequence, and actions or alternatively an A4 sheet turned on its side and ruled into columns. More recently, computer-aided recording has been implemented.

Time to complete the hazard and to disseminate the results is variable, with studies of small systems taking several days (for each group member), up to months of analysis for large systems. HAZOP must therefore be considered as being very resources-intensive.

A venue is also required where the group can work together undisturbed.

Links with other techniques

HAZOP has strong links with other risk assessment approaches such as fault and event trees (pp. 188 and 178), and techniques such as failure modes and effects analysis (pp. 184).

HAZOP for human error identification requires that a hierarchical task analysis (pp. 104) and preferably some form of decomposition analysis (pp. 95) have taken place, so that the goals and sequences of operations etc. are understood, and the information being utilized is also described.

Human HAZOP may also benefit from a talk-through (pp. 162) or a table-top analysis (pp. 155) for areas of detailed human involvement. HAZOP can also interface with any of the other behaviour assessment techniques in this chapter, as part of a human reliability assessment or more general risk assessment approach.

Advantages

- HAZOP identifies potential hazards before they become built into the system.
- It is a systematic method covering all potential hazards for all sections of the plant and system.
- It provides the basis for a list of 'actions' in order to prevent or rectify problems.
- HAZOP already takes place in many industries and can readily be extended to address human factors issues.

Disadvantages

- Ideally HAZOP requires a mixed team of design engineers, engineering managers, process managers, operators, safety personnel and human factors engineers in order to cover all aspects of the system.
- If a human factors engineer is not included, human failures are sometimes considered reasons for hazards in themselves rather than looking at the underlying reasons. If this occurs, the suggested preventative measures can entirely miss the point leading to vague and ultimately under-effective recommendations relating to procedures and training.

- Often the HAZOP is completed before full details of controls and instrumentation are available therefore possible hazards associated with operator interpretation and action cannot be fully considered.
- It is likely that the study will generate extensive information, which must be recorded and for which someone needs to take responsibility.
- A period of several days/weeks must be set aside to complete the study, dependent upon the size of the plant and the extent of the investigation.

References

Andow, P.R. (1990) *HAZOP Course Manual.* Department of Chemical Engineering, Loughborough University: Loughborough.

Comer, P., Fitt, J.S. and Ostebo, R. (1986) *A Driller's HAZOP Method.* Paper at Eurospec '86, Soc. Petroleum Eng., London.

Kletz, T. (1986) *HAZOP and HAZAN Notes on the Identification and Identification of Hazards.* Rugby: Inst. Chem. Engrs.

Kirwan, B., Cox, R.A., Embrey, D.E. and Miles. A. (1988) Lifeboat Evacuation Assessment. In *Human Reliability Assessor's Guide*, Humphreys, P. (ed.). Rept. No. RTS 88/95Q. Warrington: UK Atomic Energy Authority.

Whalley, S.P. (1988) Minimizing the Cause of Human Error. In *Reliability Technology Symposium*, Libberton, G.P. (ed.). London: Elsevier.

Whalley, S.P. and Kirwan, B. (1989) *An Evaluation of Five Human Error Identification Techniques.* Paper at 6th International Loss Prevention Symposium, Oslo, Norway.

Influence diagrams

Overview

Influence diagrams are a method of graphically representing various factors which directly influence the occurrence of a particular event. The development of an influence diagram involves defining the target event and describing the general setting and conditions which lead up to the event. The effect of each identified influence is then evaluated quantitatively, with the resulting values used to calculate human error probability estimates. For anyone who wishes to use this approach for quantification, Phillips *et al.* (1983) and Humphreys (1988) provide detailed information on applying the technique.

Application

The influence diagram approach has been derived from decision analysis, and is based on expert judgements, so it is useful for human reliability assessment, particularly in high risk situations for which there may be little, or no empirical data available. Typically, the technique is used when the system has been operational for some time, because it is preferable that some of the experts have operational experience in order to provide a more realistic evaluation of the scenario in question. The technique is essentially a table-top exercise requiring iterative calculations. Target events (e.g. system failure) and descriptions of the associated influences (e.g. quality of the design of the operator interface) together with their causative structure (see Figure 5.4), are normally defined fairly extensively, which may necessitate site visits and lengthy interviews with subject matter experts.

Influence diagrams can be regarded as causal models which are useful for the successive decomposition of events, so that the impact of identified influencing factors in a particular event sequence can be studied. The development of the diagram enables an investigation to probe beyond the surface causal contributors and to consider the underlying root causes, or influences, such as organizational and management variables. Accordingly, the technique has been used in incident analysis as well as being used to model event and accident sequences.

Description of technique

For the application of the influence diagram, a model is developed which defines the factors that influence error probability for a task (or a broad category of tasks) and the interrelationship between these factors (e.g. see Figure 5.4). Assessments of dependencies between these factors are made at each level of the influence diagram, and these dependencies are mapped onto the diagram. If required, the overall probability of success or failure of the event under consideration can be calculated by numerically assessing the effects of influences and their dependencies upon the target event.

The application of this technique involves the steps described below.

Select experts and describe a scenario

Identify the scenario of interest, such as a particular accident sequence, and define the target event (e.g. 'operator controls repressurization', or, 'manual depressurization of the reactor'). Depending on the scenario of interest, appropriate subject matter experts need to be selected, in order to describe the general setting in which the target event may occur, as well as all the conditions leading up to that target event. If possible, a broad range of experts should be involved, in order to provide different perspectives on the scenario in question. The description and statement of conditions is very important, because it forms a context for subsequent assessments, and these assessments are conditional on this context. At this stage, and in consultation with the experts, the analyst should try to elicit from the subject matter group the most important factors which might influence operator performance.

Develop influence diagram

Through a process of discussion and interaction, and guided by the analyst, the group develops a structural representation of the main influences which have an

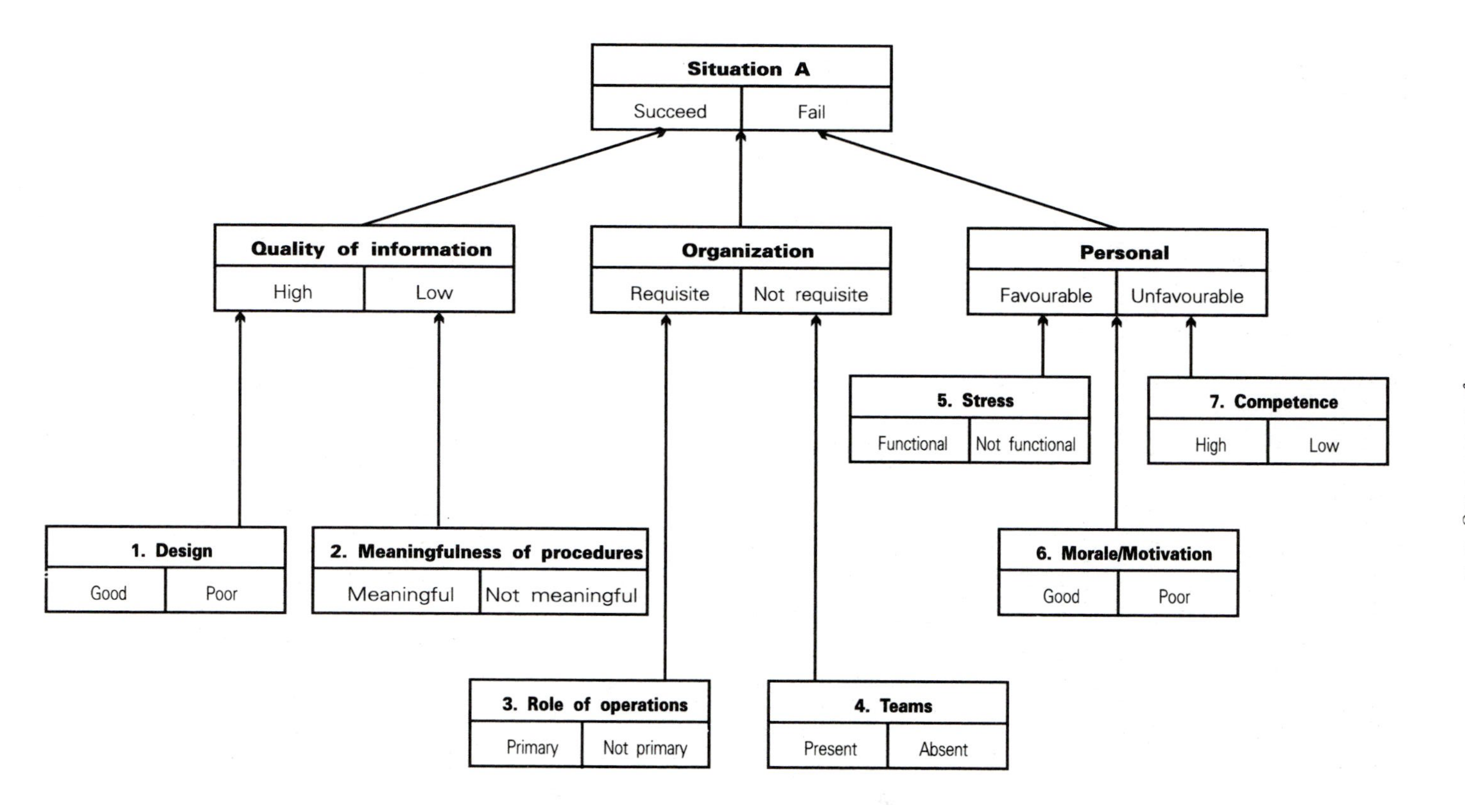

Figure 5.4 *Influence diagram model of situation A: 'What is the likelihood that the operation will successfully achieve recirculation at the correct time?'.*

effect on the probability of success in the scenario of interest. The technique requires the identification of significant influences on human performance, but does not require that all possible influences are modelled (e.g. only the most important ones are modelled). The form of the influence diagram is illustrated in Figure 5.4, from a study which was undertaken at a nuclear power plant. It can be seen that this diagram represents not only the direct influences of the various factors on the scenario of interest, but also the influences of some factors upon each other.

The analyst should ensure that the group members have a common understanding of the influences, because subsequent judgements will be based upon this understanding. For this purpose, it is essential that all the influences are operationally defined in terms of the basic dimensions which constitute each influence. The influence of a factor, such as Design is broken down into bipolar dimensions of display characteristics. For instance, good aspects of displays might be defined as consisting of mimic panels and alarms which are discriminable and relevant, whereas bad aspects of displays could be defined as non-representational displays or confusing and irrelevant alarms. An example breakdown of two influencing factors relating to the quality of information for the operators, is presented in Tables 5.7 and 5.8.

Table 5.7 Quality of information: sub-influence – Design (good or poor)

Good	Poor
Displays	
Easy to read and understand and accessible	Hard to read, difficult to interpret: inaccessible
Make sense: easy to relate to controls	Confusing: not directly related to controls
Alarms discriminable, relevant, coded	Alarms confusing, irrelevant, not coded
Mimic display	Non-representational display
Displays related to event are present, clear, unambiguous	Displays related to event are not present, unclear or ambiguous
Operator involvement	
Operators have say in modifications	Little or no say
Prompt confirmation of action	No confirming information
Automation of routine functions	
Highly automated — operators act as systems managers	Low level of automation — operators perform many routine functions

The interactive process continues until the group comes to agreement on the structure and context of the influence diagram. When the scenario has been described and influences have been identified and defined in detail, the target event definition may then need to be refined, in order to more precisely specify the conditions determining its occurrence or, non-occurrence. Diagrams which are produced in this manner are compact and contain more information than other similar approaches, such as the typical event tree structure (see pp. 178).

Table 5.8 Quality of information: sub-influence – procedures (meaningful or not meaningful)

Meaningful	Not meaningful
Realism	
Realistic: especially the way things are done	Unrealistic: not the way things are done
Location aids	
Location aids provided	Few, or no, location aids
Scrutability	
Procedures keep operators in touch with plant	Procedures do not keep in touch
Operator involvement	
Operators involved in developing procedures	Not involved
Diagnostic	
Allow unambiguous determination of event in progress	Allow inappropriate diagnosis
Format	
Clear, consistent, easily read format	Confused, difficult to read

Quantification

For the purpose of quantification, the completed diagram provides the structure in which iterative calculations can be performed in order to determine conditional probabilities and the overall joint probability of the target event.

At this stage, expert evaluations can be elicited regarding the likelihood that a particular state of the influence exists in the scenario being assessed (e.g. the extent to which the design is appropriate in the scenario under consideration). Then the effects of all combinations of the lower-level influences upon middle-level influences are evaluated. Finally, the combined effects of each of the middle-level influences upon the probability of success of the target event, are estimated.

The detailed procedures for calculation, are beyond the scope of this *Guide*. However, these procedures, along with reference to the development of the influence diagram in Figure 5.4, and the methods of facilitating the expert judgements, are comprehensively presented in Phillips *et al.* (1983), and are summarized in Humphreys (1988).

Practical advice

The validity of any error probabilities which are produced by the model depends to a large extent on the expertise of the members of the group, and also on the social processes that help to generate the model inputs.

The analyst in particular must ensure that all group members have a clear and uniform understanding of the definitions of the lower-level influences. The quality of results depend on obtaining representative experts and an analyst who is able to guide the group carefully, avoiding the biases that can occur (see 'group methods' within table-top analysis, pp. 155).

Resources Required

In most cases, particularly for complex problems, a group of experts, usually four or more, is required to provide all the relevant expertise for one, or up to several, days. In such cases, the analyst may require a technical recorder.

The subject experts usually require a practice session to help them generate the numerical assessments of how these influences combine to ultimately yield a probability of success/failure.

If quantification is required, at least a programmable calculator will be indispensable for generating on-line feedback to the group.

As an example, in the nuclear power plant case study. a group of eight domain experts took four days to quantify five target events using one generic influence diagram. This can be seen as a maximum usage of resources.

Links with other techniques

Before the group of experts is convened, it is important to identify a candidate set of the lower-level influences (particularly if quantification is intended) through structured interviews (pp. 66) with experienced operational personnel and assessment of appropriate plant and documentation. This would then have to be examined and ratified with the expert group and changed if required. The example diagram in Figure 5.4 was developed in the context of a nuclear power scenario. Whether this is a truly generic diagram containing common influences is questionable, and consequently it is important to capitalize on the uniqueness of the influencing factors for a particular event. When the influence diagram approach is used for quantification, the human error probabilities which are generated can be incorporated into fault and event trees (pp. 188 and 178).

Advantages

- Organizational and management variables which influence human reliability can be modelled and the technique can consider large task 'chunks' (i.e. any level of task decomposition can be assessed).
- Influence diagrams do not assume that all influences impinging on success, or failure, are identified within the model. Only influences which have a significant effect upon error probability are considered.
- Sensitivity analysis can be performed on the quantified influence diagram to determine where to refine the accuracy of the assessment, or where intervention could be most cost-effective.
- The structure underlying the assessment is explicit and auditable.

Disadvantages

- It is costly in terms of resources, because a variety of 'experts' will be needed for some time.
- The elicitation of unbiased probabilities in the technique is dependent upon the skill of the analyst.
- When assessing probabilities for the target event conditional on the middle-level influence, assessors may experience difficulty in dealing with three or more different influences, as well as their possible interactions at the same time.

References and further reading

Howard, R. and Matheson, J.G. (1980) *Influence Diagrams.* Menlo Park, California, USA: SRI International. *The original reference for the influence diagram method as a decision analysis tool.*

Humphreys, P. (ed.) (1988) *Human Reliability Assessors Guide* Rept. No. RTS 88/95Q. Warrington: UK Atomic Energy Authority. *Provides detailed information on calculation procedures for quantification and useful case studies which illustrate the application of the technique on a range of scenarios.*

Phillips, L.D., Humphreys, P. and Embrey, D.E (1983) Appendix D, A Socio-technical Approach to Assessing Human Reliability (STAHR). In *Pressurized Thermal Shock Evaluation of the Calvert Cliffs Unit 1 Nuclear Power Plant*, Selby, D., Research Project on DOE contract 105840RZ1400, Oak Ridge National Laboratory, Tennessee, USA *This study was a full application of the technique to an existing nuclear power plant system, and represents one of the major published studies of this technique.*

Pyy, P., Pulkkinen, V. and Vaurio, J.K. (1989) *Human Reliability Data Sources – Applications and Ideas.* Rept. No. VTT 02151 ESPOO, Helsinki, Finland: Technical Research Centre of Finland. *Examines the potential of the technique as a data source and illustrates its application for analysing and quantifying post initiating event sequences in the nuclear industry.*

Pyy P. and Pulkkinen, V. (1989) Treatment of Uncertainties in Human Reliability Analysis. In *Reliability Data Collection and Use in Risk and Availability Assessment,* Columbari, V. (ed.), pp. 385-400. Berlin: Springer-Verlag. *A detailed case study of the application of an approach based on influence diagrams for the estimation of error probabilities. Problems concerning data uncertainty are discussed.*

Management oversight risk tree technique

Overview

The management oversight risk tree (MORT) technique was developed by Johnson through the mid 1970s into the 1980s for use by the US Department of Energy. The technique investigates the adequacy of safety management. It can therefore be used either to ensure that adequate safety management functions are in place or, in an accident investigation, to determine how such functions have failed. MORT is one of the very few techniques in existence that directly addresses management effects on safety.

The principal output is a pattern of causes which have led to the event/accident, plus an identification of areas/responsibilities where the safety management structure is vulnerable.

Application

MORT can be applied late in the design process, prior to operation and commissioning of a specific plant, or at any subsequent point during the life of a system. It is primarily used for accident investigations in existing systems. MORT can be applied at any venue, provided that the assessor can gain access to company personnel to confirm any aspects that are unclear.

The major usage of the technique to date has been for accident investigation purposes (e.g. on behalf of the US Department of Energy), to determine what specific systems aspects were less than adequate, hence causing the accident. Its secondary use, along with a host of MORT-related techniques, is to consider the operational readiness of a system (i.e. whether sufficient safety analysis has been carried out, and whether sufficient safety management systems exist to mean that the plant can safely go operational).

Description of technique

Since the MORT system itself is large, only the accident investigation usage of MORT is described below.

Establishing the event sequence and barrier failures in accident causation

Accidents are usually the outcome of a sequence of events. Therefore, the MORT approach commences with the analysis of causes by constructing an accident sequence flow diagram. This diagram is constructed by detailing the direct (obvious) sequence of events, which is then built upon by adding contributory factors (secondary events and conditions). Finally, consideration is made of systemic factors (e.g. fundamental aspects of company policy) such as the quality of the original risk assessment, the level of supervision, or adopted procedures and training.

Johnson (1980) recommends the construction of an accident causation diagram to initially establish 'what went wrong'. This gives the assessor a summary of the accident. Johnson uses the term 'accident' to cover an incident resulting in either personal injury or loss and/or equipment damage for product loss – a rather broader definition than injury or death of personnel. The establishment of what went wrong utilizes the concepts of unwanted energy flow and barrier failure, and the reader is referred to pp. 169 for a description of this approach. Once the scenario event sequence has been established, the assessor should move on to the full use of MORT (the reader should note that each event in the sequence may require separate analysis using MORT).

Basic symbols and logic

The entire MORT system contains the MORT tree and a great deal of guidance on its usage, the questions it asks etc..

In essence MORT is a diagrammatic representation of potential management weaknesses, which gives an easily accessible and comprehensive overview of the situation. The MORT chart is based on the fault tree concept with the undesired event given at the top of the tree and the logic of causal events beneath. MORT logic is basically the same as that used in fault trees (i.e. it uses **AND** and **OR** gates to define its structure), but MORT is not quantitative. The top level of the MORT tree, and an example of a sub-tree are shown in Figures 5.5 and 5.6, and a task analysis of how to conduct a MORT analysis is illustrated in Figure 5.7.

Assessment

Each event in the tree is assessed for its influence on the accident. MORT works by forcing the assessor to make a series of decisions. The assessor must decide whether a particular cause shown in the tree was a relevant aspect of that particular accident; if it was irrelevant, then that particular cause plus any further breakdown beneath it is deleted. If viewed as relevant, the assessor must decide whether the condition was 'adequate', or 'less than adequate' (i.e. a likely contributory cause). If a cause is assessed as adequate then, similar to irrelevant causes, the underlying sub-tree is ignored. An additional assessment option is included; 'don't know' but this can only be an interim assessment and must be

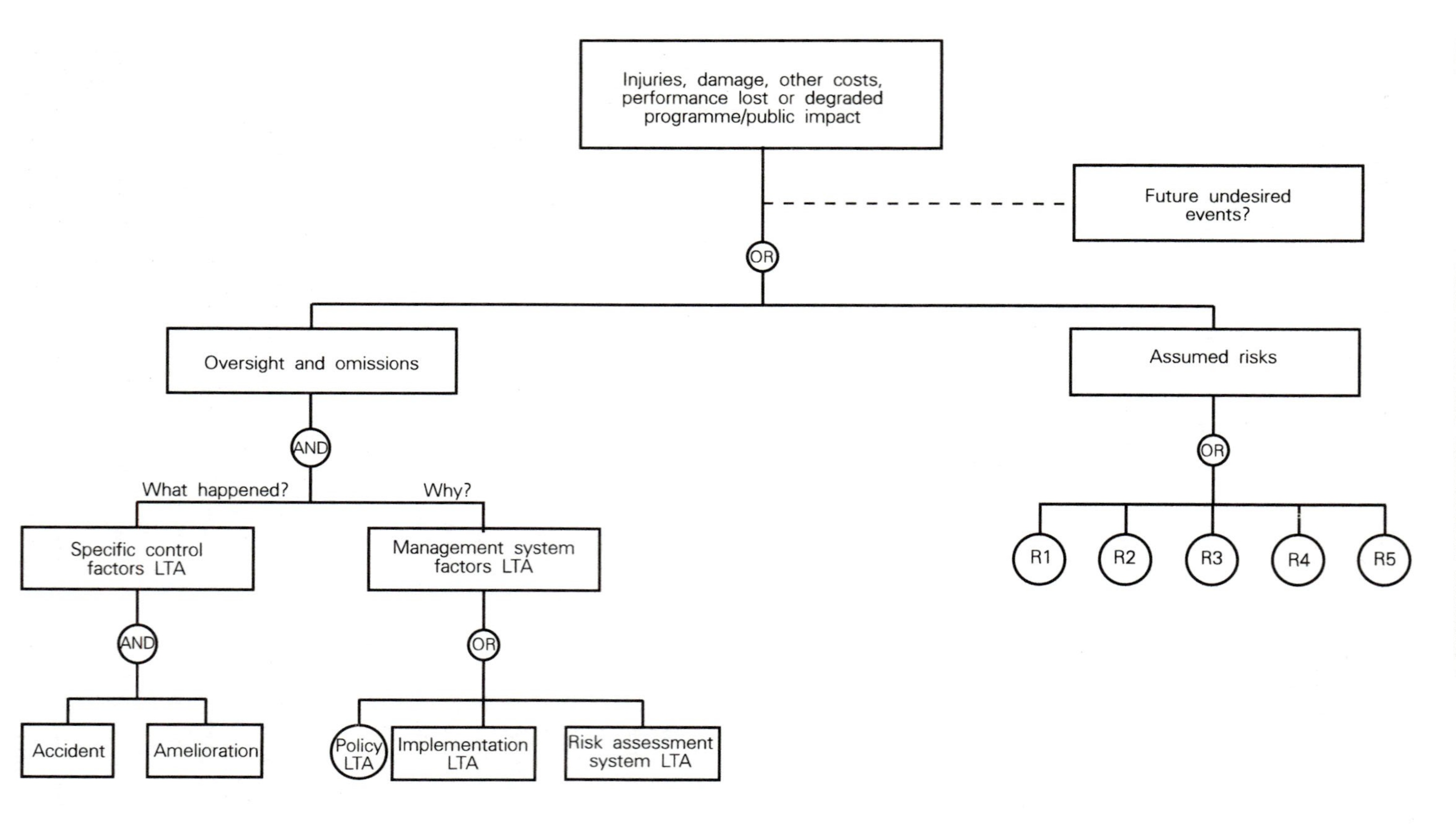

Figure 5.5 The MORT top events

converted to either 'adequate' or 'less than adequate' by obtaining additional information. The assessor will need to use judgement in some cases to decide if the answers to the questions mean that the aspect under investigation is adequate or less than adequate.

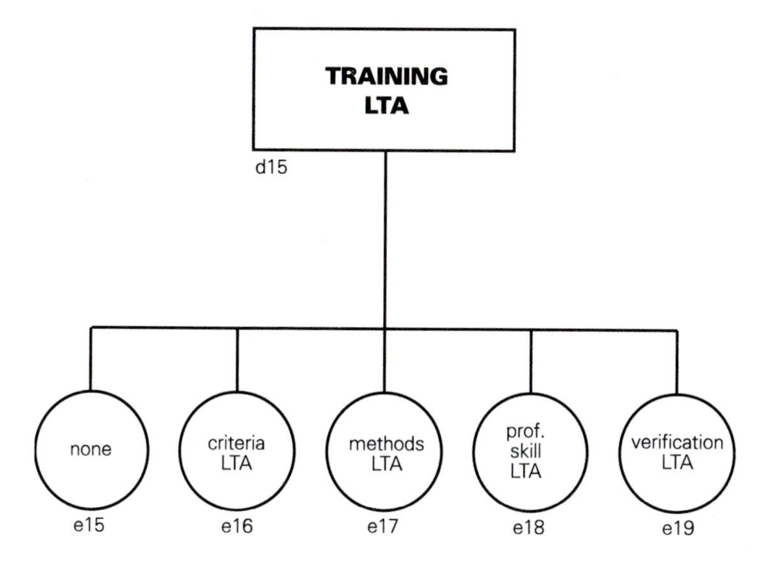

Figure 5.6 MORT training sub-tree

In the MORT chart, 'adequate, less than adequate' and 'don't know' are colour coded to facilitate a quick visual check for problem areas. This is particularly useful when the assessor presents the tree to management as a summary of the accident analysis: red = deficient (i.e. less than adequate); green = OK; blue = to be established. At the end of a MORT assessment, the assessor will have identified the principal causes of the accident, in a systematic and auditable way.

The general structure of MORT: oversight and omissions or assumed risks

The top-level event box summarizes the extent of the accident's associated losses. Once the extent of the accident is established the user arrives at the first logic gate which is an **OR** gate: only those risks which had been identified, analysed and accepted at the appropriate management level can be deemed to be assumed risks; unanalysed or unknown risks are oversights and omissions by default. It is important to remember that mistakes could have been made when initially accepting a risk, therefore the assessment should still be applied.

The next major subdivision separates 'what happened' from 'why'. The 'what happened' section of the tree considers the specific control factors that should have been in operation while the 'why' section of the tree considers general management system factors. It is the 'what happened' branch of the tree which forms the major assessment route during an accident analysis, while

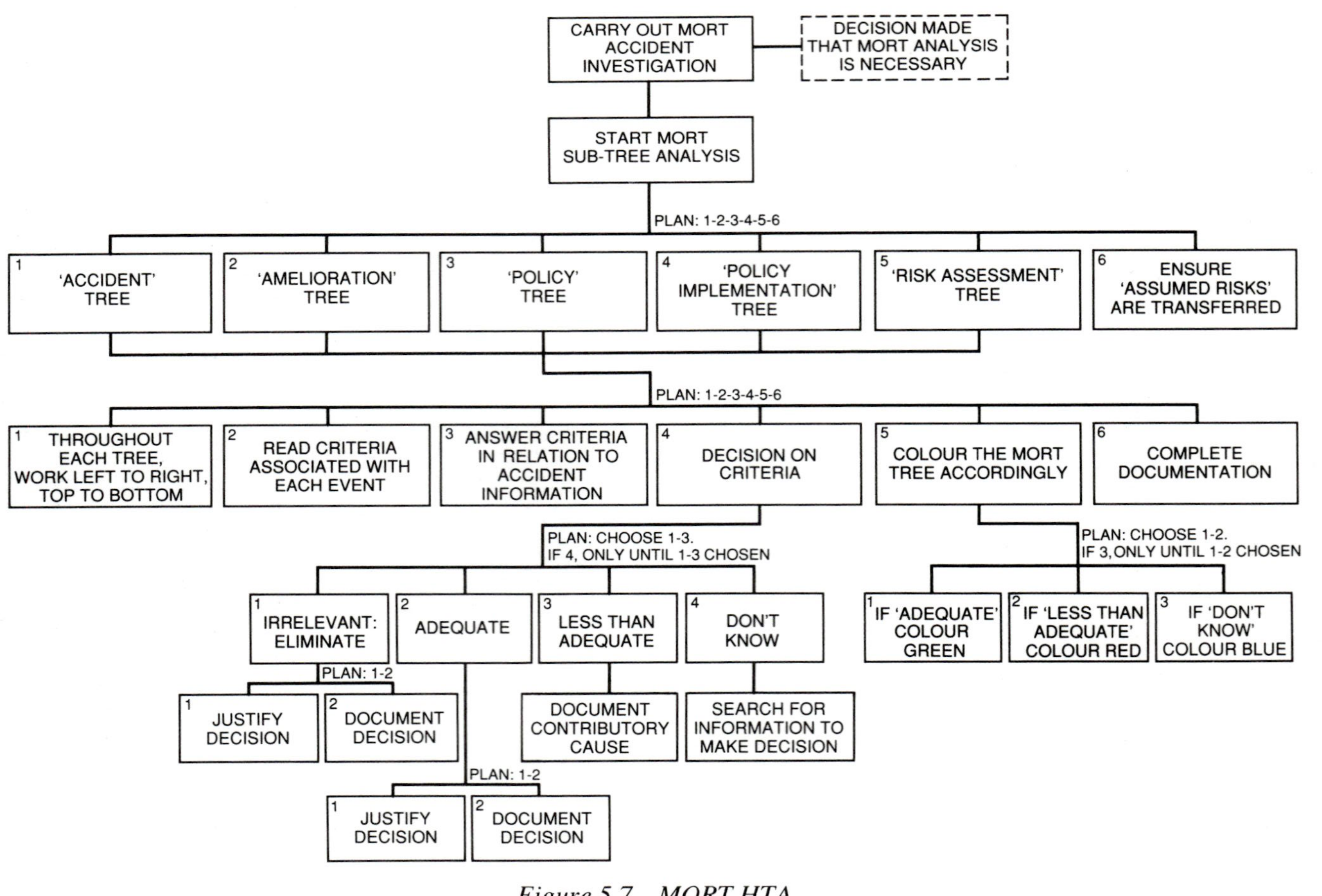

Figure 5.7 MORT HTA

management system factors are of primary importance if assessing the company's safety programme.

The major parts (sub-tree) of the MORT tree are as follows

Technical information systems
Knowledge about the type of incident; performance monitoring systems; data collection systems; hazard assessment systems. Why did these not predict the accident?

Maintenance
Was the maintenance plan or its execution contributory to the event?

Inspection
Did the inspection plan or its execution fail to aid in detecting the danger or the build-up to the accident?

Safety analysis recommended controls
Were the worksite controls placed or implemented on the facility or the equipment/personnel inadequate, and thus contributory to the accident?

Services
Did a lack of resources, research and fact-finding, or standards and directives contribute to the accident sequence?

Human factors
Was the allocation of tasks to machine and to the operator *less than adequate* (LTA), or did a failure to predict errors contribute to the accident?

In addition, other major tree sub-headings in MORT are:

- Task performance: e.g. personnel selection, training, motivation, procedures
- Amelioration: e.g. rescue, emergency response
- Supervision: e.g. performance, higher supervision etc
- Management systems: e.g. policy, policy implementation, or risk assessment
- Concepts and requirements: e.g. definition of goals and concepts of risk, safety criteria, life cycle analysis
- Design and Development Plan: e.g. energy control procedures, independent review method, general design process

Use of the MORT summary page

The first completed MORT tree provides an overview of the full assessment. Depending on the type and seriousness of the event concerned, the relevant branches can be descended to the required level of detail. Each generic event that has an associated structure of contributory events is redescribed by a lower level tree. Each of these detailed trees is referred to on the overview by a capital letter.

General information

When working with the MORT diagrams the direction of completion should be top to bottom, left to right, within each tree. Each event box has its own unique coding. The first letter in the code is an S or M, this identifying whether the event is a Specific causal factor or a Management system factor respectively. The second letter indicates the extent to which it is embedded in the tree: e.g. 'a' events are at the top, 'd' events are four levels beneath. The associated number distinguishes it from others at that level in the branch.

In addition to indicating management factors specific to the analysed accident, it is also possible to identify more fundamental management system aspects which could cause problems elsewhere. Each event included in the MORT tree is separately defined in an accompanying handbook with associated questions to consider during analysis. MORT relates the different sub-trees to different sections in the MORT handbook. These sections provide methods and suggestions for improving inadequate situations.

As an example, when considering the adequacy of the training of the individual(s) involved in the incident, the following types of questions must be answered:

d15 Training LTA – was the personnel training adequate?
e15 None – was the individual trained for the task?
e16 Criteria LTA – were the criteria used to establish the training programme adequate in scope, depth and detail?
e17 Methods LTA – were the methods used in training adequate to meet the training requirements?
e18 Professional Skills LTA – was the basic professional skill of the trainers adequate to teach the training programme?
e19 Verification LTA – was the verification of the person's current trained status adequate? Were training and requalification requirements of the task defined and reinforced?

Negative answers to any of the above five questions (e15–e19) means that an LTA response is generated for the event above them (d15). Since these questions or 'events' are beneath the intermediate event 'training' via an **OR** gate, this means that if any one of these five questions is LTA then the superordinate event 'training' is 'LTA'.

Practical advice

People within the organization may be reluctant to provide the necessary information since such a study can be seen as personally threatening – particularly within organizations requiring help. Training in accident investigation procedures and protocols is therefore desirable. When carrying out MORT accident investigations there are a number of useful supporting documents, in particular the accident investigation manual (DOE, 1985).

The output of MORT, whether in accident investigation or management audit, is concerned with people and their responsibilities, and as such can be taken rather personally. The analyst must be objective and unbiased at all times, and not jump to conclusions, based on circumstantial evidence, for example, during an accident investigation. It is an American system, and there may be cross-cultural differences between both laws and management styles/responsibilities in the UK, which an analyst needs to be aware of.

Besides energy trace analysis and barrier analysis, MORT analysis often recommends carrying out 'change analysis'. This is concerned with seeing if the accident is a result of a change in work practices and so on, which may have occurred within the last 6–12 months, for example. As noted in the section on WSA (pp. 170) accidents are far more likely following a change to the work procedures.

Resources

MORT really requires MORT training or else an extensive period of self-training with the system and its associated techniques. In its application it also requires interviews with various personnel, sometimes at fairly high levels within the organization. As with any accident investigation technique, its length of application will depend upon the accident and the availability of evidence, etc. It is likely, however, that MORT will take longer than other accident investigation tools because of its thorough nature and its structure, which is aimed at ensuring that all causes are considered during the investigation process. This of course is not necessarily a disadvantage.

Links with other techniques

MORT is highly linked to barrier analysis (pp. 169) and has used fault trees (pp. 188) to develop its structure. In the data collection phase it makes use of structured interviews (pp. 66) and the critical incident technique (pp. 47). It may also make use of event trees (pp. 178) when defining the event sequence.

Advantages

- MORT is one of the few techniques addressing management effects on safety, and is a highly structured approach (it has to be in view of its acceptance in the USA for litigation). It is also useful in the consideration of safety management adequacy prior to operation.
- MORT is actually the most prominent of a set of complementary and interlinked techniques such as job safety analysis, safety appraisal, etc., and so MORT can be used in conjunction with these other approaches.

- Energy trace and barrier analyses are useful ways of defining what physically happened prior to an accident. The presentation of the MORT chart is useful when discussing the implications of the analysis. The assessor is forced into deciding whether a situation is adequate or less than adequate and is expected to provide the reasons on which such decisions were based. MORT analyses are therefore very explicit.

Disadvantages

- MORT is complex, and while detailed guidelines are given on its application in the MORT manual, it still requires skill and training to use it in a robust manner (i.e. reliably).
- The MORT appreciation of a typical or normative safety management structure may not always be appropriate for individual organizations, particularly smaller ones.
- MORT requires a good deal of effort to implement, and requires detailed assistance from the organization in an accident investigation, or (less usually) in a safety-management audit situation.

References and further reading

DOE (1985) *Accident/Incident Investigation Manual.* 2nd Edition, DOE, SGDC-76/45-27, Department of the Environment.

Johnson, W.G. (1980) *MORT Safety Assurance System.* New York: Marcel Deckar.

Knox, N.W. and Eicher, R.W. (1976) *MORT Users Manual.* Rept. No. SSDC-4EG. Idaho Falls, Idaho, USA: EG & G.

Nertney, R.J., Clark, J.L. and Eicher, R.W. (1975) *Occupancy Use Readiness Manual.* Rept. No. ANC, ERDA-76/45-1 SSDC-1. Idaho Falls, Idaho, USA: EG & G.

Reunanen, M., Suokas, J. and Heikki, S. (1981) *Drafting a MORT Analysis of a Maintenance Organization.* VTT Rept. No. TYO/5. Helsinki, Finland: Technical Research Centre of Finland.

Suokas, J. (1988) The Role of Safety Analysis in Accident Prevention. *Accident Analysis and Prevention*, 20/1, 67-85.

Chapter 6
Task requirements evaluation methods

Ergonomics checklists

Overview

These are checklists which an analyst can use to ascertain whether particular ergonomics criteria are being met within a task, or whether the facilities which are provided for that task are adequate. The items within these checklists can range from overall subjective opinions to very specific objective checks, which may have to be made several times under varying conditions on different equipment or task elements.

Application

Checklists can be used to collect information about either existing or proposed systems. Usually checklists can be completed by the analyst observing the task interfaces (or some representation of them such as diagrams and documentation) without any additional equipment (i.e. the checklist can be a 'pencil and paper' evaluation method requiring no special tools). However, for some purposes it will also be necessary to observe the tasks themselves, and if the checklist contains very specific behavioural items, it may be helpful to have a comprehensive set of task descriptions available before attempting to fill in the checklist.

The range of issues covered by ergonomics checklists is vast. They can cover any aspect of a task, its associated interfaces, or the environment in which it is undertaken. The checklist items can vary greatly in the amount of detail which they seek, for instance an item could be applied to a complete control room, a single indicator and control panel, or even to a single instrument. Similarly, some checklist items might be answered by general subjective assessment about the acceptability of certain task features, while special measuring equipment might be necessary in order to answer other items.

Description of technique

The design, use, administration and analysis of checklists are described below.

Design of checklists

Ergonomics checklists can be used to ascertain how well a system conforms to accepted ergonomics principles, or, more specifically, to ensure conformance with particular standards or codes of practice. For the former purpose there are several checklists already in existence, but it may be necessary to specially design a checklist to ensure that other standards, or codes of practice, are being met. However, the development of checklist items can be very resources-intensive, and so apart from applications which only require a few checklist items, it will generally be preferable to use, or adapt, an existing checklist than to design a new checklist.

The first stage in checklist design is to select the criteria against which checklist assessments are to be made. These should then be worded as clear, positive statements which demand either agreement, or disagreement. All of these items should be written consistently, so that disagreement is always associated with a failure to comply with the ergonomically desirable design. To the right of each item, at least two columns should be provided, preferably with tick boxes, so that the analyst can record whether or not a particular feature conforms to the checklist criteria. In many situations it will also be helpful to utilize a *not applicable* column as well.

If the checklist items are relatively straightforward, or if they are only intended for users with detailed knowledge, the checklist items can stand on their own, because they will only be used as an *aide-mémoire* by the analyst. However, for most situations, it will be necessary to provide some background information to assist the analyst in making his checklist judgements. Such information should be provided as close as possible to the checklist items and should remain part of the checklist (i.e. not separable from it), with explanatory diagrams where possible. Ideally such information should be attached to each related checklist item.

If it is necessary to have a large number of items on a checklist, it will be helpful to group these under a limited number of sub-headings.

The use of existing checklists

There are several ergonomics checklists which can be used in part, or in their entirety, for analysing tasks and their associated interfaces. However, some of these are particularly lengthy, and so the analyst should either restrict the extent to which such checklists are used, or else regular breaks should be taken. Otherwise, the responses could be more a statement of the analysts vigilance (or lack of it), than of the state of the system.

Amongst the most notable ergonomics checklists are the following.

CRT display checklist (Blackman *et al.*., 1983)
Appendix A of NUREG/CR-3557 gives a 93 item checklist, which was intended for the subjective comparison of different ways of presenting the same information on a CRT/VDU screen. However, the checklist could also be used for the evaluation of specific CRT displays.

Human engineering design checklist (Woodson, 1981)
This is a detailed checklist which has been used extensively by an ergonomics group in industry. It appears in the *Human Factors Design Handbook* by Woodson.

Maintenance checklist (Seminara, 1982)
This checklist (EPRI NP-2360) is devoted to maintenance issues in power plants, but many of the issues are also relevant to maintenance issues in any large system. The intention of this checklist is for each item to be answered on a plant-wide basis, but some of the items could also be used to study plant areas, or particular functions. The items in the checklist are lengthy and self-explanatory, with clear illustrative photographs, close to the appropriate items, to provide guidance. The checklist is also organized so that it is fully compatible with a more detailed report (EPRI NP-1567), which can be used to clarify any items where necessary (see Seminara and Parsons, 1981).

MIL-STD 1472C
This is an extensive standard which covers most aspects of equipment design for military systems. It is written as a checklist with explanatory notes.

NUREG-0700 (Nuclear Regulatory Commission, 1981b)
NUREG-0700 is an extremely detailed checklist for control room design, which incorporates clear explanations of all the items. Most of the checklist items are associated with the design of individual instruments, and so it would be an extremely lengthy process to use this checklist in full, for anything but the simplest of systems. A shortened version of the checklist is given in Section 7 of NUREG-0659 (Nuclear Regulatory Commission, 1981a).

Short guide to reducing human error
This is composed of a series of checklist items for five distinct areas of system design (namely the interfaces, the procedures, the environment, training, and work organization), which are considered to affect the safety of complex systems. The checklist items are intended to assess the overall design of the plant, either for general use, or for assessing the feasibility of a particular task. In a more detailed guide (Ball, 1991) full explanations are provided for each checklist item.

Transport checklist (Woodson, 1981)
This is a short ergonomics checklist, which is presented in the *Human Factors Design Handbook*, following the general ergonomics design checklist (see earlier). It is very much directed towards military vehicles.

User interface software checklist (Smith and Mosier, 1986)
This is a checklist based upon extensive guidelines for VDU software user interfaces. It is currently available in a software form.

VDU checklist (Cakir *et al.*) 1980
This is an extremely detailed checklist for assessing VDU terminals and the workplaces in which they are used. Some of the items require relatively detailed technical information about VDU characteristics (i.e. details which may only be

available on request from the VDU manufacturer), and there are only a few items which are directly related to task requirements.

HSE Checklist (HSE, 1989)
This checklist entitled *Human Factors in Industrial Safety* introduces human factors and the concept of error, and human error types. Prevention and control measures are then defined in three areas: the organization, the job, and personal factors. In order to illustrate potential problems and their consequences, this checklist usefully cites many examples of error problems in real industrial contexts.

Administration of checklists

Checklists should be filled in by an analyst in a systematic manner, either by checking through each checklist item in turn, or else by checking the same item on all instruments, or parts of the system, before moving on to the next item.

Analysis of checklist information

The main concern with checklist information is with items which have failed to meet the prescribed criteria, and thus it is tempting to use the number of items which do not comply with the checklist recommendations, as a measure of the inadequacy of a system. However, because checklist items are not generally scaled in terms of their importance, and because the degree of failure is not usually recorded, it is not possible to compare the responses to different items without further information. Therefore, for every checklist item which fails to meet the prescribed criterion, the analyst should gather further information so that the significance of that non-compliance can be judged. In some cases, it may be sufficient for the analyst to make a general subjective assessment for each non-compliance, while for other purposes, more objectivity will be needed. One relatively simple method of making such an assessment is to use two numerical rating scales for each non-compliance. The first of these scales should assess the degree of non-compliance on a three-point scale, with borderline non-compliances rated as one, and large non-compliances rated as three. The other scale should assess the likely consequences of a failure due to a non-compliance, with a rating of three being given to the most serious errors. Then the significance of each non-compliance can be calculated, by multiplying the two ratings. Most effort should then be directed at rectifying the highest scoring non-compliances.

Practical advice

Extensive checklist surveys can become very monotonous and prone to error. Therefore, much effort must be devoted to trying to limit the number of checklist items which are used for any particular situation. Unfortunately, there are no empirical data available which can be used to assess the optimum number of checklist items which should be used, and in the absence of such guidance, the analyst should not expect to continue using a checklist effectively for more than an hour without a break, or a change of activity.

Prior to undertaking any checklist survey, analysts should take some time to familiarize themselves with the necessary background, or else a considerable amount of the survey time on site will be devoted to discovering this background, rather than assessing the criteria.

Resources required

The main resource which is used by checklist surveys is time. This can range from a few minutes which might be sufficient to conduct a simple checklist, to several days which would be necessary to fill out a detailed checklist, such as NUREG-0700, on many different instruments within a complex system.

Links with other techniques

Although checklists can be used with other techniques, it is most likely that they will be used as a stand-alone technique, to direct the analyst to potential problems, which can then be subjected to more detailed analysis using other task analysis techniques. However, a checklist approach (pp. 223) could be particularly useful for obtaining the information for some interface surveys. The method of checklist evaluation will often utilize observation (pp. 53). If detailed task knowledge is required, it may be necessary to carry out a hierarchical task analysis (pp. 104) and checklists can be used during the use of human error analysis methods such as failure modes and effects analysis and HAZOP (pp. 184 and 194).

Advantages

- Checklists provide an easy way to evaluate whether a system meets specific criteria, which are either mandatory, or of particular importance to an organization.
- The completed checklist can be used as evidence that the system meets a particular set of criteria.

Disadvantages

- It is often necessary to have a good understanding of the underlying ergonomics and psychological principles before certain checklist items can be effectively answered.
- Checklists do not generally make any attempt to assess the relative importance of different items, or to indicate the degree to which items may fail to meet the criteria. Thus, there is a need to undertake some prioritization of checklist failures, in order to avoid misinterpretation of the information.

- Checklists generally take no account of the context in which the tasks that are being checked are carried out. For instance, although an instrument scale marking may be slightly smaller than the legibility standard given in a checklist, this may be acceptable on a regularly-used analogue display on which the normal reading is expected to be close to the 12 o'clock position, because the operator will tend to monitor the pattern of the pointer position, rather than the actual value. In this case, therefore, the actual value does not have to be read so closely, since the operator is only monitoring its relative position.
- Checklist items are usually uni-dimensional, and take no account of important interacting variables (e.g. one item may concern the avoidance of a 'cluttered' display structure on a VDU system, and another checklist item may recommend that all relevant information should be placed together on the same page). In cases such as this, the analyst must use judgement to decide where the 'trade-off' between these potential opposing principles should lie.
- In large systems, equipment-based checklists can soon become unwieldy, and use up a considerable amount of an analyst's time.

References and further reading

Bainbridge, L., Ball, P., Eddershaw, B., Hunns, D., Kirwan, B., Lihou, D. and Williams, J.C. (1985) *Short Guide to Reducing Human Error in Process Operations.* Rept. No. R-347. Warrington: UK Atomic Energy Authority.

Ball, P. (ed.) (1991) *The Guide to Reducing Human Error in Process Operations.* Rept. No. SRDA-R3. Warrington: SRD Association.

Blackman, H.S., Gertman, D.I. and Gilmore, W.E. (1983) *CRT Display Evaluation: The Checklist Evaluation of CRT-Generated Displays,* Rept. No. NUREG/CR-3557. Washington D.C.: U.S. Nuclear Regulatory Commission.

Cakir, A., Hart, D.J. and Stewart, T.F.M. (1980) *Visual Display Units.* Chichester: Wiley and Sons.

Department of Defense (1981) *Human Engineering Criteria for Military Systems, Equipment and Facilities.* MIL-STD 1472C. Washington D.C.: US Dept. of Defense.

HSE (1989) *Human Factors in Industrial Safety.* HS(G)48. London: H.M.S.O.

Nuclear Regulatory Commission (1981a) *Staff Supplement to the Draft Report on Human Engineering Guidelines to Control Room Evaluation.* Rept. No. NUREG-0659. Washington D.C.: US Nuclear Regulatory Commission.

Nuclear Regulatory Commission (1981b) *Guidelines for Control Room Design Reviews* Rept. No. NUREG-0700. Washington D.C.: US Nuclear Regulatory Commission.

Seminara, J. (1982) *Human Factors Methods for Assessing and Enhancing Power Plant Maintainability.* Rept. EPRI NP-2360. Palo Alto, California, USA.: Electric Power Research Institute.

Seminara, J. and Parsons, S.O. (1981) *Human Factors Review of Power Plant Maintainability.* Rept. No. EPRI NP-1567. Palo Alto, California, USA: Electric Power Research Institute.

Smith, S.L. and Mosier, J.N. (1986) *Guidelines of Designing User Interface Software*. Rept. No. ESD-TR-86-278. Bedford, Massachusetts, USA: Mitre Corporation.

Woodson, W.E. (1981) *Human Factors Design Handbook*, New York: McGraw Hill.

See Chapters 9 and 16, which both utilized a checklist approach.

Interface surveys

Overview

Interface surveys are a group of information collection methods which can be used to gather information about specific physical aspects of the person–machine interface at which tasks are carried out. For these purposes the person–machine interface is defined very broadly, so that it encompasses the ambient environment, as well as the controls and displays for the system. Each survey method is limited to considering specific aspects of the interfaces, and so it will usually be necessary to select a set of these methods which can address the particular issues of interest to an analyst.

All the methods in this section require the analyst to systematically conduct a survey of the man–machine interface, in which specific features are measured (e.g. illumination levels etc.) or recorded. Brief summaries are given of the following survey methods:

- Control/display analysis
- Labelling surveys
- Coding consistency surveys
- Operator modification surveys
- Sightline survey
- Environmental surveys

Examples of the use of most of these surveys in a control room are given in Ainsworth (1985), and guidelines to help the analyst to evaluate the findings from the surveys can be obtained from most general ergonomics texts.

Application

Some of the survey methods can be used throughout the life cycle of a system, but others, such as environmental surveys and operator modifications surveys, cannot be used until there is an actual system in existence.

The information from interface surveys can be used to highlight directly poor ergonomics practice, or to help produce detailed system control and display lists which define the facilities that the system provides (or will provide) for the user. These control and display lists can then be compared to the task requirements to check whether there will be any mismatches between the information and control facilities that are supplied by the system to the operator, and those required to conduct particular tasks effectively.

Description of techniques

Each of the interface survey methods requires an analyst to record, and sometimes also to measure, specific aspects of the system. These aspects are summarized below for each method.

Control/display analysis

A control/display analysis is a detailed survey of the control and display facilities which are provided in a system, and this can then be used to ensure that all the necessary instrumentation has been provided for a particular task. Such surveys should examine all the parameters which can be displayed or controlled, and should then list these together with additional details which more fully describe each display or control. These control and display lists are particularly useful as a check during the design stage, in which case the necessary information will probably be derived from drawings, or high-fidelity mock-ups. However, the lists can also be generated by reviewing operational systems.

In order to develop a controls list, the analyst should systematically examine each control in turn, and record the parameter which is being controlled, together with some means of identifying the control (such as its location, or its reference code). The analyst should also record the type of control (e.g. push-button, rotary switch, toggle switch, etc.) and other details which might be useful during any subsequent analysis, such as the range of movement, the accuracy with which a particular control can be used, the units which are given, or its safety classification.

A displays list will be based upon the examination of the displays in a system (e.g. dials, indicator lights, computer screens, etc.), but otherwise these will be analogous to control lists. However, for some displays it may also be appropriate to note any particularly important values, such as normal and abnormal operating limits.

Once all the necessary information has been gathered, the lists should be sorted into an order which facilitates the search and location of particular parameters. The manner in which the lists are ordered will depend upon the system which is being investigated, but for most systems, a hierarchical breakdown into *system, subsystem* and then *parameter*, with alphabetical and/or numerical ordering within each level of the hierarchy, should prove sufficient. For some parameters it is likely that there will be more than one control or display, and in order to emphasize this, a horizontal line can be drawn across the list wherever a parameter changes. It may also be helpful to use a format in which the parameters concerned are listed in a column at the centre of a page, with their associated controls and displays to either side of this column, as illustrated in Figure 6.1.

Control			**Parameter**	**Display**		
Ref.	Type	Range		Ref.	Type	Range

Figure 6.1 Control/Display list

The primary use of control and display lists is to enable an analyst to check that a system presents all the information and provides all the controls which a user will need to undertake particular tasks. The information about what information and control is necessary in order to undertake a task will normally be derived from detailed task descriptions, or from decompositions of these. The lists can also be used as a source of information for checklist studies.

Labelling surveys

Incorrect, unclear, or inconsistent labelling (both on hardware and on VDU displays) can be a significant cause of confusion and error for the users, and so it is useful to perform some systematic examination of the labelling. During such a survey, the following details should be recorded, using a data collection form such as that illustrated in Figure 6.2:

Ref.	**Wording**	**Size**	**Position**	**Colour**

Figure 6.2 Labelling survey data sheet

- A reference code for the label
- The wording of the label
- The relationship between the position of each label and the item to which it refers (e.g. above, below, left or right)
- The size of the label within the labelling hierarchy (e.g. system, subsystem, component, etc.)
- Label colour

If there are any labelling omissions, these should also be recorded. The information which is obtained from these surveys should then be analysed for inconsistencies, or poor ergonomics practice, to ensure that all labels are legible and meaningful, and that they are positioned consistently. It will also be useful to check whether the wording which is used on the labels is consistent with that used on related documentation, such as procedures, or training manuals.

Coding consistency surveys

A fundamental ergonomics principle for the design of complex systems, is that the human perception and organization of information can be significantly aided by providing additional cues to the users, in the form of various types of perceptual coding. The most prominent of these is the use of colour coding, but relative position (e.g. *Start* positioned to the left of *Stop*), size and shape coding are also useful.

Unfortunately, however, coding systems are often not formally developed, but rather they 'evolve' in a piecemeal fashion. A coding survey is intended to systematically investigate the coding systems, which may unwittingly, or wittingly, have been introduced into a system. This should enable the analyst to determine what meanings have been associated with various codes, so that ambiguous or misleading coding can be identified, and it should also show where additional coding is required.

In its simplest form, the output from a coding consistency survey could just be a tally of the number of times that a particular coding was associated with different meanings. However, on its own, this will seldom provide sufficient information, and so it is suggested that there are four features which should be recorded for each item. These features are:

- An item reference, or a description of its location
- A full description of the coding which has been observed for a particular item (e.g. a red flashing light)
- Any other supplementary coding which is present (for instance, a particular control may be coded using both colours and relative position). This means that some items might be surveyed more than once
- A full description of the feature/function which is being coded (e.g. *valve open*). In some circumstances it could be appropriate to subdivide this description, so that it also gives an indication of the overall system state. For instance, in the above example, an open valve which would normally be open, might be coded differently to another open valve which was normally expected to be in the closed position

A sample data collection form from a study which collected this information at a nuclear power plant, is shown as Figure 6.3

For coding surveys of paged VDU systems, it may also be appropriate to record the level within the hierarchy, or the type of operation which is being carried out.

Ref.	Coding	Supplementary coding	Meaning
CIES valve 1	RED BUTTON	Right of 2	Close valve
CIES valve 2	RED BUTTON	Right of 2	Close valve
Valve D3	RED BUTTON	Centre of 3 (horiz.)	Stop valve movement
Valve D3	BLUE BUTTON	Left of 3	Close valve
Valve 225	BLUE BUTTON	Top of 2	Open valve
Lamp Test	BLUE BUTTON	None	Lamp test for panel

Figure 6.3 Example of coding consistency data collection sheet

It is usually easiest when conducting a coding consistency survey, to restrict the initial survey to a single type of coding, and to examine the occurrence of this type of coding systematically throughout the entire system, before looking at other coding systems. However, for some VDU-based systems, it may be preferable to systematically work through the display pages for all coding methods at once. The coding systems which could be surveyed could include:

- Colour of illuminated indicators
- Colour of controls
- Background colour
- Size of controls
- Shape of controls
- Relative position of instruments
- Relative position of specific markings (e.g. on, off, open, close, etc.) on multi-position switches
- Relative position, or orientation, of scale markings, such as zero points, on scales
- Instrument identification codes
- Auditory coding
- Highlighting methods (e.g. inverse video, flashing, etc.)

Once all the information has been collected, it can be analysed to identify how well particular coding conventions have been established, and also to show where the coding needs to be modified. For instance, as a result of a colour coding survey at Three Mile Island, Malone *et al.* (1980) discovered that red had been used to code 14 different functions or plant states, some of which were diametrically opposed to each other, and these were seen as a serious source of potential human errors.

It is good ergonomics practice to combine at least two forms of coding to represent important functions, particularly when colour coding is being used. Therefore, it will also be necessary to analyse the information in terms of different combinations of coding methods. Guidelines on coding/labelling requirements be to compared against the coding characteristics actually used, as found in the survey (see Ball, 1991; Oborne and Gruneberg, 1983; Pheasant, 1986)

In any analysis, it will also be important to consider coding, both in terms of the underlying function being encoded and the coding method which is employed. For instance, it could be just as important for the analyst to ascertain that green had been used inconsistently to represent both *open valve* and *close valve*, as it would be to discover that four different colours have been used to represent the *close valve* function. The coding survey should also note the absence of coding where it may be useful.

Finally, if a coding consistency survey has examined all the relative positions of groups of controls and displays, the analyst should use this information to determine whether groups of controls and their related displays, are related to each other in a compatible and ergonomically acceptable manner.

Operator modifications surveys

The end users of systems are the people who have to contend with problems and difficulties which were often not envisaged by the designers. A common response to such design inadequacies is the addition of temporary modifications by the users. Ainsworth (1985) has suggested that these predominantly tend to be of three main kinds

- *Memory aids* – such as dymo labels to mark safe limits on displays, or to show control positions
- *Perceptual cues* – such as makeshift pointers, or additional mimic lines.
- *Organizational cues* – such as grouping instruments together by a demarcation line

Such modifications can provide a wealth of information about the difficulties which users have experienced with systems, and these can be recorded relatively easily from a quick survey of an operational system. This information can then be used to identify both specific and general areas of the design which merit particular attention. However, because these modifications will only reflect difficulties which users have experienced, they must obviously be gathered from systems which have some operational history. Therefore, these surveys cannot be used as an aid to design, unless the surveys can be done on generically similar systems.

It must also be stressed that an operator modifications survey will only highlight problems which have been perceived as such by the users, and such perceptions could well be influenced by the users' personal biases. Thus, it is possible that features which have only a minor effect might be modified, while more serious shortcomings may be left, either because they are not perceived as being amenable to modification by these users, or because they are considered as being too much trouble to modify.

Some results from an operator modifications survey of a power plant are given in Ainsworth (1985).

Sightline surveys

A sightline is a direct line from an observer's eye position to other positions of interest. If it is not possible to draw such a line between an observer and a particular instrument, then it is obvious that the instrument will not be seen from that position, and this may seriously compromise the use of the system.

Although there may be a direct line of sight to particular instruments, it may still not be possible to read them effectively, because the reading distance is too great and because the viewing angle may not be acceptable. Therefore, it is often helpful to supplement sightline data with some measurement of the distances to instruments of interest, which can be incorporated with the sightline data for presentation. The viewing distance information can then be used to estimate the size of markings which are necessary to ensure legibility.

For existing systems, sightlines can be determined relatively easily. If the analyst is only interested in a relatively few sightlines, these can be obtained directly by simply recording whether particular instruments can be seen and are legible from specific viewing points. However, if a more comprehensive sightline survey is required, photographic or video methods will invariably prove to be the most effective means of gathering data. In either case, it is usual to take the viewing position as the eye height (standing or seated) for the smallest likely users (5th percentile). If a photographic method is being used, a camera should be placed at the required 'eye' position, and then panned round to record the view for later analysis. Thus, the collection of sightline information can often be achieved very rapidly.

For proposed systems, sightlines can be measured from drawings or diagrams, or else a computer can be used to generate them.

There are four main ways that sightline information can be presented

Text
For simple systems, or where the legibility of only a few instruments and/or labels needs to be determined, sightline information can be presented as plain text.

Layout plans
Scaled layout drawings which show the limiting sightlines from selected viewing positions will be sufficient for many purposes, providing that these limiting sightlines do not vary at different heights. Viewing distances can be shown on these drawings as contours based upon particular viewing positions.

Simulated views
The clearest way to present sightline information is as simulated views. These can be drawn in perspective, or can be based upon panel layout drawings, as in

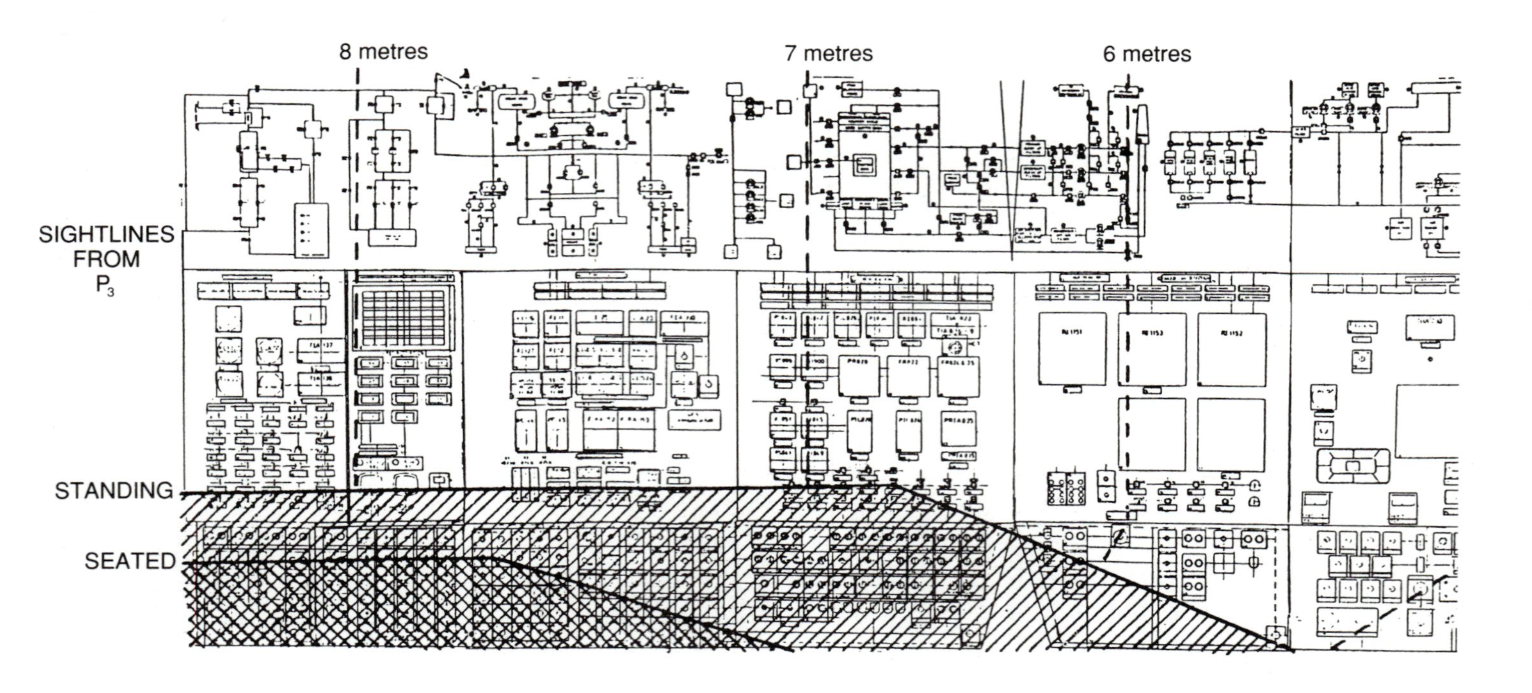

Figure 6.4 Sightline survey of a power plant control room from the operators desk. Shaded areas show the view for a 50th percentile male when sitting or standing at this position. Adapted from Ainsworth , 1985.

Figure 6.4. Viewing distance contours can be superimposed on these diagrams if necessary. Simulated views can also be shown on some computer-aided design packages.

Photographs
If photographic methods have been used to gather the sightline information, the photographs themselves provide a very convenient means of presenting the sightline information, and if necessary, distance contours can easily be superimposed on them.

Environmental Survey

Environmental surveys measure the state of the ambient environment (i.e. noise levels, illumination levels, temperature and humidity) directly within systems, and thus they can usually only be undertaken in fully operational systems.

Within this report it is only possible to give some general guidelines about conducting environmental surveys, and therefore any reader who is not familiar with environmental measurement techniques, is directed towards equipment handbooks and standard environmental texts, such as the IES Code (Illuminating Engineering Society, 1977) or Oborne and Gruneberg (1983), to learn how to use the specialized measurement equipment which is necessary.

Apart from technical advice about using environmental measurement equipment, there are two important guidelines for conducting environmental surveys, which must be mentioned.

- Measurement points must be chosen which are appropriate to the tasks which have to be undertaken. For instance, illumination should be measured at points where there are critical instruments to be read, not at floor level. The position of such points must also be clearly defined, in order to ensure repeatability, and to assure users of the information that the measurements were taken at appropriate points.
- The nature of the ambient environment often varies according to the time of day, the time of year, and the tasks being undertaken. Therefore, several measurements should be taken at predetermined times, under typical operating conditions.

Practical advice

For each of these survey methods, the most important practical advice to users is that in order to ensure that all items in a complex system are fully considered, a very systematic approach must be adopted for gathering the information. For instance, in a large control room, each section of the control panel should be scanned from top left to bottom right for a limited number control/display of features in turn. The importance of adopting a systematic approach will be particularly important for complex VDU systems, where there could be a large amount of information to be recorded, and much of this could be repeated on different display pages.

Resources required

The information for each of these interface surveys can generally be collected by a walk-round inspection, or by the examination of system drawings, and so they only require a limited amount of an analyst's time. However, for very extensive systems, there could be a very large number of items to be considered, and in such cases, the workload involved could become very high.

It is advisable to prepare special data sheets in advance for each survey, but this can generally be accomplished relatively easily, by adapting existing data sheets, such as those shown in Figures 6.1, 6.2 and 6.3.

The only other resources which may be necessary, are specialized environmental measuring equipment, which will be required for environmental surveys.

Links with other techniques

Prior to the construction of a system, control and display lists which have been developed from interface surveys can be compared to lists of the control and information requirements for particular tasks (e.g. task decomposition analysis methods – pp. 95), to identify shortcomings which might render these tasks untenable.

The information which is derived from interface surveys defines the context in which tasks have to be carried out. Therefore, it can be drawn upon by all other task analysis techniques which need to determine whether the human–machine interfaces support the user to undertake particular tasks (e.g. ergonomics checklists, pp. 217). The interface surveys can also be performed on simulators/mock-ups (pp. 150) or panel layout drawings, to provide exploratory investigation of these interfaces.

Advantages

- Each survey method is relatively easy to administer.
- The methods can be employed on existing systems with little, or no, interruption to the normal users of such systems.

Disadvantages

- Although the surveys are relatively simple, in some situations, the sheer volume of items which have to be recorded could make this a somewhat onerous task.
- Some of the survey methods can only be used when a system is already in existence.

References

Ainsworth, L.K. (1985) *A Study of Control Room Evaluation techniques in a Nuclear Power Plant.* Applied Psychology Rept. No. 115, University of Aston. *Gives examples of most of the survey methods being used to study a control room.*

Ball, P. (ed.) (1991) *The Guide to Reducing Human Error in Process Operations.* Rept. No. SRD R484. Warrington: Safety and Reliability Directorate.

Illuminating Engineering Society. (1977) *The IES Code for Interior Lighting,* London: IES. *An excellent practical summary of illumination and light measurement, which presents a comprehensive list of recommended illumination levels for different situations.*

Malone, T.B., Kirkpatrick, M., Mallory, K., Eike, D., Johnson, J.H. and Walker, R.V. (1980) *Human Factors Evaluation of Control Room Design and Operator Performance at Three Mile Island – 2.* Rept. No. NUREG/CR 1276. Washington D.C.: US Nuclear Regulatory Commission. *Illustrates some of the coding deficiencies which were found at Three Mile Island.*

Oborne, D.J. and Gruneberg, M.M. (1983) *The Physical Environment at Work.* Chichester: Wiley and Sons. *Provides background information about the effects of the ambient environment (lighting, noise, heating and ventilation) upon performance, health and well-being.*

Pheasant, S. (1986) *Bodyspace* London: Taylor and Francis. *Provides general ergonomics background information.*

See Chapter 8, which used a form of interface survey technique.

PART 3
TASK ANALYSIS CASE STUDIES

Introduction to Part III

Part I of the *Guide* defined task analysis and the task analysis process, and discussed where and when task analysis can be used. Part II detailed 25 task analysis techniques. The purpose of Part III is to demonstrate how some of these task analysis methods have been used in real applications to improve systems. There were four main criteria for selecting case studies: it was desirable to have one or more case studies deal with each of the six human factors issues raised in Part I; a representative range of industries was desirable; case studies at different system life cycle stages were also searched for; and last, but by no means least, the case studies were to show a diverse range of task analysis techniques. Therefore, within these constraints, a variety of case studies were chosen to demonstrate the use of task analysis in solving problems relating to the human factors issues discussed in Part I:

- Person specification
- Allocation of function
- Staffing of job and organization
- Task and interface design
- Skills and knowledge acquisition
- Performance assurance

All of these application areas have been covered in these case studies, apart from *person specification* because no suitable case study of this application area could be found in a form which would be useful to demonstrate the use of task analysis.

In addition, there was a general lack of available suitable case study material which could be published, for reasons of confidentiality, and as a result there has been a focus by some of the case studies on one heavily published area, that of the nuclear power industry. There are also, however, case studies from the military domain, chemical and offshore industries, and the problems addressed by all of these case studies may arise in other industries, and can be solved using the same sorts of task analysis methods or a combinations of such methods.

As noted above, Part III was also intended to include case studies which have used task analysis in different stages of the design life cycle. This has been achieved with case studies representing interventions at various system life cycle stages, ranging from the early stages (Chapters 7 and 8), to the detailed design stage (e.g. Chapters 9 and 10), to the later stages of commissioning and operation (e.g. Chapters 13 and 14. The final stage, that of decommissioning, is not represented as no relevant task analysis case study of this life cycle stage could be found.

The ten case studies incorporated in this part of the *Guide* span a wide range of techniques. Hierarchical task analysis (HTA) features widely throughout the case studies as it is a fundamental way of analysing and describing how subtasks fit together to achieve the overall goal, thereby clearly representing the task elements and operations, which can then be more usefully examined.

Structure of the case studies

The case studies each outline the problem or reason for the study and as far as possible they provide the background information necessary to understand the process of analysis. The objectives of the study are discussed and the method(s) of task analysis described, noting any particular modifications made to the standard task analysis technique for the specific purposes of each study. The case studies discuss the resourcing of such projects, although it was not always possible to detail the number of person-days involved, and inevitably this will have depended to some extent on the experience of the assessors in that particular industry and their familiarity with the assessment method.

The case studies indicate the results of the analysis and, where possible, show how they impacted upon the design or assessment of the system. If relevant, recommendations and/or conclusions which arose out of the analyses are cited, and lessons learned are discussed where appropriate. Some attempt has been made to indicate the cost-effectiveness and impact of using the task analysis method(s) by discussing the perceived benefits to the organization. It is not possible to state how much money is saved by assisting system design in this way. The savings are likely to be in terms of a reduction in down-time and an increase in safety, both of which are difficult to quantify.

What the case studies do not attempt to do is instruct the reader how to perform the particular task analysis methods (Part II of the *Guide* is more appropriate for this), but they do provide some useful examples of where, when and how task analysis methods can be used, by themselves or in conjunction with others, to achieve a particular purpose.

Summary of case studies in Part III

Below, each of the case studies is briefly summarized in terms of its nature, industry, life cycle stage and the task analysis methods used.

1 Balancing automation and human action through task analysis

This case study reviews the assessment of a specific nuclear power plant operation to ascertain whether the proposed staffing levels were adequate. This operation had a high safety significance, and followed a variety of faults. The case study utilized hierarchical task analysis and the functional analysis system technique (FAST) approach, and timeline analysis was undertaken to generate workload estimates for the task, resulting in changes to the proposed staffing arrangements.

2 A preliminary communications system assessment

This case study documents a task analysis of communications on an offshore platform where a drilling crew are to be integrated into what was originally a production platform. Although the platform existed, this was a design study to expand the platform's capabilities. The study highlights the utility of carrying out the study early on, as a number of potential problem areas and items for consideration were identified by the study. This study in particular shows the impact of task analysis on a system being redesigned.

The techniques used were interviews, task decomposition, controls/displays survey and informal talk-through analysis.

3 A plant local panel review

This case study is concerned with a large nuclear chemical plant, and in fact shows the detail of part of the case study documented in chapter 16. It demonstrates the depth of analysis that can be required to fully assess a system, in this case a set of local panel instrumentation at the late detailed design stage. Much of the detail of the task decomposition is included, as well as the human error analysis which was used to explore the consequences of errors and their rectification. The study contains a 'before-and-after' design of a panel, showing the improvements implemented on the basis of the task analysis. Techniques used were hierarchical task analysis, task decomposition and a form of failure modes and effects analysis.

4 A staffing assessment for a local control room

This case study again shows detail from case study 10, this time focusing on a workload assessment for a local control room. The assessment, used to estimate the number of staff required to operate the system, occurred in the earlier stages of detailed design, and used a fairly straightforward approach to predict workload. The results suggested extra staff support would be required for one of the workstations in the local control room. The assessment methods used hierarchical task analysis and timeline analysis.

5 Simulation to predict operator workload in a command system

This case study in the naval defence area was again concerned with workload. However, this time a computerized system called Micro-SAINT was used to predict whether operators could perform their tasks successfully within the available time. GOMS was used to generate some of the data required for input into Micro-SAINT. The main result was a workload simulation model of the tasks, and further confidence in the interface design for the operators during these tasks.

6 Operator safety actions analysis

This case study details a task analysis which was carried out for the Sizewell 'B' pressurized water reactor which was still at the design phase. Specifically, the task analysis concerned a set of operator actions which were necessary following a particular reactor fault. In particular, this case study demonstrates the usage of a specific format for summarizing the results of such an analysis in a manner which is useful for relevant assessment and design personnel.

7 Maintenance training

This study concerns the development of a scheme for supervising and managing the training of mechanical fitters for an existing chemical installation. Hierarchical task analysis was the principal method used, and resulted in operating procedures, detailed statements of the ranges of skills of the fitters and supervisor responsibilities, and a statement of the interaction between maintenance staff and line personnel.

8 A method for quantifying ultrasonic inspection effectiveness

This case study involved the investigation of likely inspector performance and errors in ultrasonic non-destructive testing (NDT) operations. The techniques of hierarchical task analysis, and event and fault tree analysis were utilized, and the latter two were quantified in order to highlight the relative significance of different types of error. In particular, the actual scanning part of the task was seen as an area of performance difficulty.

9 Operational safety review of a solid waste storage plant

This study of an existing operational plant concerned the problem of investigating the operability of a particular subsystem. A set of task analysis techniques including structured interviews, hierarchical task analysis, walk-throughs, task decomposition, timeline analysis and interface surveys revealed a number of problems resulting in operating difficulties. However, more importantly, the results of the study were used to render the system operable through a variety of suggested modifications to procedures, training and hardware aspects of the system.

10 A task analysis programme for a large nuclear chemical plant

This case study from a nuclear chemical plant details an entire task analysis programme for a large design project at the late detailed design stage. The task analysis programme made significant use of hierarchical task analysis, as well as timelines, a failure modes and effects analysis-type of approach, ergonomics checklists, and fault and event trees. The task analysis programme was part of a human factors and human reliability evaluation of the design of the plant, and the case study shows the interrelationships between the various task analysis approaches.

Chapter 7

Balancing automation and human action through task analysis

A. Fewins – *NNC Ltd*, **K. Mitchell** – *NNC Ltd*, **J.C. Williams** – *Nuclear Electric/DNV Technica*

Introduction

This paper describes the application of a practical methodology to assess the implications of carrying out a specific plant manoeuvre with various assumed staffing levels. The exercise was carried out during the design phase of the Sizewell 'B' pressurized water reactor (PWR). The manoeuvre selected for analysis was a normal reactor coolant system (RCS) cooldown and depressurization to the Cold Shutdown State. This manoeuvre was chosen because it was known to involve significant operator actions which might have an impact on automation requirements and staffing levels. Since the manoeuvre must also be performed following a variety of faults, for example a steam generator tube rupture (SGTR), it was perceived as having a high safety significance. An additional advantage was that the information obtained could form the basis of similar future assessments of post-fault recoveries. This exercise involved the integration of several task analysis techniques to ensure the overall objectives of the exercise could be achieved.

Objectives

The primary objective of the exercise was to assess whether automation provisions within the design system would enable a specified plant manoeuvre to be adequately carried out given the defined minimum main control room (MCR) staffing complement of one supervisor and one desk operator.

Additional objectives were to identify any important requirements for the following:

- The Man–Machine interface (MMI)
- System design
- Work organization
- Training
- Procedures

Information sources

Information was obtained from the following sources:

- Standard nuclear unit power systems (SNUPPS) procedures
- Expert knowledge from different design and operations related teams
- Expert knowledge from the training discipline
- Detailed information on plant components and operations

Method

The methodology employed in this exercise was developed by the study group and is described in detail in Williams (1988). This involved the integration of several human factors techniques to ensure that the overall objectives of the exercise could be achieved. These techniques included:

- Hierarchical task analysis
- Timeline analysis
- Workload assessment

Hierarchical Task Analysis

Since detailed Sizewell 'B' operating procedures did not exist at the time, the first requirement was to identify the 'genuine' operator tasks to be carried out as part of the normal RCS cooldown and depressurization manoeuvre. This was done by performing a hierarchical task analysis (HTA) utilizing the function analysis system technique (FAST; see Creasy, 1980).

Using FAST enabled the HTA to be developed in a logical and consistent manner. FAST is a method for arranging functions to answer HOW and WHY questions. Working upwards through the hierarchy enables the analyst to determine WHY a task is performed, in functional terms (i.e. what system functions it achieves). Working downwards through the hierarchy enables the analyst to determine HOW a function is achieved in operational terms. In addition, the use of conventional **AND** and **OR** gates allowed a very robust and logical HTA to be developed. Two different HTAs were produced. A function-based HTA was developed in parallel with, but slightly in advance of, an operations-based HTA. The functions-based HTA described how the provisions in the plant design supported the performance of this manoeuvre and ensured all analysts had a common knowledge-base. It also served as an important aid in the development of the operations-based HTA.

The operations-based HTA represented the operator actions required for successfully carrying out the plant manoeuvre. The HTA was developed in sufficient detail to allow a workload and MMI assessment to be made while avoiding unnecessary detail. This resulted in 56 separate tasks being defined which formed the basis for the timeline and workload assessments discussed below.

Figure 7.1 shows a small part of the operations-based HTA and serves to illustrate the features described above.

Timeline analysis

Timeline analysis is a method of identifying the density of tasks to be performed. It graphically represents the relative timing of different tasks and the duration of individual tasks, and hence identifies where parallel tasks are required to be performed.

The operator tasks identified from the operations-based HTA were represented in the timeline analysis by using:

- The combined operational experience of the study group
- The relationships between tasks and functional goals within the HTA itself
- The specific cooldown and depressurization trajectory assumed in this analysis (used to accurately locate tasks required to be performed at given RCS pressures or temperatures)
- The assessment of individual tasks (see workload assessment below)

Figure 7.2 shows a portion of the timeline analysis which was developed, and provides a clear example of the advantages of this type of graphical representation in allowing the ready identification of periods of high task density.

Workload assessment

The workload assessment was performed in two stages. Firstly, each individual task was assessed by the study group and rated in terms of difficulty, complexity, accuracy, special skills/knowledge requirements, time constraints and critical cues (e.g. when to start/finish the task) and criticality (for safety, and/or for plant availability). Secondly, a summated task rating was produced for each 12 min interval throughout the cooldown and depressurization manoeuvre (i.e. aggregating the rating 'scores' for difficulty and criticality with a range of 0–6, with 6 representing very high difficulty and criticality, see Williams, 1988, for detail).

To assist in the task assessment and rating, a task description worksheet was developed. This worksheet concentrated on the assessment of the tasks in terms of their difficulty and criticality for both safety and system availability. It combined the virtues of being simple, resource-efficient and capable of identifying potential MMI shortcomings, while containing a rating scheme which could allow a rapid indication of the task workload to be obtained. The relationships between tasks revealed while performing these task assessments, also assisted in the iterative process of the development of the timeline analysis.

The reliability of the assessments was enhanced by adopting a systematic process of assessment by one study group member a and review by another. A small number of tasks were assessed by all members of the study group and the ratings were found to agree approximately, indicating a fair degree of inter-analyst consistency.

In order to derive an overall assessment of the potential workload, a workload indicator was produced based on time-weighted task ratings summated over 12 min intervals throughout the manoeuvre. The summated task ratings were plotted in the form of a histogram to give a comparative indication of the potential operator workload throughout the manoeuvre. The workload indicator produced is shown in Figure 7.3 (the time before the manoeuvre begins is represented as negative time. In this figure, the summated task rating may go well above the value of 6, as several tasks can be competing for the operators' attention at once.

In order to assess the workload against the operating staff capabilities, it was necessary to calibrate the workload indicator. This was achieved by using the well-known preliminary phase of the manoeuvre where the hot shut-down state is being maintained and the required shut-down margin is being established. This period was assessed by the study group to be within the capability of a single desk operator and the supervisor, with some spare capacity, for example, to respond to alarms.

Evaluation

MCR Control location

To assess the implications of the task groupings shown on the timeline analysis, a stand-alone evaluation of the MCR control location was made at various points throughout the manoeuvre. The aim of this evaluation was to identify any coincident tasks requiring control from different sections of the existing main control suite concept. Where potential problems were identified, MMI layout changes were considered, irrespective of the assessed workload at that time.

Workload

By plotting the RCS cooldown and depressurization trajectory on the same time axis as the workload indicator, it was possible to see where the workload might be affected by key features of the manoeuvre. Following calibration of the *workload indicator* it was possible to determine where a single desk operator appeared to be insufficient and to investigate the sensitivity of the overall workload to various assumptions about the degree of automation.

Before changes in automation were considered as a means of reducing operator workload other possibilities were explored. These included:

- Where local peaks in workload or a large number of simple but coincident tasks occurred, the workload was evened out by re-ordering the tasks or simply relocating a task with respect to time, where this was technically feasible

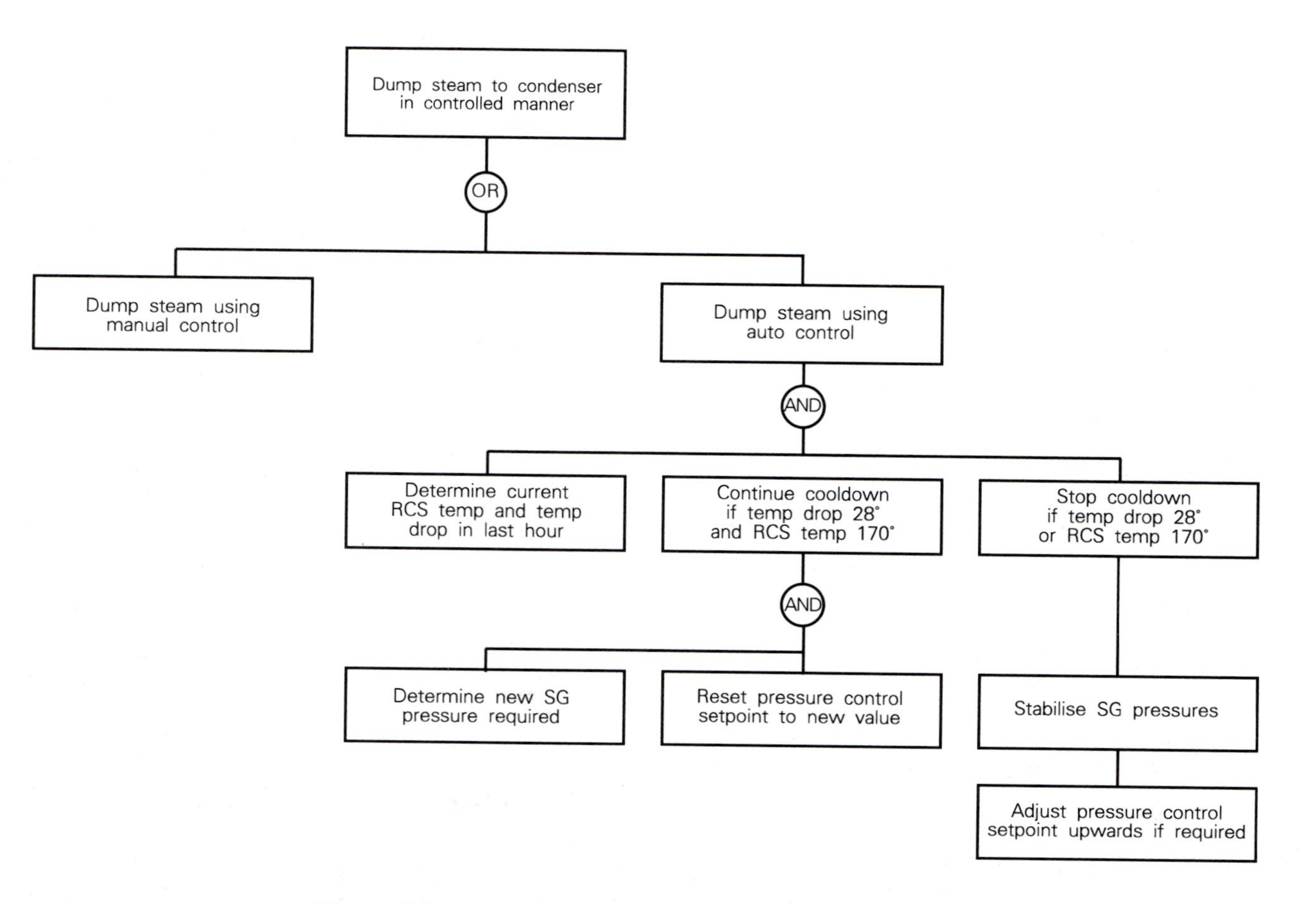

Figure 7.1 An operations based hierarchical task analysis

- Where a high difficulty rating for a task was found to contribute significantly to the workload, an investigation into possible MMI or system design changes was made to attempt to reduce the assessed workload, by simplifying one or more tasks
- Where two tasks occurring closely in time were considered incompatible with performance by a single operator, MMI changes were again investigated

Having attempted one or more of the above, if the workload was still high, automation was investigated in relation to various possible staffing levels.

Resources

Because of the range of objectives of the exercise a multi-disciplinary study group was assembled. Each member of the group had considerable expertise in one or more of the following areas:

- Human factors
- Operator reliability analysis
- PWR systems + safety
- PWR operations
- Operational experience of nuclear power plant
- MMI development
- Procedures development
- Training development

The time taken to complete this study was estimated at approximately 44 person weeks. It must be expected that much time will be spent ensuring that operator actions are correctly recorded, as any inaccuracies in that part of the study would render the rest of the analysis meaningless.

Results

Staffing level

A significant increase in workload was identified during the cooldown and depressurization to the intermediate shut-down state with demanding tasks on the feed and steam control required to achieve the desired cooldown rate.

The workload during this period was assessed to be within the capability of two desk operators and the supervisor, provided that limited developments to the automation provisions were adopted. The workload was assessed to be too high for the minimum staffing complement without widespread automation, which was not considered practicable because of other safety implications (as well as cost). Therefore this analysis recommended a revision to the proposed staffing

TASKS	BORATION (-1 to 0)	COOL DOWN & DEPRESSURISATION TO INTERMEDIATE SHUTDOWN (0 to 4)	PLACING RHRS IN OPERATION (4 to 6)	RATING
COMPLETE ACS & SO CHEMISTRY SPEC. CHECKLIST BEFORE STARTING COOLDOWN				4
VERIFY BORATION COMPLETE				2
RESET BLEND TO MAINTAIN MAKE-UP AT THE REQUIRED BORON CONCENTRATION				4
WAIT 30 MINS AFTER BORATION TERM'N BEFORE STOPPING PRESSURISER SPRAY				2
ADJUST PRESSURISER SPRAYS IN MANUAL CONTROL AS REQUIRED				4
ADJUST PRESSURISER HEATERS IN MANUAL CONTROL AS REQUIRED				4
STOP REDUCING PRESSURE WHEN REQUIRED PRESSURE REACHED				4
DUMP STEAM TO CONDENSER USING AUTOMATIC CONTROL				6
DUMP STEAM TO CONDENSER USING MANUAL CONTROL				6
VE TO P11/P18				4
BYPASS LOW ROS TEMPERATURE INTERLOCK				3
CONTROL MAIN FEED FLOW IN MANUAL CONTROL				4
REPOSITION AUX. FEED DOWNSTREAM VALVES TO CHANGE SO LEVEL				4
ISOLATE ACCUMULATORS				4
ENSURE HHSIS LOCK OUT BEFORE OPENING RHRS TO RCS				3
OPERATE EXCESS LETDOWN				3
COMPLETE LCO OHEOXLIST APPLICABLE TO RHRS HEAT TRANSFER MODE				3
SAMPLE ROS BORON CONCENTRATION				2
OHEOX OOW ALIGNMENT				2
CARRY OUT PRE-OPERATIONAL CHECK OF RHRS ALIGNMENT				2
OPERATE RHRS TO OVCE VALVE TO BACKCHARGE RHRS LOOPS				2
OPERATE RHRS PUMPS FOR FIVE MINUTES				3
SAMPLE RHRS BORON CONCENTRATION				2
OPEN RHRS SUCTION VALVE				2
OPERATE RHRS TO OVCE LETDOWN TO BORATE ROS				2
OPERATE AAA DIVERT VALUE IN MANUAL CONTROL TO PREVENT RCS DILUTION				3
CHECK OPEN RHRS SUCTION VALVES				2
ENSURE OOMS ARMED ON RHRS SUCTION VALVE OPENING				3
CLOSE HX BYPASS & DISCHARGE VALVES IN MANUAL CONTROL				2
CHECK OPEN OOLD LEG INJECTION VALVES				2
START RHRS PUMP				3

RHRS = RESIDUAL HEAT REMOVAL SYSTEM

Figure 7.2 Timelines analysis of a cooldown and depressurization

philosophy, requiring a second desk operator to be available at relatively short notice.

Automation

Automation was considered as an acceptable option for reducing operator workload in one specific scenario, during the period when the plant is being depressurized and cooldown is being achieved by steam dump.

The tasks which dominate the assessed workload during this period, between time zero and 4.6 hours (see Figure 7.3), were found to be:

- Control of steam dump to the condenser in automatic (or manual) control
- Control of main feedwater flow in manual control
- Operation of pressurizer sprays and heaters in manual control

These were all identified as high priority tasks that must be performed continuously. Additional automation was therefore seen as providing maximum benefit in reducing the workload during this period.

Automation desirable to support MCR staffing levels

To ensure that the workload remained within the capability of the two desk operators, it was judged that one major control function should be automated specifically for this manoeuvre (depressurization and cooldown via steam dump). This would reduce the summated task rating given by the workload indicator to a value which was assessed as compatible with this staffing level. Candidates for automation development were steam generator (SG) level control (main feedwater), SG pressure control (steam dump), and RCS pressure control (pressurizer sprays and heaters). For economic reasons discussed below, extended automation of SG pressure control via steam dump to the condensers (the normal means of SG pressure control during this manoeuvre) was recommended.

Automation desirable to support safe, reliable operation

The control of SG pressure is achieved in two ways: via the steam dump control system to dump steam to the condensers; or via the SG power-operated relief valves (PORVs) to dump steam to the atmosphere.

The existing automation of the steam dump to the condensers allowed control of the SG pressure to a fixed steam pressure set point. To achieve an RCS cooldown required frequent small downward adjustments of this set point. Extended automation was recommended by the group so that the system would be capable of comparing the actual cooldown rate with a pre-set desired cooldown rate and continuously adjusting the steam dump valve(s) setting to equalize the two. This additional mode of control would eliminate the requirement for continuing operator action, and hence avoid errors arising through

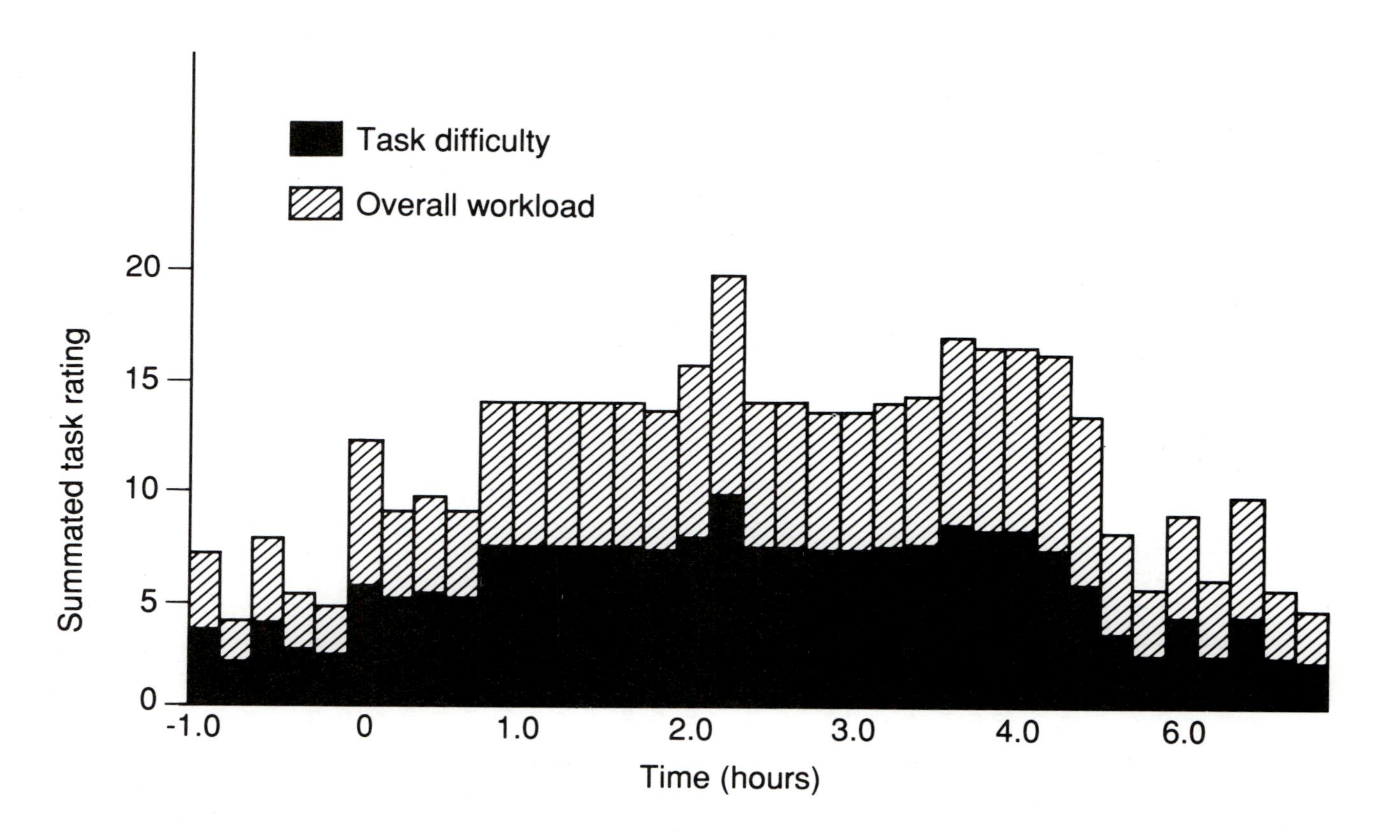

Figure 7.3 Workload assessments during the cooldown and depressurization

failure to compensate accurately for other sources of cooling such as additions of feedwater.

This extended automation, specific to the cooldown process, would reduce the likelihood of excessive cooldown rates occurring and thus avoid significant economic damage to major RCS components.

While the SG PORVs are not the normal means of cooldown, automation was considered desirable to ensure compatibility of operation when cooling down with the condensers unavailable, for example, following certain faults.

In addition, substantial operational benefits could also be derived by extending the main feedwater control system to enable automatic control during shutdown conditions. During this manoeuvre manual slug feeding was required in order to avoid continuous low flow conditions which were judged to have led to feedline cracking problems on some operating PWRs. This task required considerable and continuing operator attention, and this requirement would be removed by automatic feedwater flow control (to carry out slug feeding under low or zero power conditions).

Other considerations

Other factors that were considered during this study included MMI, procedures, training and systems design considerations.

Assessment of possible workload reductions through MMI changes led to the investigation of the effects of task performance location, recommendations for presentation of information to the operators via the data processing system, use of the station computer for operational calculations, and computerization of task checklists.

Procedural considerations led to the need to pay particular attention to parallel activities and to the identification of the need for specific *cautions* in the procedure. For example, a caution might be recommended if a task was identified as particularly critical for safety or availability, or if the task assessment indicated that particular care was required in carrying it out.

Training considerations centred on the positive benefits of the systematic approach to the exercise which clearly identified and documented the following: the tasks required, their interrelationships, their difficulty and criticality, their sequence, timing and density; and an indication of potential problems of high workload.

Various systems design requirements were also identified, primarily as a result of the group discussions that took place, supported by operational experience and operator error analysis, rather than as a direct output of the analytical approach.

Discussion

As a result of this assessment it was recommended that:

- The staffing strategy should be revised to recognize the specific need for at least two desk operators in the MCR, during a cooldown and depressurization manoeuvre
- One major control function should be automated specifically for this manoeuvre, in order to limit the staff required to two desk operators. The preferred control function should be the control of SG pressure via the steam dump control system
- It was desirable that SG pressure control via the SG PORVs be automated for cooldown and that an extended main feedwater control system suitable for shut-down operations should be provided

Perceived benefits to the organization

- One of the main benefits to the organization is that a methodology has been developed and utilized which will provide a rigorous and systematic way of ensuring that an adequate level of automation and manning is provided for future designs.
- By producing graphical representations of the progression of the manoeuvre, and by comparing this with the progression of operator actions, the workload of the assumed number of operators could be calculated. It is possible to modify these graphs accordingly to different numbers of operators and different levels of automation to ensure that the 'best' decisions are made during the design and subsequent phases.
- The use of this methodology has not only provided useful recommendations for the automation of parts of the control process, but also for man–machine interface design, and procedures and training. This can be done at an early design stage to provide the greatest benefit at minimum cost, and due to the systematic nature of the assessment it is highly auditable. By providing a system which is designed to maximize safety and operability, human error will be minimized.

References and further reading

Creasy, R. (ed.) (1980) Problem Solving the FAST Way. *Proc. Society of American Value Engineers International Conf., 173-175. Irving, Texas USA.*

Mitchell, K. and Fewins, A. (1990) *Balancing Automation and Human Action Through Task Analysis.* Paper at IAEA Int. Symposium on Balancing Automation and Human Action in Nuclear Power Plants. Munich, 9-13 July.

Williams, J.C. (1988) *Human Factors Analysis of Automation Requirements – A Methodology for Allocating Functions.* 10th Advances in Reliability Technology Symposium. Warrington: UK Atomic Energy Authority.

Chapter 8

A preliminary communications systems assessment

J. Penington – *WS Atkins*

Introduction

A safety study was undertaken to examine the effects of introducing drilling on an offshore platform from which previously only production had taken place. The purpose of the study was to highlight any potential difficulties and solutions before commencing drilling operations, so that any necessary modifications could be made to ensure safe operation. The HAZOP identified that effective communications between the drilling crew and the production crew would play an important part in safety. This would not only mean introducing new tools and equipment, as well as extra members of personnel, but would also entail the integration of another team of workers into the already existing platform personnel structure. It would also therefore entail consideration of the potential for conflicting goals and objectives and perhaps different sets of safety priorities.

It was decided therefore that a human factors assessment was needed to consider the communications aspects in more detail to highlight areas of potential conflict and make recommendations to reduce error potential.

System description

The offshore platform on which the study took place is a floating production vessel (FPV) in the UK outer continental shelf. Production takes place via sub-sea wells, some of which are satellite wells and are tied back to the main template by flow lines. Produced fluid, export crude oil, injection water and gas-lift gas will pass between the vessel and the template through flexible risers. The FPV contains the necessary facilities and systems to monitor and control the recovery of crude oil reserves, and to stabilize oil for export.

This production vessel consists of five different levels. At 8m above the main deck are the helideck, drill floor (where the drilling activities are performed and the drill cabin is located) and radio room, and at 4.5m above the main deck are the pipe deck, the wheel house, a number of offices, an accommodation block and the gas turbine house. The main deck houses the gas compression and separation area, the gas turbine room, the crane handling area and the hospital. The next level, the so-called 'first tween deck', houses the mud rooms/pits, various workshops and an accommodation block. The second tween deck contains the production control room and dive control area and further accommodation blocks.

The platform is manned by the following people in the following staff structure. The offshore installations manager (OIM) is the overall man in charge of the vessel (the captain of the ship, so to speak), but beneath him the production team and the drilling crew tend to operate autonomously. Within the drilling team, the drilling representative is a representative from the oil company itself and provides advice as and when necessary to the drilling part of the operations on the platform (See Figure 8.1).

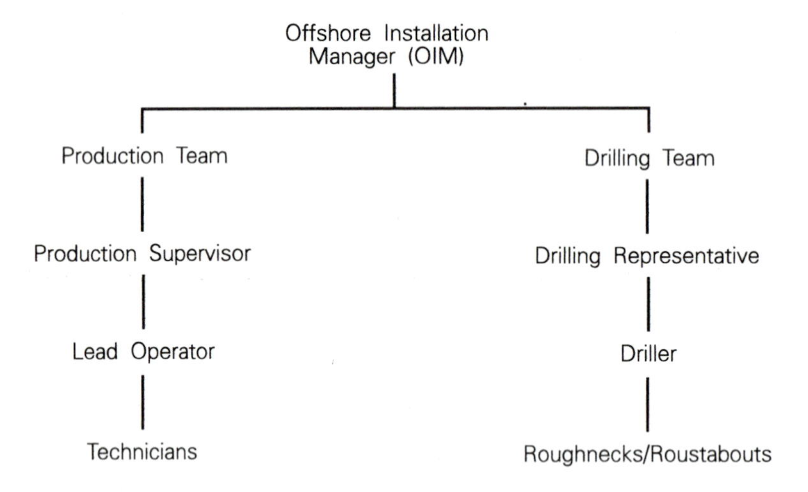

Figure 8.1 Lines of authority

Communication facilities

The communications facilities which exist on the FPV are outlined in this section for each of the main production and drilling offices.

Production control room (PCR)

Public address system

There is a dual circuit public address system. The circuits are A and B, and each has full redundancy (i.e. if one fails the other automatically takes over).

Indicator lights, located above the public address buttons on the communications panel, display the system condition: green lights show healthy operation, and conversely a red light indicates system inoperative or failure of either system. The communications panel occupies a section of the control desk in the PCR.

The public address control buttons allow the user to talk to different areas: e.g. the *ALL* button means that the public address sounds in all areas except the two accommodation areas where the nightshift may be asleep; the *GENERAL ALL* also sounds in the corridors of all accommodation areas; and the *EMERGENCY ALL* sounds in all areas including cabins.

In addition to the PCR, the public address (PA) system can be accessed in the following areas, which are listed in order of their priority of access to the system:

- The radio room
- Wheelhouse
- The drill floor
- OIM's office
- Ballast control room
- Drilling/sub-sea office
- PCR
- Telephone access (Private automatic branch exchange – PABX)
- Lifeboats (PA)

If one of these areas is using the public address system, only higher priority areas can interrupt in order to gain access to the system.

General alarm and abandon platform alarm have priority over any usage of the public address system.

UHF radio system

A UHF radio system utilizing portable handsets is used to talk to operators out on the FPV. The PCR issues the portable UHF handsets for use by personnel. In addition to the PCR there are base stations on the UHF system in the radio room, wheelhouse, etc. However, there are reception problems in certain areas resulting in low quality communications. Several operators indicated to the investigator that for this reason they do not bother to carry the handsets. It is understood that a study to alleviate this problem is planned by the operating company in the near future.

Internal FPV telephone set

There is a telephone linked to the internal FPV telephone system located in the PCR. However the set is not in an immediately convenient position for use by the PCR operator. The investigator observed that the operator tends to use an alternative cordless handset instead. The telephone is therefore available for use by other personnel who may be present (e.g. OIM), although both sets use the same line.

Satellite communications

Three satellite telephone systems are available in the PCR for external communications. There are two INMARSAT telephones which, though in practice are not often used, could be used if other facilities failed. The third is via C-SAT, which is linked to British Telecom onshore and can be used for data transfer and speech communications.

VHF radio
The VHF radio is used for short range communications (e.g. to the standby boat in the event of an emergency). It is a shared facility with the radio room having top priority. The PCR (and the wheelhouse) have the second level of priority. There are also portable handsets available on this system, which are issued from the radio room.

Intercom
A full inter-office intercom facility is provided which links all of the major FPV offices.

Other telephones in the PCR
In the PCR, there are two other internal FPV telephone lines, as well as a hotline to the emergency control room at the headquarters of the operating company and the red phone to receive emergency calls. There is also a hand-cranked telephone which is linked to the drill cabin and the tool pushers office (as well as other areas).

Drill cabin

The doghouse has the following communications facilities available:

- Telephone linked to the internal FPV system
- Public address system (with the same facilities as the PCR)
- UHF radio:
- VHF radio: portable handsets with headphones are available for use in the drill house
- Intercom linked to the production and sub-sea control room (PSSCR)
- Hand-cranked telephone (also known as 'sound powered')

There are four different levels of emergency shut-down (ESD) in use on the platform.

- ESD 0 – Complete shut-down and evacuation
- ESD 1 – Complete shut-down and assemble at muster points
- ESD 2 and ESD 3 – Shut-down of the part of the platform requiring evacuation from a specific area or areas
- ESD 4 – a fire or gas situation which may require the shut-down of part of the equipment until the situation has been rectified
  There are ESD 1 and ESD 2 buttons available on the drill floor.

The Drilling/Sub-Sea Office

This office contains the following communication facilities:

- Telephone linked to the internal FPV system
- Public address system (with the same facilities as the PCR)
- VHF radio
- Intercom linked to PCR
- Hand-cranked telephone

In addition, in the drilling/sub-sea office there is an ESD 2 button.

Objectives

The aim of the study was to identify points in the procedures of both the drilling crew and the production team where communication would be necessary, in both normal and emergency procedures. The assessment was to note not only inadequacies in the physical communication facilities, but also to consider the lines of authority and responsibility. In addition, any inadequacies in the interface design and emergency procedures were to be noted.

Method

The human factors analysis of the communications systems initially required in-depth discussions with members of the drilling crew and the production team to identify the particular procedures for which they felt communication with the other team was necessary. Informal interviews were used to obtain more detailed information about particular procedures where communication is or would be required, how this communication is implemented and why it is important. It must be noted that the study could not have been completed by performing analytical task analysis alone; the one-to-one interviews were essential to provide information about how the tasks are actually completed, rather than how the procedures suggest they are completed.

Information on the interface was noted on a proforma (see Table 8.1). The contents of the table are exemplary to show how the information can be recorded. This, combined with the task analysis and communications analysis enabled inadequacies to be identified

A task analysis and communications analysis were completed and the information was recorded in a tabular format (see Tables 8.2 and 8.3). The task analysis detailed the subtask elements, the information available to decide upon the correct action and the resulting feedback to the operator. The communications analysis recorded between whom the communications took place, their locations, the facilities available to them and any related performance shaping factors (PSFs). These are defined as factors which affect human performance, and may be of an external nature (e.g. equipment design, procedures, the environment, etc.) or of an internal nature (e.g. skill, motivation, experience, etc.). From this information recommendations were made for the improvement of communications.

The identification of PSFs was an important part of the task analysis as it identified constraints upon the performance of communications systems such as noise, which can make the difference between ineffective and effective communications. Another predominant PSF was the potential conflict between safety and production. This provided a stressor which will affect the decisions about the corrective actions to be taken in abnormal or emergency situations. It can be seen from the three tables (8.1 – 8.3) and the recommendations that these different types of assessment identified different types of inadequacies and, hence, different types of recommendations. The table recording the details of the controls and displays and the task analysis table were completed at the same time. The former clarified the details of the task analysis and enabled the identification of any hardware inadequacies. The task analysis then served to provide the basis of the communications analysis by breaking down the tasks into subtasks. This indicated where communication may be necessary, either to complete the task effectively or to provide the other team with valuable information.

In addition to assisting the investigator in performing a thorough and detailed study, the task analysis techniques used provided an auditable trail of the route to the recommendations, which was useful to the organization concerned.

Resources

This particular study was completed by a single analyst experienced in the use of these assessment methods, but who had no experience of working in the offshore industry. A preparation period of approximately 3 days was necessary to read appropriate documentation, and preliminary discussions took place with a member of the drilling team. Four days of intensive information gathering took place on the offshore platform itself, and the study then took a further 2 weeks to complete. It can be estimated that the study took a total of approximately four person weeks.

Results

Generic problems

- Drilling crew and production team had limited appreciation for each others activities
- Lines of authority and command were not fully understood

Table 8.1. *Details of controls and displays used.*

Task Step	Control	Purpose	Method of Implementation	Feedback	Adequacy of feedback	Comment
1. To stop the drill	Brake (foot control)	To halt progress of drill	Depress	Gauge	Acceptable	
	Handle	To stop rotary table	Turn the handle	RPM shown on digital and analogue displays	Acceptable	
	Clutch pedal	Pick drill string off bottom	Depress	Tactile (In = Up)	Acceptable	
	Handles	Shut off pumps	Turn handles	Position indicators	Acceptable	
PROCEDURE: TO CONTROL AN ESD 4 SITUATION: FIRE AND GAS						
1. To formulate action to control incident and maintain safe conditions		To inform appropriate personnel to take effective action as soon as possible	Production team assess importance themselves and if in drilling area contact drill team	No information available to drill team of whereabouts of incident	Unacceptable	Some direct feedback should be provided for the drill team

ESD – Emergency shut down

Communications facilities

- Levels of noise in some areas made public address systems ineffective
- The pagers were made ineffective by background noise
- There may be some congestion of the telephone lines when number checkers all call the PCR
- Some staff were not aware of the priority system of the public address system

Information feedback

- There was a lack of feedback in the drill cabin of the whereabouts and severity of a fire on the platform
- The PCR is visually isolated from the rest of the platform
- There was no direct feedback to the drill floor when a hydrogen sulphide release had occurred
- There were a number of inadequacies noted associated with the alarm information and alarm handling in the PCR

Emergency shut-down (ESD) information

- There were various different levels of severity of an ESD (0–4), each of which demanded a different response. The lines of authority for the operation of each ESD level were not always clear
- In an ESD scenario the drill crew were expected to muster on the drill floor. Alternative muster sites were not suggested, however, if the drill floor was rendered unsuitable by the situation

Procedures

- There were no procedures to detail by whom and under what circumstances the ESD 1 and ESD 2 buttons in the doghouse should be activated. There were significant reservations expressed among the production staff as to whether these buttons should be used at all. The view was that the drill team would not necessarily have the information to make the decision to instigate an ESD 1 or 2. The senior tool pusher considered, however, that in certain situations the driller should use these buttons
- Specific points in normal and emergency procedures require communication between the drilling crew and the production team. These are outlined as (e) to (m) on the next page.
- Interviews with both the production and the drilling team revealed that there were inadequacies with the permit-to-work procedures

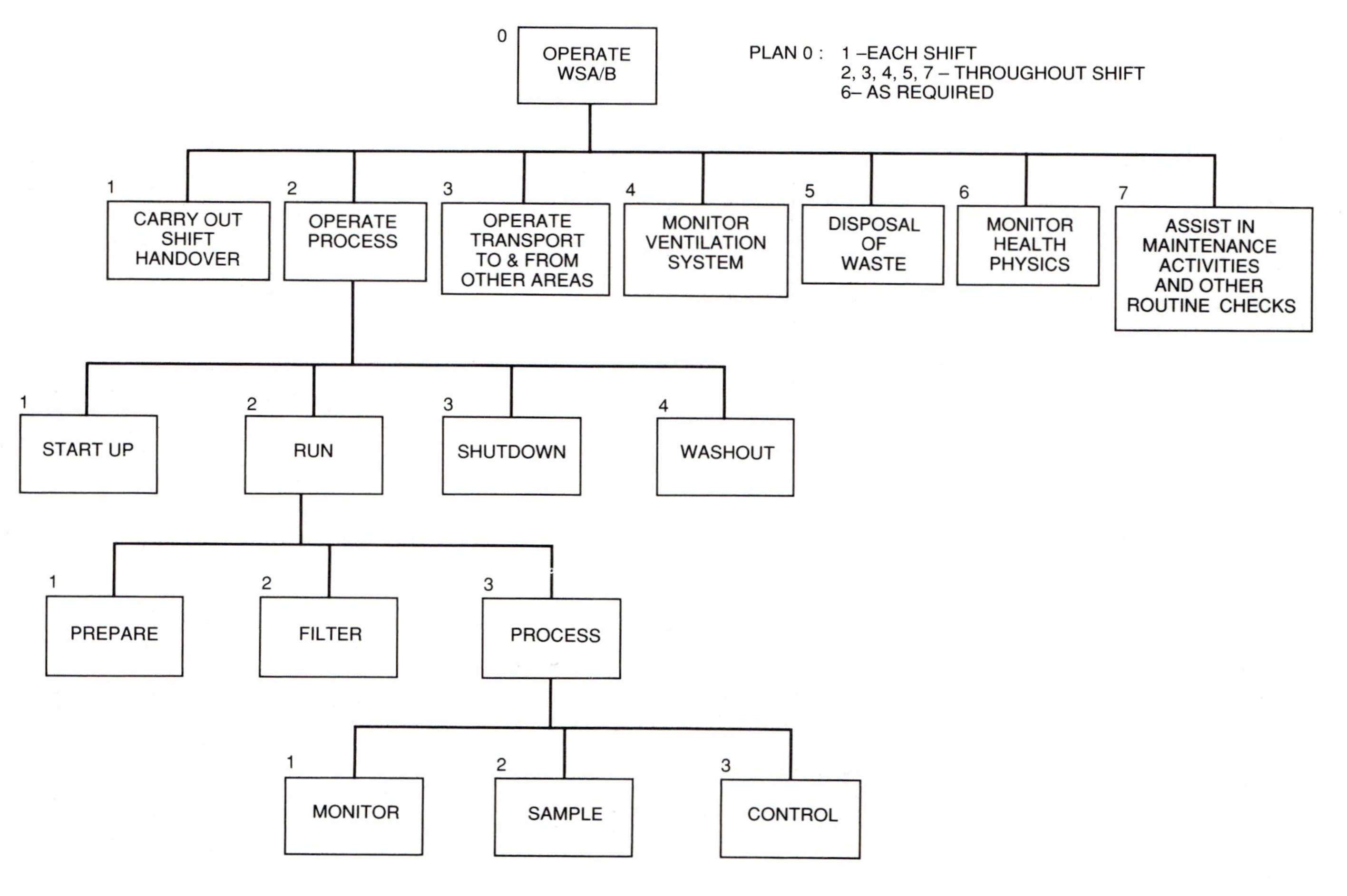

Figure 10.2 HTA of the process operation

Results

There was very little useful guidance in the human performance literature on the subject of what are and are not acceptable levels of workload for process operators. Experts on workload were consulted and suggested that beyond a 75% loading operators would make more errors and be under significant stress, and below 50% it was suggested that operators may be bored, which is undesirable. The boundaries of 50%–75% for the purposes of this study were considered acceptable and, although a fairly gross criterion (which does not take account of the cognitive complexity of tasks), it was all that was available for use at this time.

Initial results: workload estimates for WSA and WSB

The WSA workload was just below 50%, but the calculation relates to workload in normal running conditions. In an abnormal situation (i.e. one which requires operator action), the workload will be higher and in such situations assistance may be required. A supervisor will provide this assistance and therefore this workload can be considered acceptable.

The percentage workload for a WSB operator was calculated as 101% without consideration of the transportation of materials to and from other plant areas. This was obviously in excess of the 75% criterion. Therefore the facility of using two operators for WSB was considered, as described below, to see how this would impact on workload.

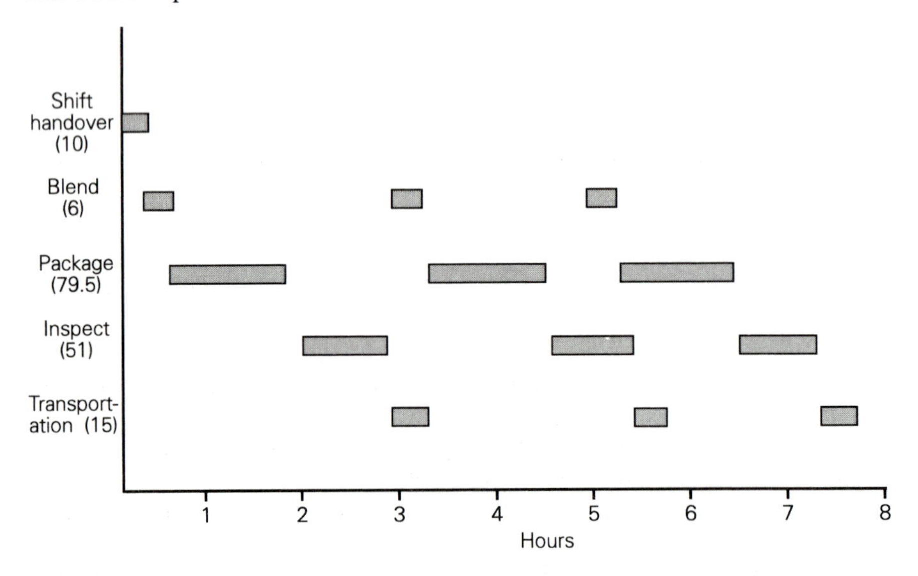

Figure 10.3 Modified timeline of WSA operations – 1 operator

Workload assessment for WSB with an additional operator

All operations have to be completed three times within an 8 hr shift. Figure 10.3 shows how these tasks fit into a simplified timeline if only one operator is available. In this particular operation however, it is not possible to initiate another part of the sequence until the former is complete.

Figures 10.4 and 10.5 show that it is possible to fill three vessels within a shift if the operator initiates the next filling operation after he has arrived at a certain task step, namely inspect vessels. The WSB process will allow the operator to do this. This means that another operator will be required to complete the remainder of this sequence plus the additional transportation operations.

As a result of this observation, calculations could be completed for two separate operators, using the same formula as used before for WSB.

Operator 1

Operator 1 completes actions when finished with the product packaging phase, before the can is inspected.

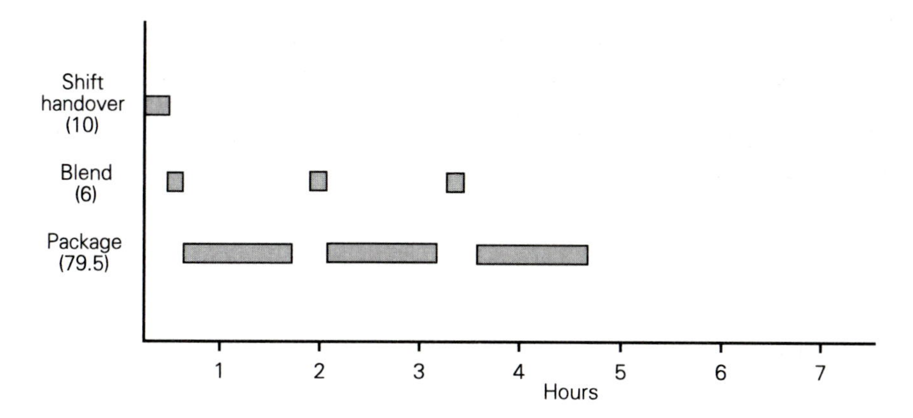

Figure 10.4 Modified timeline of WSB operations — Operator One

Operator 2

Operator 2 commences actions at the start of inspection and completes the store movements.

This workload has been calculated on the assumption that the operator will transfer three vessels to the main store and three to a position in preparation for export. These workloads for operator 1 and operator 2 are considered to be acceptable.

Figures 10.4 and 10.5 show the tasks plotted on a timeline after the introduction of a second operator who would take over operations after the packaging phase and complete the rest of the tasks, including the store operations for which information was not available at the time of the study.

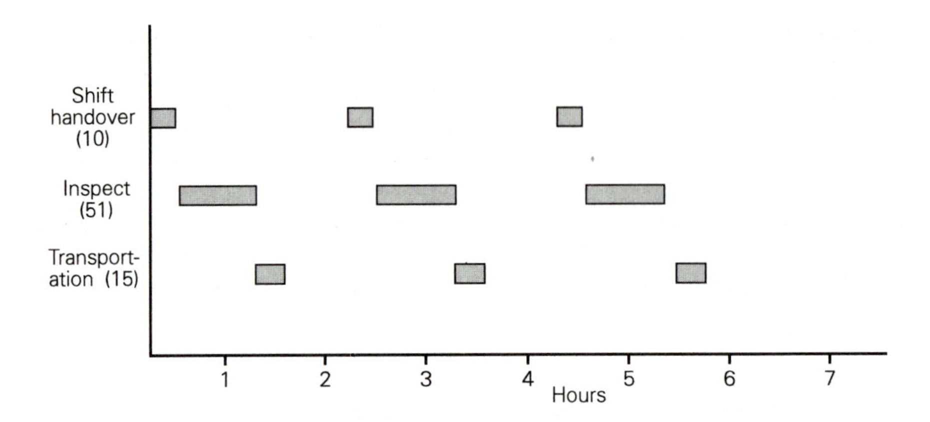

Figure 10.5 Modified timeline of WSB operations — Operator Two

Discussion

It was evident from the workload analysis that two operators would be required to control the WSB operation since process times and targets could not be altered. It would not solve the problem to designate some of the WSB tasks to the WSA Operator as the workload may increase for the latter at any time if an abnormal situation arose.

It was, however, possible to modify the workload calculations for the WSB process to see the effect of adding an extra operator. This resulted in percentage workloads in the middle of the acceptable bounds. These WSB workloads would not be likely to increase as, in an abnormal situation, the process would stop: the sequence would not continue until the problem was rectified and therefore decreased productivity would be the only effect.

Productive and efficient coverage of operations in the control centre would be achieved by two WSB operators, one WSA operator and the availability of a supervisor. This study gave a meaningful estimate of the required staffing levels for this particular control centre. The results are, however, only preliminary: levels will be finalized by the operations group during commissioning.

The assessments above have quantified workload for 'average' operators. In reality there is uncertainty attached to the workload assessment process, both in terms of the individual task times and the assumptions of how the operators will schedule the activities whose timing is under their control. If more than a gross estimate is required, uncertainties can be modelled by estimating maximum and minimum anticipated task times for all the operations, and factoring these into the assessment. The result will be a confidence interval for the workload estimates (i.e. boundaries within which the workload is likely to fall).

In addition, in any workload assessment it is worth considering the sensitivity of the workload predictions to the sequence of events. If for example, a particular operation occurring late causes a cascade of process-time-dependent slippages, a 'bottle-neck' situation may occur. This is only likely to happen if the system is 'tightly-coupled', but the sensitivity of the system to such events should be assessed. Such assessment will obviously require input from systems/operations personnel who know a great deal about the process.

Confidence estimates, and assessment of the sensitivity of the estimates to process disturbance or stoppages, are therefore important additions to the workload assessment, and do not require significant extra resources. As the plant progresses from the detailed design phase to the commissioning phase, the workload estimates can be updated, and if necessary, modelled by a computerized workload assessment package.

Perceived benefits to the organization

The benefits of performing a workload assessment at this design stage were that the task-related information was already available, and staffing provision decisions could be made at a stage when any related design effects could be modified accordingly. This assessment could therefore ensure that neither productivity nor operator efficiency would suffer when the plant was actually running. It was difficult to quantify the benefits of this assessment in terms of potential cost savings. The analysis formed a useful basis for early manning decisions.

The workload assessment took approximately 3 months to complete. This may seem a high resources cost for the workload estimates, but the rewards gained from the HTA developed during the assessment were significant (e.g. the HTA could provide the basis for communications analysis, detailing the communication needs between plant areas and between specific members of personnel). This breakdown of operator actions can also be used later in the design process to form the basis of training and operational procedures. Furthermore, the information about operator actions at the detailed level of the task analysis can be compared with the operator interfaces to assess the compatibility of the tasks with the operators information requirements (i.e. displays, etc.).

Finally, the workload assessment derived was useful as a model of the tasks to be completed by the operators in a particular system, and if any design or personnel changes arose in later stages of the project, these could be fed into the workload assessment and their impacts assessed. Therefore the workload assessment was considered a useful process, and even with its uncertainties, was seen as preferable to having simply an intuitive feel (backed up by design experience) as to whether or not the workload would be adequate.

Acknowledgements

The authors acknowledge the following design and operations personnel at British Nuclear Fuels plc. for their assistance in the completion of this study: John Deakin, Phil Egington, Ray Bryan and Steve Hall.

Chapter 11

Simulation to predict operator workload in a command system

R.W. McLeod and B.M. Sherwood-Jones – *Yard Ltd.*

Introduction

Increasing emphasis is being placed on the integration of human factors considerations throughout the design and development of complex systems, so that users perform their tasks successfully. The human factors specialists of Yard Ltd have focused on attempting to identify problems at the earliest stage in the system development. An advantage that modifications can then be made with minimum impact on project costs or time-scales. In 1988, the UK Ministry of Defence and the Department of Trade and Industry jointly published a set of guidelines (MoD (PE) and DTI, 1988) for incorporating human factors at all stages of a systems life cycle. The sponsors intend that these guidelines should eventually become mandatory standards in the procurement of computer-based systems.

An important indication of the acceptability of the design of a man–machine system is that of the physical and mental workload of the operators. Physical workload in this context refers to the time for which an operator is physically busy, as opposed to physiological capacity or endurance. Operator workload is widely recognized as having a potentially major effect on human reliability; high levels of workload may lead to stress and hence to an inability to perform the operations required to the necessary standards in the time available. If the workload is too low, reliability can be affected through boredom or reduced operator vigilance.

Techniques for addressing workload issues have until recently involved timeline analysis and lengthy assessment of operator performance using simulators, or prototypes of the design conducted at an advanced stage in development. This can be extremely labour intensive, and problems identified late in the design cycle can be expensive to solve. Techniques are now becoming available which will allow workload issues to be addressed earlier. Although two other case studies are included in the *Guide* which deal with workload assessment, this particular case study seeks to demonstrate how workload may be predicted at a relatively early stage, using a simulation computer package. This case study in particular demonstrates the use of Micro-SAINT (system analysis of integrated networks of task).

A task in Micro-SAINT can be modelled as a representation of the time taken to perform the real-world task being simulated. The analyst conducting the simulation specifies the type of statistical distribution from which the time taken to perform the task is sampled during model execution, and the mean time and the standard deviation or minimum time estimated to be taken in performing the task, are then generated by running the simulation. The simulation includes the facility to modify the assumed task performance times during the simulated task to represent the possible effects of stressors such as fatigue.

This particular application looks at a naval command system, namely one single operator within a naval command team (other studies have been undertaken by Yard Ltd which have examined a team of operators and the interaction between team members). The growing complexity of naval command systems has increased the potential for overloading the operator and in naval operations the performance capability of the force is dependent on the ability of the command systems to maintain an accurate tactical picture, so it is important to ensure system efficiency.

The case study describes an initial workload prediction exercise which examined operations room personnel under a range of specified operational scenarios. These scenarios represented situations of high workload both over short periods of time with extreme time demands (such as responding to a missile attack) and over longer periods with high levels of steady-state activity. The aim was to conduct an initial assessment of the workload likely to arise with the proposed human-computer interface design, to identify problem areas to make any necessary modifications resulting in a proposed design which will provide the required levels of operator performance.

Objectives

The main objectives of this study were as follows:

(a) To find a method of validly predicting operator workload so that any design modifications necessary can be made at an early stage
(b) To enable training and procedural requirements to be identified at an early design stage to optimize the final design of the interface
(c) To reduce the risk that unacceptable operator workload may impair system safety, performance or efficiency
(d) To reduce the risk that the system which is developed does not meet the criteria established to satisfy the users
(e) To provide an auditable system whose inputs can be altered according to system or manning modifications

Method

The study involved three stages:

- Task analysis
- Generation of human performance data on task element timing
- Computer simulation

The task was a naval command task (similar to an air traffic control task) which consisted of monitoring the system and detecting signals, followed by decision-making and classification tasks, and finally by interaction with the computer system.

The intention was to predict the maximum proportion of a 6 hr watch in which one single operator would be actively involved. Although the method concentrates on physical demands, allowances were made for the time taken by the cognitive components of tasks.

The baseline scenario (case A) included the requirement to detect the following:

- Tracks of various types air, surface, sub surface, link-received, radar received, including new tracks appearing through the watch and problem tracks
- Passive and data link bearings
- Own forces (helicopters, ships-in-company)

Case B involved the display of half the number of signals and case C double the baseline number of signals.

Six hour runs were completed for each of the cases to see the effects of externally imposed demands upon workload.

Task analysis

A hierarchical task analysis was completed to a level where detailed activities involved in performing each operation using the human–computer interaction (HCI) facilities, were identified (this was a standard console – i.e. a VDU, keyboard and the trackerball).

For each task, the analysis identified:

- The conditions which would initiate the task
- The task or tasks which would follow it
- The priority of the task
- The HCI operations involved in performing the task
- Cognitive activities (such as decisions or judgements) involved

Generating performance data

The simulation requires the input of times taken to complete the different types of subtasks which make up the total task, and then combines them and modifies them according to various stressors to predict the workload for the overall task. The validity of the workload predictions depends not only on the accuracy and comprehensiveness of the tasks represented in the model, but on the times assumed to be taken to perform the modelled task.

In this case study the predictions of human performance times were generated theoretically by using the goals, operators, methods, selection modelling approach of Card *et al.* (1983). GOMS is a method of representing routine operations on a computer-based system as a basis for predicting performance times. As with hierarchical task analysis it breaks down tasks from a general level to progressively more specific subtasks with the most basic action unit being at the bottom of the goal hierarchy. If performance times are available for each type of operation represented, it is possible for GOMS to predict overall task performance times.

The GOMS model assumes that the time required for an activity comprises three components: a perceptual processing component (Tp); a cognitive processing component (Tc); a motor component (Tm).

Estimated Times Range are Tp = 100 (50–200) micro sec; Tc = 70 (25–170) micro sec; Tm = 70 (30–100) micro sec.

Thus the typical time for an operation that involves forming a perception, translating the perception to a motor response and making the response is calculated as:

$$Tp + Tc + Tm = 100 + 70 + 70 = 270 \; micro \; sec$$

The upper limit (i.e. the sum of the top of the range of estimated times) would be:

$$200 + 170 + 100 = 470 \; micro \; sec$$

Where operations involve significant movement components, values were calculated principally by using Fitts Law where:

$$\text{TIME TAKEN} = lm \log_2 (D/S + 0.5)$$

Where D = distance to target; S = target size; lm = 100 (70–120) micro sec/arm movement and 220 micro sec/cursor movement. (The latter figure is based on empirical data with an assumed time range of 154–264 micro sec)

The GOMS approach was used to generate estimates of times likely to be taken to perform a range of simple operations involved in manipulating the HCI (e.g. moving the hand around the console or making computer key selections). Estimates for slightly more complex activities, such as moving the hand to the trackerball and selecting a track on the display, were generated by combining the lower level operation timing estimates. Estimates for more complex activities were generally based on the upper limits of the predicted times and were used as the basic input data for the workload model.

To validate the theoretically predicted times, empirical data were taken from a range of simple tasks, performed by four experienced naval command team personnel, using a demonstrator. These operators were given a 45 min introduction on how to use the facilities and the method of operation. They were then allowed 1 hr of practice, followed by a series of nine controlled tasks, each requiring the use of one or two interface devices (e.g. trackerball, touch screen) from which data were obtained.

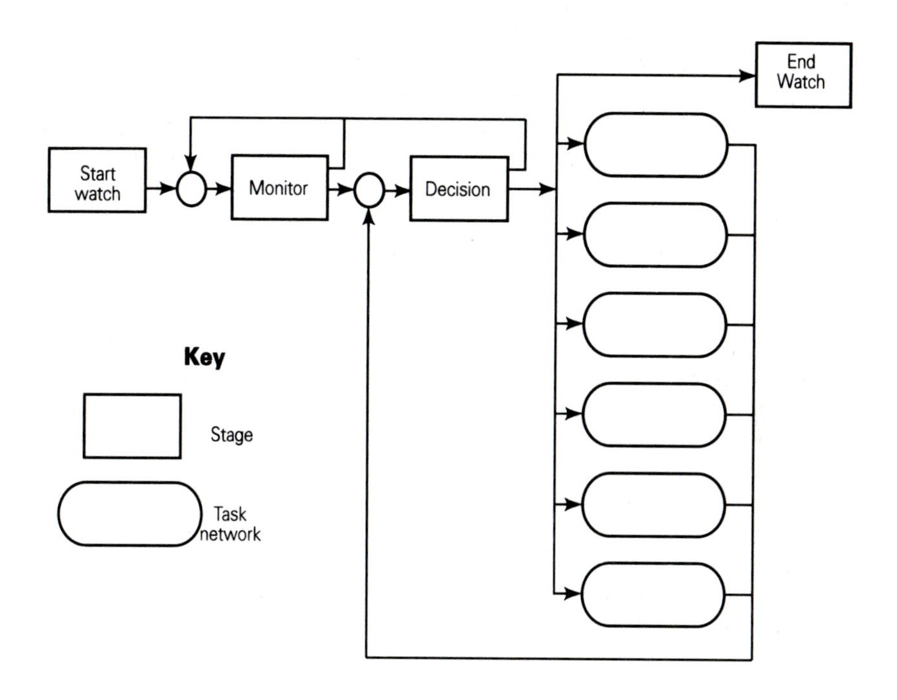

Figure 11.1 Conceptual framework of workload model

Tasks were recorded on video tape for analysis. In the analysis up to 40 measurements were taken for each simple action of interest for the four operators. The data were believed to represent the upper range of times which could be expected due to the small amount of training provided.

Table 11.1 lists some of the operations with typical times predicted from the GOMS model, the mean times measured from the demonstrator (mean, standard deviation, and number of instances analysed) and those used in the workload model (with their standard deviations).

Computer simulation

The computer simulation was developed using the commercial software package Micro-SAINT (developed by Micro Analysis and Design: Laughery, 1984).

Using performance data (as above), Micro-SAINT simulates the operator activities in responding to events. These events can be:

- In the external world (such as detecting a new track)
- Associated with the command system (e.g. losing an auto-track)
  Generated by the operator (e.g. reducing display clutter)

Table 11.1 Times to perform interactive operations

Operation	Predicted	Measured (SD/n) [mean (SD/number of instances analysed)]	Adopted
Home to keyboard, 1 press	2	1.3 (05/15)	1.3 (0.5)
Keyboard to keypad, 1 press	1.35	1.5 (0.45/14)	1.5 (0.5)
Home to bezel, 1 press	1.1	1.1 (0.4/41)	1.1 (0.4)
1 touch panel selection	1.0	1.3 (0.3/26)	1.0 (0.3)
2 touch panel selections	2.0	2.5 (0.55/10)	2.0 (0.6)
Typing rate (unskilled)	1 – 2 char	0.7 char (0.2/24)	0.7 (0.2)
Data entry	0.6 – 1/number	0.4 numbers (0.1/12)	0.4 (0.1)
Cursor and select	1.1 – 5	0.35 (2.25/17)	1.0 (0.25)

All events are assumed to be independent. The simulation included the facility to modify the task performance times during the simulated watch to represent possible effects of stressors (e.g. the effects of fatigue can be represented by including dependencies of the task time on the time through the watch in which the task is performed). Figure 11.1 illustrates the conceptual framework of the model.

The *Start Watch* and *End Watch* stages represent the hand-over periods at the start and end of a watch and are modelled as 2 min periods during which the operator does not respond to events. After hand-over, the operator monitors the system until an event is detected. The time spent monitoring with no event waiting is accumulated over the watch as an indication of *Not Busy* time.

When an event is detected, the model passes to the *Decision Stage* which identifies the task required to process the event. Where necessary the highest priority event waiting to be dealt with is responded to first. On completion of the task control passes back to the *Decision Stage* which either selects the next highest priority task, if there is an event waiting, or else activates the *Monitor* task, incrementing the *Not Busy* counter. When the *Decision* process detects the end of the watch, control is passed to the *End Watch* stage and simulation execution ends.

Assumptions were made to allow for errors and thinking time (i.e. one error requiring three sec recovery was included in every task involving data input, and up to 10 sec were allowed for non-observable tasks such as thinking, reading and decision-making). The simulation allows events to occur either at a constant rate or allows the scenario parameters to be altered at run time, thereby generating workload predictions for a wide range of operational scenarios.

On completion of a specific task, Micro-SAINT allows for three types of decision-making; *simple* in which the same specified task always follows the one completed; *probabilistic* in which the probability of each of two or more following tasks is specified; or *tactical* in which the following task is determined by the current state of actions. All three types of decision process were used in the simulation.

Resources

The absolute costs of conducting either individual workload activities or an integrated program, will vary greatly between projects. The object of predicting workload at an early design stage is to achieve a major reduction in the costs which would be incurred to ensure acceptable levels of workload late in the development and during system operation (including training and manning costs). Work can be done early on, re-used and modified, with considerably less effort than would be required if separate workload assessments were completed to address individual problems.

As an example YARD Ltd have conducted a workload prediction exercise involving computer simulation and attentional demand estimation (another workload assessment technique focusing on cognitive aspects of tasks) involving 9 man-months of effort over a 3-month period. This level of effort allowed detailed quantitative analyses of a number of a worst-case multi-incident scenarios to be conducted. Results allowed assessments based on a design mock-up to focus on predicted critical activities within each scenario. The overall outcome provided a significantly higher level of confidence in the proposed design than would have been achieved based on only separate passive assessments for a wide range of scenarios.

Results

Sample results are shown in Figure 11.2. The measure of workload used was derived as:

$$WORKLOAD = [\frac{T_{not\ busy}}{T}]/100$$

Where T is the measurement period. Samples were taken at 9 min intervals throughout the 6 hr watch simulated.

Results for the baseline scenario (A) showed a fairly consistent level of workload for this particular operator. Given the worst-case assumptions made, such as the frequency with which the tasks were assumed to be required, the number of operator errors included and the times used for the human computer interactions, the data were judged to provide some confidence in the interface design.

In (B), all components of the scenario were reduced by 50% (e.g. half the number of tracks to be recorded). This only reduced the workload by 10%, reflecting that there were a number of tasks which were independent of the scenario. Many of these tasks were however rated as *low priority* and therefore would not necessarily be performed in a worst-case watch.

A further set of data was gathered, to investigate possible effects of variations in operator performance times (e.g. due to fatigue) during model execution depending on the time through the watch. Sample data are shown as (C) on Figure 11.2.

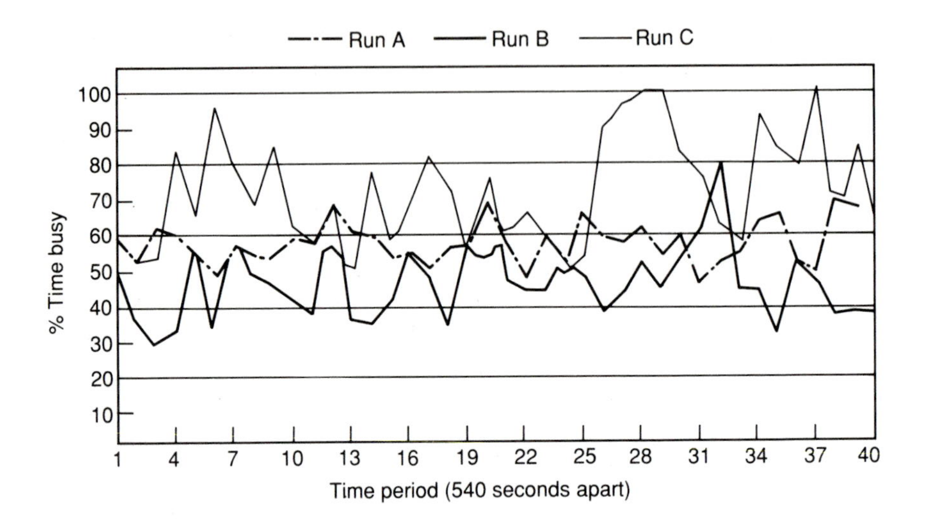

Figure 11.2 Workload graph

Discussion

Workload prediction based on a paper design can provide a cost effective means of addressing operator workload issues early in system development. Results generated in this case study are tentative and need to be interpreted with caution, bearing in mind the assumptions made and the constraints and limitations of the data. However, predictions can be progressively improved and validated by incorporating relevant results from operability tests scheduled through development of the system.

Computer simulation can provide an extremely powerful and flexible tool for predicting operator workload. An increasing number of computer software packages are available which can be used to develop simulations, but probably the most widely used commercially available package is Micro-SAINT.

Some of the principal advantages of using computer simulation are the facility to generate estimates of the statistical workload estimates, the facility to simulate alternative ways of performing tasks; the ability to generate potential measures and indicators of workload; the facility to simulate procedural errors (such as slips and lapses) and error recovery, and the ability to examine potential areas of difficulty or to explore alternative design solutions.

Some of the principal disadvantages of simulation are the effort required to produce accurate simulations, the number of parameters it is necessary to specify and the number of assumptions which need to be made. As with timeline analysis, simulation techniques are currently poor at covering mental aspects of workload.

To achieve the greatest benefits and cost effectiveness, workload considerations should be integrated with other aspects of the system development. Integration ensures:

- Consistency of workload assessments – minimum replication of effort. – maximum use of information
- Timely input of design detail into the workload assessment and development program
- Timely identification and feedback of recommendations for design
- Overcoming problems of communication between personnel involved in different aspects of the development through a mutual understanding of each other's priorities and constraints

Perceived benefits to the organization

At the start of the development process the risk is high due to the multiplicity of factors, which can, if not carefully controlled, make the system unusable by its operators. This case study illustrates the reduction in workload risk using a conventional approach compared with a computer simulation. At the end of the process, both approaches must reduce the perceived risk of the operators being unable to cope to a level considered to be acceptable.

Yard Ltd are developing an integrated approach to the considerations of operator workload extending throughout the development process. This involves the use of computer simulation as well as other techniques such as subjective rating and attentional demand analysis.

An integrated programme can achieve a reduction in risk during the requirements specification stage by using computer simulation to explore the implications of design options. Results can be fed directly into the design to reduce or remove potential 'hot spots' with minimal impact on development time or costs. During the later design stages the existing computer simulations can be refined to reflect the detailed design, and so more accurate predictions can be made. Simulations can also be used to complement workload assessments conducted on active mock-ups and prototypes by simulation situations and making assumptions which cannot be represented within other assessments.

The benefits to the organization are that a system can be produced which is safe, efficient and meets the specified requirements without undetermined costs and design changes due to late implementation.

Acknowledgements

The work relating to this case study has been carried out with the support of the Procurement Executive, Ministry of Defence and was undertaken while under contract to Dowty-Sema Ltd as part of an MoD Command System contract. The authors express their thanks to Dowty-Sema Ltd for permission to publish. The views expressed are those of the authors.

References and further reading

Card, S.K., Moran, T.P. and Newell, A. (1983) *The Psychology of Human-Computer Interaction.* New Jersey, USA: Lawrence Erlbaum.

Laughery, K.R. (1984) Computer Modeling of Human Performance on Micro-computers. *Proc. Annu. Meet. Human Factors Soc.*, 884-888.

McLeod, R.W., and Sherwood-Jones B.M *Predicting Operator Workload in Naval Command Systems. Glasgow: YARD Ltd., Charing Cross Tower, Glasgow G2 4PP.*

MoD (PE) and DTI (1988) *Human Factors Guidelines for the Design of Computer–Based Systems.* London: HMSO.

Chapter 12

Task Analysis of Operator Safety Actions

G. Pendlebury – *Nuclear Electric*, **L.K. Ainsworth** – *Synergy*

Introduction

The Pre-construction Safety Report (PCSR) for Sizewell 'B' nuclear power station explicitly identified over 50 operator actions which might have to be undertaken to ensure that the plant could be operated safely, or could be safely shutdown following a fault. This case study reports on how these safety actions were analysed using a uniform task analytical approach, and it also demonstrates the utility of using a standardized reporting format which enables quick and easy information transfer to potential end-users of the analysis. To achieve this, much effort was directed towards developing a method of reporting the analysis which, as far as possible, ensured that a diverse group of end-users could easily determine which parts of the report were particularly relevant to them.

A complete and virtually unabridged example of a report of one of these analyses of a PCSR-listed Operator Safety Action is attached for illustration (see Annex to this chapter).

Objectives

The prime aim of each task analysis was to examine a task, in order to determine whether the facilities which are to be provided (such as the procedures, hardware, software or training) would match the users' requirements and thereby enable them to perform this task effectively. Where a specific design had been developed, the possibility of undertaking the task was assessed against the design specification, and recommendations were made for overcoming any features which were considered likely to make the task particularly difficult or prone to human error.

Where the design had not yet been specified in detail, recommendations were given for the basic features which were considered necessary to ensure that the task could be accomplished. Typical examples of such features included some procedural issues, the format of the final operating instructions, the labelling of equipment on plant and training issues.

Method

The method of analysis of each task involved the following three phases:

- Information collection
- Task description
- Task decomposition

Information collection

A short description of the function which had to be achieved by the operator action was derived from the information which was given in the PCSR, and this was then used as a basis for the analysis. An example of such a description is provided in subsection 1.3 of the attached example report (Annex to Chapter 12).

Additional information about this function, and how it could be achieved, was then obtained by consulting technical information sources, including:

- Control panel drawings
- Drawings of proposed VDU formats
- System functions documents
- System descriptions
- Preliminary operating instructions (if available)
- Piping and instrumentation diagrams

Where this information was either unclear or incomplete, clarification was sought by informal discussions with technical experts. These discussions were generally limited to a small number of questions, most of which had been prepared in advance by the analyst. To ensure that there had been no misunderstandings, each discussion ended by the analyst giving a précis of his understanding of what should happen during particular scenarios, and then, if necessary, the technical expert corrected any misinterpretations.

Task description

From the information the analyst attempted to identify the main steps which an operator would have to undertake to successfully achieve the required function. A short talk-through of the proposed steps was then undertaken using the appropriate control panel drawings using a person with operational experience of reactors, to ascertain whether all necessary actions had been listed. Each of the steps was then re-described to a level of detail which indicated how each step could be carried out, and what equipment was necessary. Examples of these task descriptions are given in section 2 of the attached example.

These task descriptions were then checked by the analyst in an accurate control room mock-up.

Task decomposition

Each step was then decomposed using a set of six categories (as used in the subsections of section 3 of the attached report and as listed below) which had been selected to isolate behavioural/cognitive aspects of the task:

- Initiating cues
- Control actions
- Decisions
- Communications
- Sustaining cues (feedback)
- Termination cues

This decomposition of the individual steps was necessary to ensure that the analyst had all the necessary information about the task. However, it was felt that this level of detail was not necessary for the end-users, and so within the report of each analysis, only a summary of the findings of these decompositions was provided.

The analyst then identified points at which there was a mismatch between the information/control which was currently available in the design, and that which was required to successfully undertake each step. The type of features assessed included:

- Inadequate initiating cues or alarms
- Insufficient information in the procedures or on the instruments/VDU displays, to enable the operator to quickly and accurately locate the correct information or control
- Incompatibilities between the labels or units which were used on the various instruments (including VDU formats), or between these and the wording of the procedures
- Steps which provided either insufficient information, or which relied too heavily upon the operator's memory or training
- Difficulties associated with the operator's processing of information
- Incompatibilities between the demanded accuracy and that which was likely to be achieved
- Lack of adequate feedback
- The potential for omission errors
- Poor grouping of displays and/or controls (including functional relationships and use of mimic layouts)
- Instruments which were too widely separated (this included situations where information which had to be used together was shown on different VDU pages)
- Points at which the wrong display or control might be used

Although these features have been presented here as a list, during the analysis these decisions were made by using the analyst's professional judgment, rather than by exhaustively applying a checklist approach. However, if the analyst had not had a sufficient ergonomics background, it might have been worth developing a checklist.

Where the analyst identified a mismatch, an assessment was made of the potential of this leading to a failure of the safety action. Mismatches which were considered highly likely to result in a failure to adequately fulfil a safety action were given the highest priority (*safety-related*). Mismatches which were judged to increase the difficulty of the operator's task only without having a direct effect upon the ultimate achievement of safe conditions, were categorized as being *Ergonomically undesirable*. On their own, mismatches of the latter type were likely to increase operator workload, or to delay (but not prevent) the recovery to safe conditions. However, it must be stressed that the accumulated effect of different *ergonomically undesirable* features could also lead to a more significant failure.

For each of these mismatches, the analyst attempted to determine the behavioural mechanisms which were involved, and to recommend potential remedies. These were then described in the appropriate recommendation sections of the report (see sections 5 and 6 of the example), using terms which were analogous to those used in the prioritizations. Although it was considered that these recommendations would remedy the problems, it was also accepted that there were other possible solutions which could overcome these behavioural/cognitive mismatches. Therefore with each recommendation, references were provided to the part of section 3 where the underlying problem was defined.

Resources

The resources required to task analyse an operator safety action was mainly dependent upon its complexity and the availability of information. However, as an illustration, the example operator safety action took four man-days to complete by a fully skilled and experienced human factors analyst. The talk-through with an operator took only about 15 min. The analyst also consulted five technical experts on about a dozen different occasions ranging from 5 min to 30 min.

Results

Reporting of analysis

Sizewell 'B' is a very large and complex industrial project, which is composed of over 150 diverse electro-mechanical systems. To ensure that there is an effective and efficient design process, the project is managed on a 'systems' basis, with a specific engineer who is responsible for the design of one, or more, of these systems. In addition to this, the performance of any task could involve the use of control panels, a VDU-based system known as the data processing system (DPS), alarm cues, procedures and trained staff. There are other groups of engineers who deal with each of these areas.

Therefore, it was considered to be particularly important that the reporting of this task analysis exercise was done in a manner which maximized its use to all the personnel who were responsible for implementing any recommendations, without imposing too great a burden upon them. In other words, any engineer who might be required to act upon the recommendations, should be able to determine rapidly whether any particular analysis is of relevance and if it is, should be directed straight to the relevant parts of that analysis. The attached example illustrates how this was achieved.

Although these task analyses were aimed at safety issues, for some of the tasks it was noted that behaviour could be elicited which was perfectly safe, but which could have another impact upon the performance of the plant (for instance the use of a corrosive spray before it was required would be regarded as 'safe' behaviour, but this could cause severe economic damage to the plant). Wherever there was a conflict between safety and non-safety goals, the analysis always concentrated upon ensuring safe behaviour. However, in circumstances where there was no such conflict, the analyst thought that it was appropriate to air these considerations. Accordingly, any such operational considerations were summarized in subsection 1.4 of each analysis report, with further detail if necessary being given in section 4. The example safety action which follows was specifically chosen as being one in which some operational concerns were mooted by the analyst,

Discussion

Follow-up of the analysis

When a report of an analysis had been completed, copies were sent to all the engineers who might be involved in implementing the recommendations. They were then required to respond within a prescribed time, indicating how each recommendation was to be dealt with. In most cases, the recommendation which was given in the report was implemented, but in some cases other suggestions were made by the engineers. These were then discussed with the analyst, and generally they were found to provide a feasible alternative means of eliminating the mismatch. Where they did not do this, discussions continued until an acceptable solution was found.

It was also decided not to implement some of the recommendations, and where the analyst felt that this was acceptable, it was agreed not to pursue these recommendations any further. However, it must be stressed that this course of action was only taken for the lower priority (*ergonomically desirable*) recommendations.

Perceived benefits to the organization

The task analysis of each operator safety action has enabled recommendations to be made that improve task performance for these vital operations. Because the analysis was undertaken early on in the project, the recommendations that have been made can be implemented at little cost.

The reports of the analysis are sent to the licensing authorities to give them confidence that each operator safety action has been adequately assessed.

The standard format for reporting each analysis has ensured that the personnel responsible for implementing the recommendations could easily scan each report to determine what, if any, action they had to take. They could easily obtain the relevant background information and explanation so that they could understand the reason for the recommendation. It is believed that by making this task easier for the engineers, and by supporting the recommendations with the underlying reasoning, the recommendations were implemented more readily. Also, the problems of misunderstandings on the part of the engineers, thereby leading to the inappropriate implementation of some of the recommendations, was completely avoided.

Example task analysis report

In the next sub-section there is a representative sample of a task analysis which was documented in the standardized format which has been previously described.

It can be seen from this example how easy it is for an engineer to rapidly ascertain from the first page what systems are involved and where any *safety related*, or *ergonomically desirable* recommendations have been made.

It should be noted that figures and diagrams referred to in the example are not included for reasons of confidentiality. Reference to such figures is included, however, to show the level of documentation that was deemed necessary to transmit effectively the results of the analysis.

As an example of the use of the document in the following analysis, the person who is responsible for implementing particular parts of the training programme could rapidly ascertain that there were some *safety related* recommendations in 5.7 and some *ergonomically desirable* ones in 6.7. Each of these recommendations is a clear statement of what is required, together with a cross-reference to the task decomposition. The reader could also decide whether to read just a short summary of the task in 1.3, or to read a more detailed set of task descriptions in section 2.

Annex to Chapter 12
TASK 23 – Re-alignment of auxiliary feedwater to town's water reservoir

1 Summary

1.1 Systems involved

- AL Auxiliary feedwater system (AFWS)
- AP Condensate storage and transfer system.
- KT Town's water system.

1.2 Summary of recommendations

Recommendations have been made in the areas shown in Table 12.1.

Table 12.1 Summary of recommendations

Safety	Ergonomics	Interface area
None	Yes	Auxiliary feedwater panel
None	Yes	VDU system
Yes	None	Alarms
Yes	Yes	External to main control room
Yes	Yes	Operating instructions
Yes	Yes	Other procedures
Yes	Yes	Training

1.3 Summary of safety actions

Following a reactor trip, if main feedwater is unavailable, the auxiliary feedwater system should automatically start to provide decay heat removal, and this feed-water would normally be returned to the condensate storage tanks via the main condensers. However, if for any reason the condensers are unavailable, the condensate storage tanks will become depleted. If it is not possible to use the residual heat removal system (RHRS) to remove the remaining decay heat, the operator must obtain additional supplies of auxiliary feedwater from the town's water reservoir, by re-aligning valves at the reservoir and on plant.

1.4 Summary of operational implications

The water in the town's water reservoir will be relatively impure and so its use could lead to serious steam generator tube corrosion problems, which could impact upon safety at a later date. Therefore, the operator should attempt to avoid getting into a position in which the use of the town's water is necessary.

1.5 Assumptions and caveats

It has been assumed that procedural cues will be derived to prompt the operator to undertake the necessary re-alignments at the appropriate time.

This analysis assumes that prior to re-aligning the feed to the town's water reservoir, there are no significant leaks along the route from the two condensate storage tanks, and the valves to the condensate storage tank are all open. However, if it is necessary to interrupt the supply from the condensate storage tanks before re-aligning to the town's water supply, it will be necessary to take some precautions to prevent cavitation.

2 *Task descriptions*

2.1 Initiating cues

If the main feedwater system has failed, the operator should be monitoring the rate of depletion of the condensate storage tanks, using analogue meters (AP-LI0075A and AP-LI0077A) on the auxiliary feedwater panel in the main control room. When either of these approach specific action levels, which are to be defined within the procedures, the operator should undertake the actions which are necessary to re-align the auxiliary feedwater system to the town's water supply.

2.2 Providing a route from the town's water reservoir to auxiliary feedwater system

The layout of the pipework from the town's water reservoir to the auxiliary feedwater system is shown on piping and instrumentation diagrams, and this is illustrated schematically as Figure 23 (for reasons of confidentiality this figure is not reproduced).

Before the town's water can be used as a source of auxiliary feed, a route must be opened up from the reservoir to the auxiliary feedwater system. This involves manually opening two gate valves on the 300 mm pipes at the reservoir (KT-V0022 and KT-V0023), which are both normally left locked closed. Therefore, the operator must direct someone to get the necessary key and then go to the two valves and open them.

2.3 Getting town's water to the auxiliary feedwater pumps

When the town's water supply reaches the auxiliary feedwater system, it splits into four routes which each run to an auxiliary feedwater pump (two motor and two turbine driven). Prior to each pump, there is a locked closed gate valve, similar to those at the town's water reservoir. In order to get the town's water to the auxiliary feedwater pumps, the operator must send someone to these valves with the necessary key, and then instruct them to unlock and manually open particular valves. Although all four valves may be opened, it is likely that sufficient feed could be provided by only opening those to one, or both, of the motor-driven pumps.

2.4 Terminating cues

There are two sets of terminating cues, namely the configuration of the valves and the continued presence of a flow through the auxiliary feedwater system.

At present, feedback about the valve configurations will come from verbal confirmations that they have been opened. There are flow meters on each auxiliary feedwater line, and so the operators can easily determine whether there is still a flow through the auxiliary feedwater pumps, but this can only confirm that the town's water supply has been correctly routed after the condensate storage tanks have been fully depleted (i.e. no recovery possible from this cue if an error has been made).

3 Task feasibility

The actual task of re-routing auxiliary feedwater flow from the town's water supply, is relatively straightforward, but the success of the task depends upon initiating it within the appropriate time period. This in turn, depends upon monitoring two analogue dials for several hours, while other tasks are being undertaken. Under such conditions, there is a high probability that the operator will fail to recognize that the re-alignment must be initiated, until it is too late. Therefore, under current conditions, the feasibility of the task must be questioned, and it is concluded that additional cueing must be provided in order to ensure that the task is feasible.

Furthermore, once someone has been sent to open the necessary valves, he will only be able to do so if he can locate and operate them relatively easily.

There is no evidence which could be found concerning these issues, and so some specific recommendations (5.4A, 5.6A and 5.6B) have been proposed in order to ensure that the difficulties of finding and operating the valves are minimized. Unless these are also implemented, the task feasibility will still be in some doubt.

3.1 Initiating cues

The main cue for prompting the operator to initiate the re-alignment comes from monitoring the condensate storage tank levels on two analogue meters. This cue could be improved by marking the critical levels on each of these meters (see 6.1A). However, because this monitoring has to be sustained for several hours while other tasks have to be undertaken, it is reasonable to anticipate that monitoring alone could be insufficient, and so some additional cueing must be provided to warn the operator that the condensate storage tank is approaching depletion.

This should be provided as a *LOW CONDENSATE TANK LEVEL* alarm for each condensate storage tank on the VDU system, which is set at a level that would provide the operator with sufficient time to get someone to open the appropriate valves before the condensate tank was depleted (see 5.3A). In addition to this, an alarm based upon flow rate should be used (see 5.3A). (For further discussion, also see 3.3.)

3.2 Control actions

For the purposes of this analysis, it has been assumed that a maximum of 30 min should be allowed for the operator to take someone from other duties and direct them to either of the valve locations with the necessary key. At the town's water reservoir valves, a further 10 min has been allowed for opening the two valves; giving 40 min from initiation of this task element to its completion. For manually opening the valves to the auxiliary feedwater pumps, 15 min has been allowed; giving a total of 45 min for this task element. If these two tasks were undertaken sequentially by the same person, the total time could be reduced by at least 15 min because they could move directly between the two locations, and so a total time of approximately 70 min could be anticipated.

It is reasonable to assume that these two task elements could, if necessary, be undertaken by different personnel, and so the operator should be prompted to ensure that both tasks have been initiated, at least 45 min before condensate depletion is anticipated (5.3A).

However, although it is believed that all the necessary re-alignments can be undertaken within 45 min, there is a wide time-window during which most of the re-alignment could be done. Therefore, it is felt that by utilizing this time effectively, the reservoir valves could be opened during a period of lower workload, while possibly avoiding the need to go out in adverse weather conditions, or in darkness, under which errors would be more likely.

The actual control actions associated with this task are considered to be perfectly feasible, provided that an adequate maintenance programme is implemented (5.6B), that the procedure provides some guidance as to the location of the valves (5.6A), and that the valves themselves are clearly labelled (5.4A).

3.3 Decisions

The only significant decision which should be necessary in order to effectively undertake this task is at what time the operator should initiate the re-alignment. Therefore, the feasibility of the task will be dependent upon suitable criteria being developed to enable the operator to decide when to initiate this task. Such criteria could be based either directly upon condensate levels, or, on the rate of condensate depletion. In the former case, if a specific level within the condensate storage tank was to be designated an *action level*, the maximum flow rate would have to be assumed, and hence if the re-alignment was initiated at this point, the town's water would probably be used prematurely, with all the associated corrosion risks (see 4). On the other hand, there is the danger that predictions based upon flow rate could be invalidated by any intervening rise in flow rate, leaving insufficient time for the necessary re-alignments. In order to achieve a balance between these two criteria, it is felt that a condensate storage tank level criterion should be used to define the latest time at which the re-alignment would have to be started at full flow, with a VDU system alert alarm being given at this point (see 5.3A).

In order to minimize the risks of unnecessary use of the relatively impure town's water supply, the procedures could be written in such a way that personnel on the plant were required to establish a flow route from the town's water reservoir to the final valves before the feedwater pumps, and then be on standby, able to open these valves when specifically requested to do so by the Supervisor (see 6.5A). In such a case, the decision to open these valves must be taken at least 15 min before anticipated depletion, and should be supplemented by a VDU trend display of the condensate storage tank levels (see 6.2A).

3.4 Communications

If the final valves are only to be opened when the Supervisor specifically requests it, the task will not be feasible unless there is a two-way communications link between these positions and the operator (see 6.4A).

3.5 Sustaining cues (feedback)

Until the town's water has actually been called upon, the only feedback which the operator appears to have about the position of the town's water reservoir valves (KT-V0022 and KT-V0023), will come from a verbal report by the person who has been sent to open it. Although it is considered that with adequate procedural steps, this would probably be adequate, it does represent a danger, which could be eliminated by providing a valve position indication on the VDU system, using the valve position indicators which are already available from all the valves between AL-V0150 and the auxiliary feedwater pumps (see 6.2B).

Another, more direct, way that feedback could be provided to confirm that flow was available from the town's water supply, would be to temporarily cut off the flow route from the condensate storage tanks by either unlocking and closing valve AP-V0015 (to the motor-driven pumps), or, valve AP-V0050 (to the turbine-driven pumps), and then to check the appropriate flow meters. However, this would increase workload and introduce an additional point at which an error could occur, namely, in re-opening these valves. Therefore, it is felt that a better alternative would be to increase the margin available for recovery from an error in opening the valves. This could be achieved relatively easily by ensuring that the steam generators were at their maximum levels prior to the changeover (see 5.5B).

It would only be considered feasible for the operator to have someone on stand-by to open the valves to the feedwater pumps, if there were two-way communications between the operator and the personnel on plant (see 6.4A).

3.6 Termination cues

At present, prior to the condensate tanks being depleted, the only confirmation which the operator has that the appropriate valves to the feedwater pumps have been opened will come verbally from the personnel who have been sent to open these valves. This does not provide any independent check that the correct valves have been opened, and so it is not considered to be adequate feedback. Therefore, it is recommended (see 6.2C) that these valve positions are shown on a VDU display.

The final check on the re-alignment will come from the flow meters which are provided on each auxiliary feedwater line. However, these will not differentiate between flow from the town's water reservoir, and flow from the condensate storage tanks, which means that they can only confirm correct alignment after the condensate storage tanks have been fully depleted. The main use of these flow meters is, therefore, to warn the operator that the town's water supply has not been correctly aligned, rather than to warn him in advance that the alignment is incorrect. This situation could be remedied by incorporating flow meters into the town's water lines, but, for this particular application it is believed that if all the appropriate valve positions are shown on a VDU display (see 6.2B and C), and, if procedures and training stress the importance of checking these (see 5.5A and 5.7A), the task could be managed without these additional flow meters.

4 *Operational considerations*

Since the town's water reservoir contains relatively impure water, which could rapidly generate serious steam generator tube corrosion problems, it is considered that it would be extremely undesirable from an operational point of view, to use this water while there was any remaining condensate. Therefore, the operator should endeavour to optimize his use of the condensate, and only when this was virtually depleted should he have recourse to using the town's water supply. Operational staff should also endeavour to either make condensate available again, or, to get to intermediate shutdown conditions and bring the RHRS into service (see 6.5C) and this approach should be reinforced in the appropriate training programmes (see 6.7A).

In order to permit the operator to delay using the town's water for as long as possible, the procedures should require that the valves at the reservoir should be unlocked and opened as soon as practicable, but the other valves should be kept closed until a final decision has been made that the use of the town's water supply is inevitable (see 6.5A).

The risks of not returning all the associated valves to their normal positions after maintenance, or of any other inadvertent use of them, should be minimized by adopting strict administrative procedures, such as described in 5.6B, 5.6C and 5.6D.

The procedures should also assist the operator to use his system in a manner which conserves condensate (see 6.5B). For instance, he could be advised to:

- Supplement condensate from the other condensate storage tank
- Reduce the amount of feed by stopping pumps, or, closing valves
- Stop more reactor coolant pumps

5 *Safety related recommendations*

5.1 Control panels

No essential recommendations are proposed for this aspect of the task.

5.2 VDU system

No essential recommendations are proposed for this aspect of the task.

5.3 Alarms

A. A VDU system alert alarm should be given when the level within either condensate storage tank falls to an amount which would be depleted within approximately 45 min at the maximum flow rate (see 3.2 and 3.3).

5.4 External to main control room

A. All the valves must be prominently labelled, so that it is easy to locate them and distinguish between those to the motor-driven and those to the turbine-driven pumps (see 3.2).

5.5 Operating instructions

A. The operating instructions must stress the need to check the valve alignments on a VDU display (see 3.6).

B. The operating instructions should stress the importance of ensuring that the steam generators are all at their maximum levels prior to the changeover to the town's water supply, so that there is a maximum amount of time available to correct any errors in aligning the town's water supply to the auxiliary feedwater system (see 3.5).

5.6 Other procedures

A. The relevant plant operating instructions should provide guidance to personnel to assist them to locate all of the valves (see 3.2).

B. A regular maintenance and operability check should be instituted to ensure that all of these valves can be unlocked and opened relatively easily (see 3.2).

C. All the valves shown on Figure 23 should be locked in the appropriate position, and strict administrative controls should be instituted to ensure that after maintenance, they are returned to their correct positions (see 4). This could be achieved by using the log books to record all maintenance of the valves, and by instituting an independent post-maintenance check. Such checks would be much more effective if the valve positions could be shown on a VDU display (see also 6.2B and C).

D. The keys for these valves should be clearly labelled, and kept in a key press within the Main Control Room, under the administrative control of the Supervisor (see 4).

E. There should be some administrative procedure to ensure that if the valves at the reservoir were opened, this would be recorded in the control room Logs (see 4).

5.7 Training

A. The training programmes must explain the dangers of relying upon the flow meters on the auxiliary feedwater lines for confirmation of flow from the town's water supply, and must stress the need to check the valve alignments on a VDU display (see 3.6).

6 *Ergonomically desirable recommendations*

6.1 Control panels

A. A warning zone should be marked on each condensate storage tank level meter, at the point at which the tank would be depleted within 45 min at the maximum flow rate (see 3.1).

6.2 VDU system

A. Trend displays showing the rate of depletion of the condensate storage tanks, should be provided on a DPS display, in a format which allows the operator to project the trend forward for at least forty five minutes, and these should be sufficiently accurate for him to reliably predict when there is only fifteen minutes supply remaining (see 3.3).

B. The current positions of the town's water reservoir valves should be shown on the auxiliary feedwater pages of the VDU system (see 3.5 and 3.6). If this is not possible, see 6.6B.

C. On one of the auxiliary feedwater pages of the VDU system, the position of each of the four valves between the town's water supply and the auxiliary feedwater pumps, should be indicated (see 3.6). If this is not possible, see 6.6B.

6.3 Alarms

There are no other recommendations proposed for this aspect of the task.

6.4 External to main control room

A. If recommendation 6.5A is implemented, two-way communications must be provided between the main control room and the location of the feedwater valves (see also 3.4 and 3.5).

6.5 Operating instructions

A. In order to reduce the risks of unnecessarily using the town's water reservoir, the procedures should be written in such a way that the route from the town's water reservoir to the valves prior to the feedwater pumps, is prepared in advance, and then someone is placed on stand-by at these valves to open them when directly requested to do so by the supervisor (see 3.3 and 4). If this is done, recommendations 6.2A and 6.4A will also have to be implemented.

B. The operating instructions should be written in a manner which encourages the operators to conserve condensate, and which suggests alternative actions to do so. They should also be warned of the consequences of using town's water (see 4).

C. In a *caution* at an appropriate point in the procedure, a warning should indicate that when using auxiliary feedwater as the decay heat removal route, the consequences of losing the condensers and not being able to transfer to RHRS when at intermediate shutdown conditions, will be that the town's water has to used, with all the attendant corrosion problems (see 4).

6.6 Other procedures

A. The plant operating instructions should clearly state that opening the valves prior to the feedwater pumps, will admit town's water to the auxiliary feedwater system, and should ask the plant personnel to confirm that this is what the Supervisor wants, before proceeding to open them (see 4).

B. If it is not possible to show the positions of any of the valves on the VDU displays (as recommended in 6.2B and 6.2C) the plant operating instructions should require that the local-to-plant operator should give direct feedback to the operator as each of these valves is opened.

6.7 Training

A. The training programmes for operational staff must fully explain the implications of using the relatively impure town's water, and should stress an approach which attempts to establish conditions in which the RHRS can be used, or if this is not possible, auxiliary feedwater should be conserved. However, this training must also ensure that when the town's water must be employed, the operators will not hesitate to use it (see 4).

NOTE: In the original task analysis report, references were given to internal documentation and panel drawings.

Chapter 13

Maintenance training

A. Shepherd – *Loughborough University*

Introduction

Background to the project

The project on which this case study is based was concerned with developing a scheme for supervising the training of mechanical fitters. The project was carried out at one of the sites of a large chemical company. Management at this site recognized a general problem, namely how to ensure that mechanical fitters undergoing apprenticeships or development on plant gained adequate experience of different types of work. The company considered that it was unreasonable for fitters to be trained on everything and wished to maximize the use of relevant opportunities occurring during operations to allow on-job training. New fitters were therefore moved around the site to gain experience. As they moved around the plant however, the foreman had no detailed record of what the fitter had already experienced. Without such a record indicating the kinds of experience the new fitter needed, there was always the temptation to let experienced fitters do the more complex jobs; new fitters often just carried out routine jobs that were already familiar to them.

The project was carried out on an existing plant, though there is no reason why this approach could not have been commenced at any earlier stage in the plant life cycle. Indeed, there is considerable benefit in starting this kind of project as early as possible in the design/development cycle, since it would facilitate the establishment of a coherent strategy for maintenance supervision to be established at the outset of a plant's life, without the need to confront existing practices and change attitudes.

The project was carried out primarily using hierarchical task analysis (HTA) and therefore serves as a case study on the application of this technique. Of particular importance, however, is the fact that HTA on its own would have been of less value without devising a framework in which it could be applied and its findings used. Furthermore, task analysis activities are best treated as part of managerial activity to improve company performance, and hence this case study shows how such activities should fit into a coherent management methodology.

Objective

The overall objective was to develop a scheme for supervising/managing the training of mechanical fitters.

Method

The principal recognized task analysis method used in this project was HTA, although a number of other techniques were adapted or devised to collect information. These included structured and unstructured interviews with various personnel, inspecting plant diagrams, observing workplaces and walk-through/talk-throughs

It is necessary to view an application of HTA within the overall project strategy and not as something that stands or falls on its own merits. Proper operation of any task analysis method should form part of a general planned approach to a problem. While task analysis methods applied *ad hoc* may be helpful in gaining insight into a problem, if they are to be used systematically for any kind of design purpose, a broader strategy for action must be devised by the analyst into which task analysis methods fit. Such a broader strategy generally needs to address how human factors contribute to broader organizational performance and how organizational resources may constrain choice in dealing with human factors solutions. Thus, in the case described here, HTA is only relevant to the extent that it helps in making decisions for a sensible approach to maintenance training that will meet company requirements.

A major problem that had to be dealt with at the outset was that fitters, like most craftsmen, are supposed to do anything they are called upon to do: there is apparently no one task to analyse. This problem was dealt with by recognizing that the fitters tasks can be examined at two levels. At the first level, which will be referred to as the *organizational* level, fitters carried out some general procedures which the organization stipulates, must be followed for all jobs. These were mainly concerned with observing safe procedures and dealing with process personnel in an appropriate manner to get the job done. At a second level, the *equipment* level, the fitter is engaged in dealing with plant and equipment. The two levels are both essential, and must be dealt with effectively. There is a danger in analysing this sort of situation that focus is placed on the *equipment* level, since this seems to define most clearly why a fitter is different to other craftsmen. Failure to examine the *equipment* level in the context of the *organizational* level, however, can lead to oversights. In the case described here, for example, only addressing the *equipment* level would have overemphasized the procedures associated directly with items of equipment. The *organizational* level draws attention to the manner in which the fitter must interact with plant personnel to identify hazards, for example, and sets out the range of maintenance duties that the fitter is expected to undertake. If this type of detail had not been known, then the analysis would have failed to uncover crucial safety aspects of the task and would have failed to reveal the full range of training needs and the opportunities for fitters to gain experience.

Initial discussions

Initial discussions were held with managers and supervisors to establish the scale of the project. This established a broad range of duties maintenance fitters had to cover, how their work was organized and how the specific placements they were given focused their activities. It emerged that fitters were expected to work anywhere on the site and that, dependent upon their current placement, they were required to carry out a range of different types of maintenance, from carrying out repairs on equipment locally to overhauling equipment in the workshop. For each plant there was a shift foreman responsible for getting the work done. As maintenance jobs arose, a work-card was made out and placed on a rack. Fitters were required to take the next card from the rack when they had completed their previous job. It was acknowledged, however, that often more experienced fitters were given some of the more difficult jobs out of turn, to ensure that the work was done quickly and to a satisfactory standard. This left the routine jobs for less experienced personnel. It meant that opportunities for essential experience were being denied the less experienced fitters, and those new to the job. In discussing the range of jobs that might need to be done, it emerged that there were five major plants that had to be covered. There were obviously a great many common items, but there were still potentially many tasks to consider.

As a way into the problem, an analysis was carried out on a common and representative task, 'Install plant and pipework'.

Hierarchical Task Analysis of 'Install Plant and Pipework'

The analysis was carried out using an experienced fitter as an informant. A common mistake when carrying out analysis of a particular task is to make too many assumptions beforehand regarding what the analysis will reveal. In maintenance tasks, it is natural to assume that the task description will focus upon craft skills, but there is a danger in making these conclusions too readily. It is important to be circumspect and judge the task on its merits as the analysis progresses – what do people have to do from the start of a job to its end? This task analysis focused on what the fitter had to do from the point of obtaining a work order to the point when the plant was handed back to 'process'; that is, it was to encompass both the *organizational* and the *equipment* levels as discussed above.

Figure 13.1 shows how the task was represented. Note the top level contains the *organizational* level detail; *equipment* details emerge at a lower level. A fuller representation of the HTA is presented in tabular form as an Annex to this chapter. While the tabular form is less easy to grasp at first glance, it is generally more useful and easier to follow for the person undertaking the analysis and using the results to help with design work.

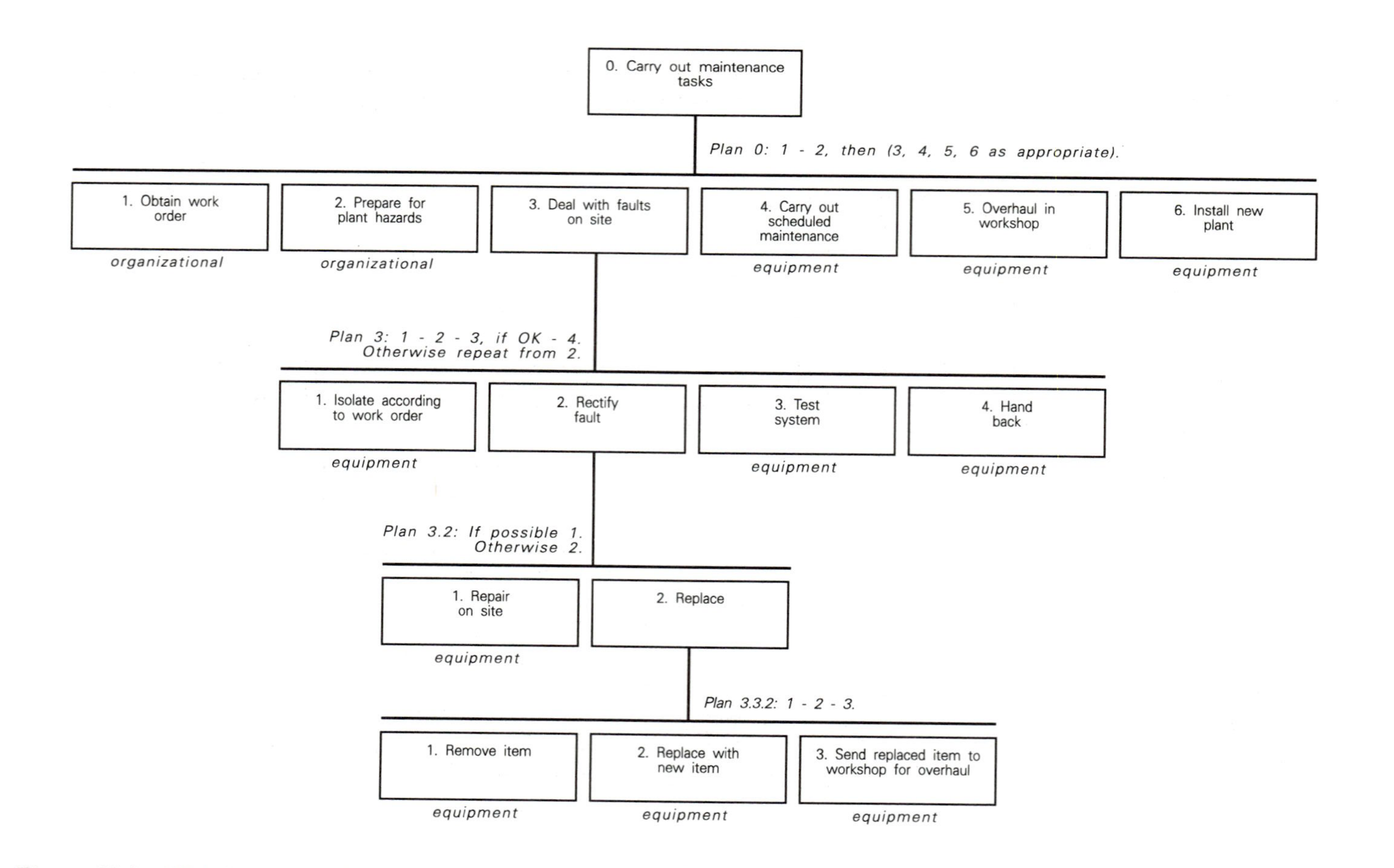

Figure 13.1 HTA for a typical engineering maintenance task from the issue of a work order to completion. © Shepherd, 1990.

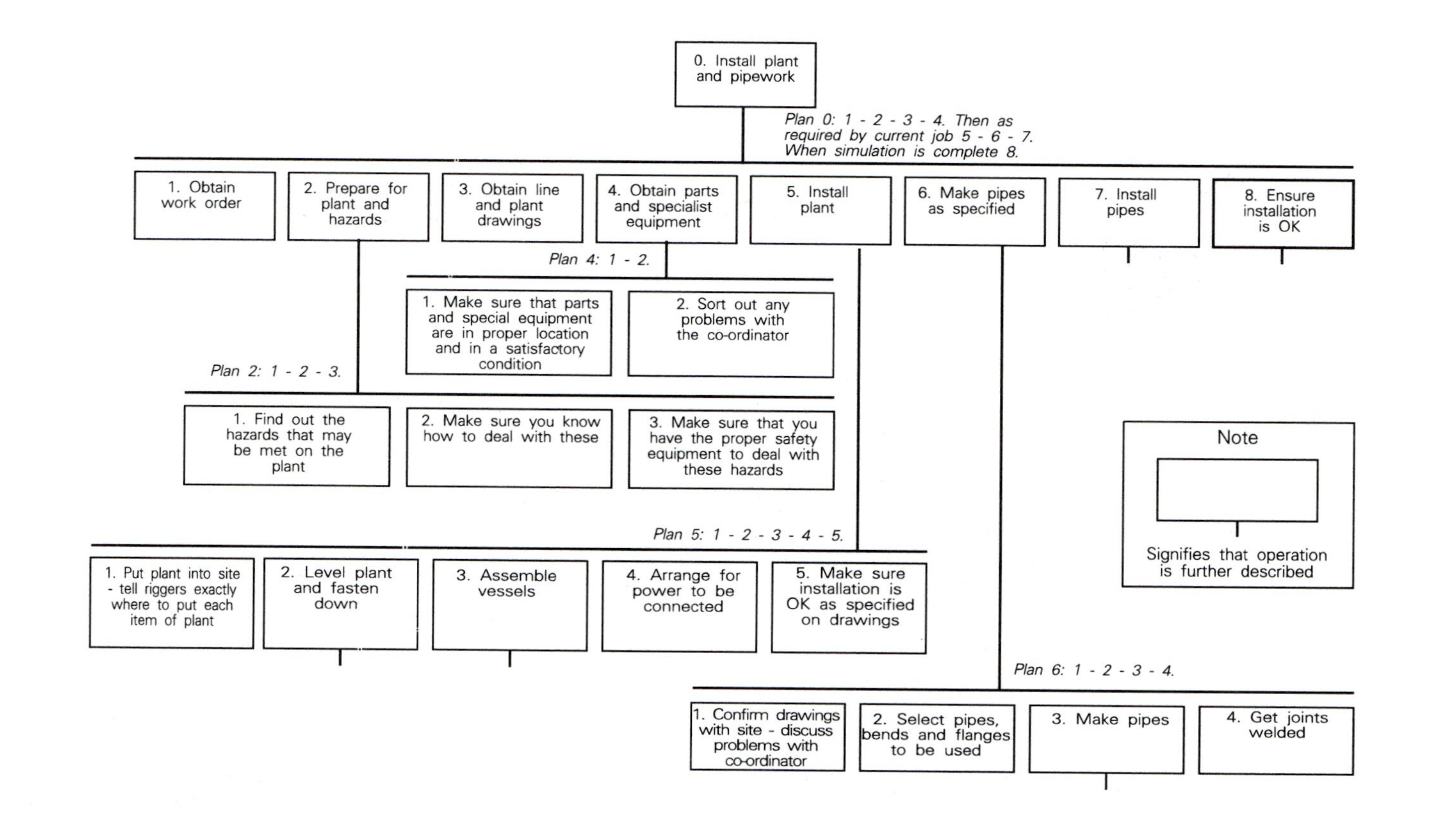

Figure 13.2 HTA of the 'organization level' engineering maintenance task for the dealing with common activities of the maintenance task. From unpublished data by Shepherd, 1990.

The diagrammatic form, (shown in Figure 13.1), is useful for representing tasks to the people with whom the work must be carried out and who are unfamiliar with tables (i.e. informants or clients) because people unfamiliar with task analysis are often happier with a diagram, although they need an expert to guide them through it. However as a working document for the analyst and as a medium for recording detailed comments, the tabular form is recommended (see pp. 112). Using a table, design issues and problems can be explored at length by recording ideas in the notes column. The tabular format has a logical structure which enables navigation through the document to be systematic, without affecting the relative importance of different parts of the task.

The Organizational-level task

Figure 13.2 shows the analysis of the *organizational* level task. To achieve this stage of analysis, further discussion, structured by the HTA method, took place with managers and supervisors. This *organizational level* task represents the general duties of any fitter on the site: on any placement, the fitter would deal with only a sub-set of the operations under 3, for example, according to the nature of the job (e.g. site maintenance versus workshop duties). Note, each operation in Figure 13.2 has been designated *organizational* or *equipment*. The *equipment* level tasks need to be examined for specific items of equipment that need to be maintained.

Identification of representative tasks

The *equipment* elements in Figure 13.3 can only be analysed with specific types of equipment. It was necessary to identify the items of equipment that needed to be dealt with. This was achieved by interviewing plant engineers and maintenance foremen. The plant flows were traced and every different type of equipment or subsystem of plant was noted. To make data easier to handle, generic types of equipment were considered. For example, there are many types of valve, and while all valves are similar for maintenance purposes, there are some critical differences. Thus, if the maintenance task was ‘replace a valve’, it is irrelevant for the task operation which sort of valve is being replaced (provided, of course, that the right type of valve is being installed), but if the task were ‘overhaul the valve’, the differences had to be observed. By opting for more generic items, 36 items were included in the list.

A matrix was prepared, with rows defined by these 36 different equipment items and subsystems, and columns defined by the eight different specific tasks from Figure 13.2. This matrix is illustrated in Table 13.1 (Table 13.1 is actually a representative sample of equipment types and subsystems – only 24 have been included). Each cell represents a separate task, potentially requiring further analysis.

Table 13.1 Plant/task matrix

	3.1 Isolate	3.2.1 Repair on site	3.2.2.1 Remove item	3.2.2.2 Replace with new item	3.3 Test item	5 Overhaul in workshop	6 Install
Valves	XXXXX	XXXXX					
Pumps	XXXXX						
Gear units							
Machine tools	XXXXX	XXXXX	XXXXX	XXXXX	XXXXX		XXXXX
Weigh scales							
Lifting tackle		XXXXX		XXXXX	XXXXX		
Hoists						XXXXX	
Pressure vessels						XXXXX	
Steam traps						XXXXX	
Filters							
Dryers							
Separators							
Glassware							
Compressors	XXXXX				XXXXX		
Condensers/heat exchangers	XXXXX						
Turbines				XXXXX		XXXXX	
Generators	XXXXX	XXXXX		XXXXX	XXXXX	XXXXX	
Crushers							
Cold box							
Pipework					XXXXX	XXXXX	
Lube oil system			XXXXX			XXXXX	
Cooling system			XXXXX			XXXXX	
Solid and oil fuel system			XXXXX			XXXXX	

Reducing the list of tasks

While there were in theory 288 separate tasks to be dealt with in the full matrix, in practice many of these could be ignored (e.g. a cooling system would not be taken into the workshop for overhaul). Another reason for ignoring a task was that several were the responsibility of other (i.e. non-fitter) personnel. Process personnel, for example, were responsible for testing several items of equipment, such as pumps and valves; electricians maintained fans, etc. To identify the items that could be ignored, each cell was considered in turn with a maintenance manager and supervisor. The rows of Xs in Table 13.1 indicate where tasks did not have to be considered.

Analysis of representative tasks

Table 13.1 could forthwith be used as a record for analysing tasks. The number of cells to be considered still seemed formidable. However, in practice the work was not nearly as extensive as it seemed. Many of the cells shared parts of the HTA with other cells. Three factors operated to enable this economy:

(a) Consider the tasks concerned with maintaining pumps. As one moves across the matrix considering each cell in turn, it soon becomes apparent that each task shares procedures with other tasks. Repairing on site for example, entails stripping the pump down until a repair can be affected. Overhauling in the workshop entails similar procedures, usually going further to replace other worn, but not yet failed, parts. By dealing with a row at a time, an item could be dealt with relatively quickly.

(b) Some of the tasks in the same column are treated identically. A particularly case is 'remove item', where identical procedures are followed for many different items of equipment.

(c) Basic items of equipment, such as pumps and valves, tend to constitute sub-routines of the more complex items and the subsystems in the matrix. As one moves *down* the matrix these economies can be recognized: HTA enables each of these economies to be dealt with very easily.

The different varieties of equipment were catered for in Table 13.1 by the introduction of *choice plans*, such as plan 4 or plan 5 (see Figure 13.3). The detail recorded here can be used for a variety of purposes, including developing procedural guides, or performance checklists. The same analysis can be used as the basis for reviewing procedures and analysing likely error sources.

Information sources

Information was taken from a variety of sources (e.g. the fitters) to serve the needs of the HTA. To analyse the *organizational* level, it was necessary to

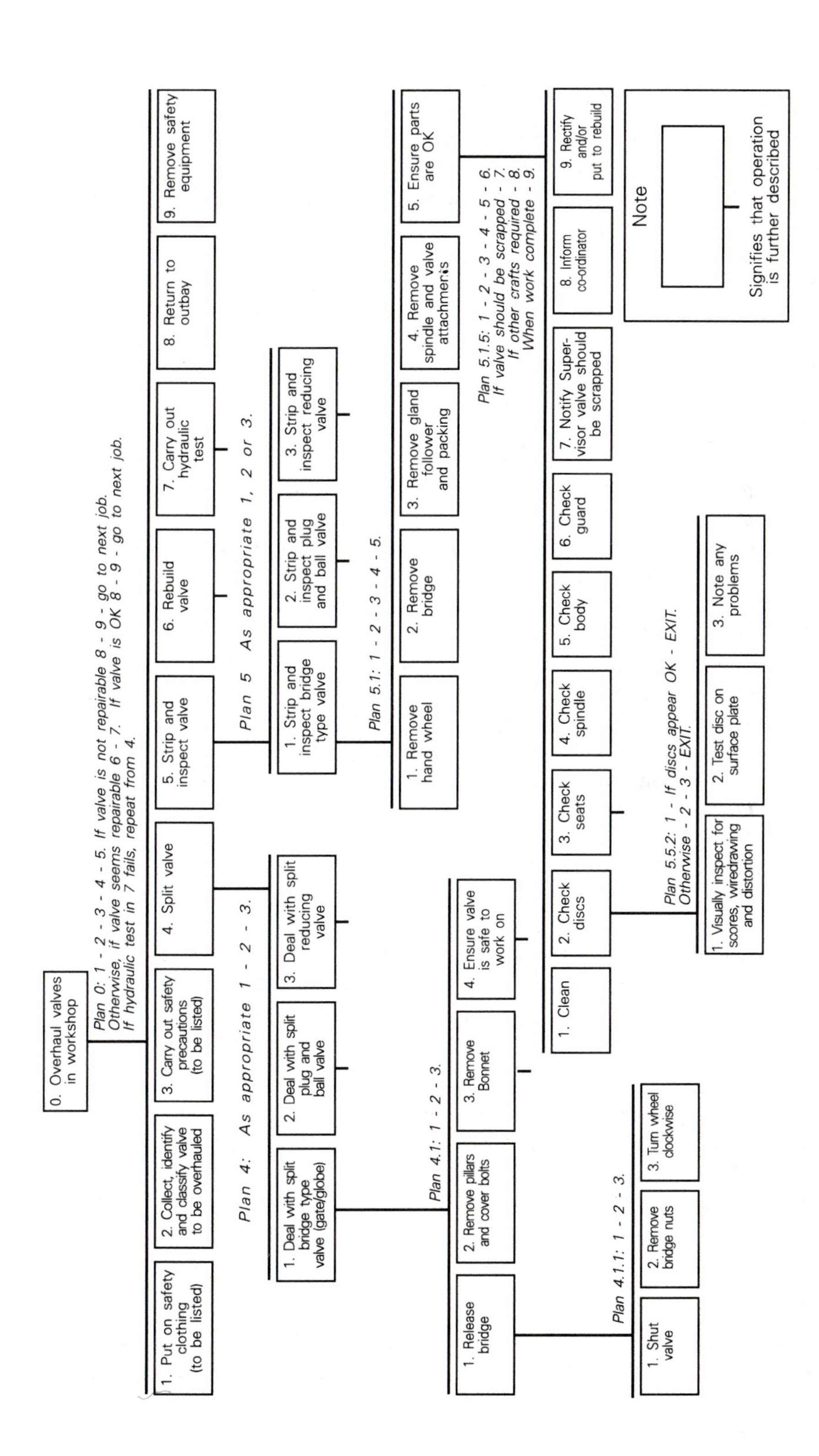

Figure 13.3 HTA of an 'equipment level' engineering task, dealing with equipment-specific activities and fitting into the 'organization level' task. From unpublished data by Shepherd, 1990.

discuss company procedures and practices with managers and supervisors. In addition existing standing instructions and record-keeping procedures were examined.

To examine the tasks at the *equipment* level it was necessary to carry out interviews with foremen and fitters, to observe some tasks being carried out and to examine written procedures.

Resources

Resources needed were time, paper and co-operation/collaboration. The general structuring of the project, including the analysis of the *organizational* level took place during discussions over some weeks, but an estimate of the collapsed time is about 5 days.

Analysis of the *equipment* level tasks depends on the quality of collaboration and the experience of the analysts. Moreover, as with any task analysis method, it is impossible to give more than a broad estimate, since, by its very nature, task analysis is looking to see what a task involves and this is uncertain at the outset. However, as a guideline, it would be reasonable to expect each row to take from half a day to a day. Some rows will take less time than this as they involve few options and may benefit considerably from the analysis that has gone before.

This study took 5 weeks in total (i.e. including the *organizational* level).

Results

The results from the project were a comprehensive set of task analyses at both the *organizational* and *equipment* levels, which could be used for a variety of purposes, including specifying safer procedures, monitoring experience and providing training guides and written operating procedures.

Specific 'deliverables' achieved during the project were as follows:

- Operating procedures (instructional guides) for the different fitting tasks
- An experience record (in the form of a personal booklet) for all apprentice fitters
- A statement of the expectations of the range of skills of fitters
- A clearer statement of the responsibilities of supervisors in allocating duties to apprentice fitters
- A clearer statement of the interaction between maintenance staff and line personnel

All of these points contributed to the development of a more effective training system.

Recommendations

The analysis provided the basis for organizing a means of assessing apprentices, and on-job training as intended. It also enabled suggestions to be formulated regarding safe practices and provided a basis for written operating procedures.

Perceived benefits to the organization

The exercise provided opportunities for improving maintenance efficiency and safety and optimizing training. The manner in which this project was conducted and the use made of HTA enabled full co-operation between all parties concerned with maintenance. This promoted goodwill towards the work and enhanced and ensured effective communication.

Annex to Chapter 13

Plans derived from hierarchical task analysis

(0) Install Plant and Pipework

Plan 0: 1 – 2 – 3 – 4. Then as required by current job 5,6,7. When installation is complete – 8:

1	Obtain work order	or can he refer to an operating instruction?
2	Prepare for plant and hazards *Plan 2: 2.1 – 2.2 – 2.3*	
2.1	Find out what hazards might be met on the plant *Plan 2.1: 2.1.1 – 2.1.2*	
2.1.1	Read the back of the work order	
2.1.2	Attend briefing meeting with manager	
2.2	Make sure you know how to deal with these hazards *Plan 2.2: do 2.2.1 – are both parties confident that the fitter will be able to cope with safety pressures? If No, then 2.2.2 – 2.2.3*	This plan incorporates the supposition that fitters should be able to specify their own weaknesses, although pressures of work and stigmas involved may mean that these decisions will not be forthcoming

2.2.1	Check with the foreman that you have been trained to cope with these hazards sufficiently recently	This was a suggestion which emerged during the analysis. If a procedure such as this is to be adopted, the foreman will need a record of each item of safety training that each fitter has received, including the date of receiving this instruction. The foreman could also be supplied with a checklist to rapidly assess the finer current knowledge and skill regarding these safety features. Developing these tools may also require a standard way for plant managers to specify hazards on their plant, rather than just using their own idiosyncratic expressions
2.2.2	Arrange with foreman for suitable training	In order that the foreman can easily indicate a course of safety training for a fitter that would be both economical and convenient to mount, training modules for each hazard type could be developed. Such modules would be appropriate to all craftsmen, not just fitters
2.2.3	Undergo training	
2.3	Make sure that you have proper safety equipment to deal with these hazards	Are checklists available so that fitters can confirm what they will need?
3	Obtain line and plant drawings	
4	Obtain parts and specialist equipment *Plan 4: 4.1 – 4.2*	
4.1	Make sure that parts and special equipment are in their proper location and in a satisfactory condition	
4.2	Sort out any problems with the co-ordinator	
5	Install plant *Plan 5: 5.1 – 5.2 – 5.3 – 5.4 – 5.5*	
5.1	Put plant *in situ* – tell riggers exactly where to put each item of plant	
5.2	Level plant and fasten down	
5.2.1	Check for level	
5.2.2	Fasten as appropriate	Nearly all fastenings are bolts into prepared positions and are straightforward. All that is required to keep a record of experience is a list of different sorts of fastenings. The fitter can then be directed to tasks where experience has so far not been gained.
5.2.3	Adjust jacking screws	

5.2.4	Shim	
5.3	Assemble vessels	
5.4	Arrange for power to be connected	
5.5	Make sure installation is OK as specified by drawings	
6	Make pipes as specified *Plan 6: 6.1 – 6.2 – 6.3 – 6.4 – 6.5*	
6.1	Confirm drawings with site – discuss problems with co-ordinator	
6.2	Select pipes, bends and flanges to be used	
6.3	Make pipes *Plan 6.3: 6.3.1 – 6.3.2*	Much of this will be superseded with standard fittings
6.3.1	Estimate the sizes of pipes to be made	This is a complex skill. There are a number of ways that it can be done and it would be wrong to be too prescriptive. However
6.3.2	Cut pipes to length	This entails considerable craft skills
6.4	Get joints welded *Plan 6.4: 6.4.1 – 6.4.2 – 6.4.3 – 6.4.4, then repeat sequence if there are more joints to be welded*	
6.4.1	Bevel ends of pipes	
6.4.2	Notify welder that some joints have to be welded	
6.4.3	Set up joint for welding and tack *Plan 6.4.3: As specified in drawing*	
6.4.3.1	Set up butt welds	
6.4.3.2	Set up centres for flange welds	
6.4.3.3	Set up template for flange welds	
6.4.4	Ask welder to weld joint	
6.5	Transport to site	
7	Install pipes	
8	Ensure installation is alright	

Chapter 14

A method for quantifying ultrasonic inspection effectiveness

J.C. Williams – *DNV Technica Ltd*

Introduction

It has become standard industrial practice for the welds in high-integrity pressurized vessels to be inspected for evidence of defects, and this is often achieved by manual or semi-automated ultrasonic inspection. This case study shows how human reliability assessment can be used to quantify the effectiveness of such ultrasonic inspection, for the identification and location of faults.

A fault is defined as a flaw or discontinuity in the substrate which is equal to, or greater than a specific size. This criterion size will also depend upon the nature of the flaw and its shape. Therefore, it is necessary to scan each weld comprehensively with one or more ultrasonic probes in order to derive a three dimensional map of the flaws within a weld to assess their characteristics as accurately as possible. The requirement for a systematic, thorough and accurate completion of this task makes it a particularly demanding task. Over the past two decades work has been performed on the development of automated methods for scanning material, and some effort has been devoted to developing an understanding of the interactions between the acoustic properties of the substrate and the flaws which may be present, but little is known about the reliability with which inspectors can identify weld faults from displays of ultrasonic signals.

Past studies have tended to lack validity because they have been based on the examination of small test pieces possessing deliberately embedded flaws at an artificially high probability of occurrence, by highly motivated teams and without time constraints. These studies indicated a level of *capability* of the ultrasonic system under ideal conditions, rather than a level of *reliability* which could be realistically expected under field conditions. Human reliability in this task must also be a function of parameters such as extent of adherence to procedures, vigilance, correct equipment usage, motivation, time, reward and job satisfaction, as well as of the technology itself.

The main research programme, of which this study is a part, was carried out by the Safety and Reliability Directorate of the UK Atomic Energy Authority and is reported in general outline terms in Williams and Featherstone (1984).

One part of the programme was designed to analyse the tasks associated with the manual ultrasonic inspection of pressure vessel welds. This was undertaken to determine how tasks were performed and to assess whether there were any potential difficulties associated with such tasks.

This particular study concerned itself with the assessment of the likelihood of failure of a total inspection system to detect and identify defects in the circumferential welds of a high-integrity pressure vessel. Event tree and fault tree analysis were used to model the actions necessary to complete the task and the possible human errors to give an overall probability of failure.

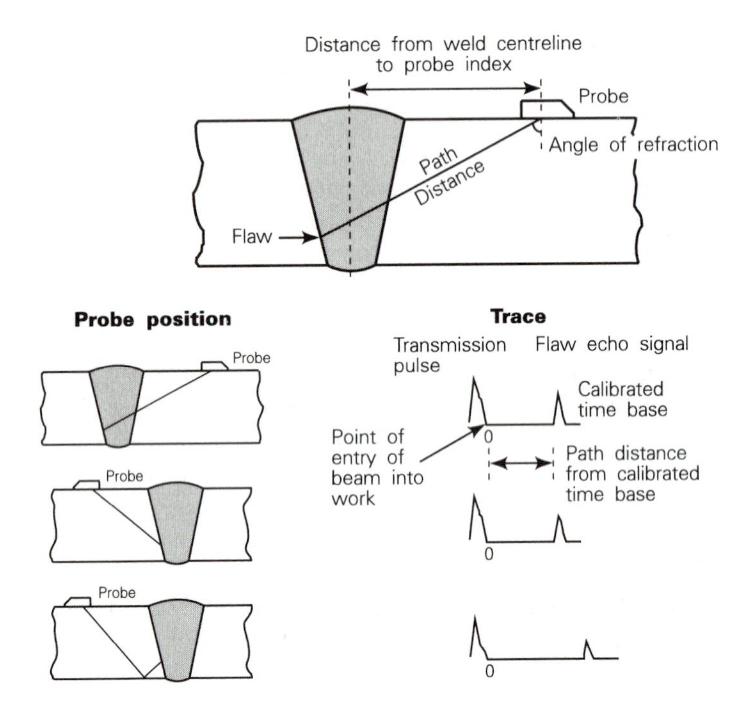

Figure 14.1 The ultrasonic inspection of welds

Figure 14.1 provides some basic examples of how ultrasonic beams might travel through welded material and also illustrates the ultrasonic traces which might be generated. A typical welding process consists of the following stages:

1. Welding
2. Ultrasonic testing
3. Heat treatment
4. Ultrasonic testing
5. Pressure testing

6. Ultrasonic testing
7. Transport of welded section to site
8. Ultrasonic reference sizing of flaws for monitoring purposes

Objective

The objective of this study was to examine the procedure used in the ultrasonic detection flaws, to assess the likelihood of human error. The overall aim was error reduction by modelling the procedures in the form of fault and event trees, so that possible identified improvements could be re-modelled to show the effect on the overall probability of human error.

The study aimed to assess the probability of failure to detect a particular defect in a weld of a particular size and at a specific depth. The study modelled the first six stages of a welding process in terms of fault detection. The last two stages were not assessed, as the transport of the welded section to site was not considered part of the detection process. The sizing of flaws (characterizing the flaws in detail for size and shape) was considered a specialist skill and therefore outside the scope of the study.

Information sources

The main information sources for this study were discussions with ultrasonics experts to find out precisely what the task entailed and where the problem areas lay. There were brief procedures to follow which formed a basis for the subsequent discussions.

Method

Three methods of task analysis were used in this human reliability assessment:

- Hierarchical task analysis
- Event tree analysis
- Fault tree analysis

Hierarchical task analysis

Task analysis was used to gather information about the operator actions involved in this task. This was represented in a hierarchical form to break the tasks down to a sufficient level for detailed analysis.

Event tree analysis

An event tree was used to model the sequence of events of operator actions and to identify the consequences of these actions. The event tree described the first six stages of the welding process (see Figure 14.2 and pp. 347–348).

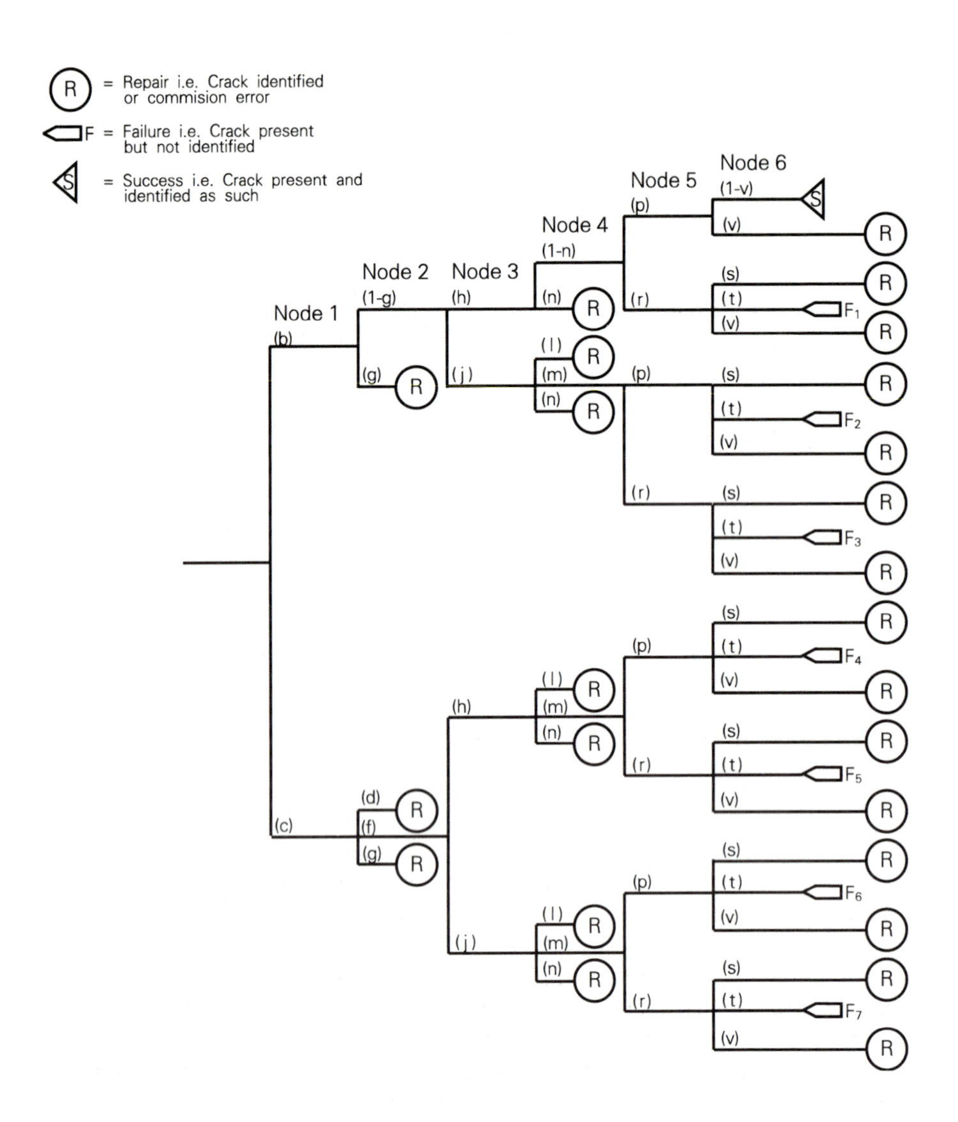

Figure 14.2 Event tree to show the first six stages in weld inspection/repair

Fault tree evaluation

A fault tree was used to model the ultrasonic inspection process and to provide a means of quantifying its effectiveness. The ultrasonic inspection process consists of a number of initial '*passes*' (through-wall inspections) to locate cracks, followed by a final evaluation inspection to size and characterize the cracks.

Table 14.1 Index to Figure 14.2

Inspection stage	Cause of crack	Symbol on Fig. 2	Error
First NDT test	Crack creation during welding	b	Probability of a crack-free weld (1-C)
		c	Probability of a crack being created
		d	Probability = 1 - (f+g))
		f	Probability of not detecting crack
		g	Probability of a commission error
Second NDT test	Crack creation due to heat treatment	h	Probability of a crack-free weld after heat treatment (1 - j)
		j	Probability of a crack being created by heat treatment
		m	Probability of not detecting crack
		l	Probability = 1 - (t + v)
Third and fourth NDT tests	Crack creation due to pressure testing	p	Probability of a crack-free weld after pressure testing (1 - r)
		r	Probability of a crack being created by pressure test
		t	Probability of not detecting crack
		v	Probability of a commission error
		s	Probability = 1 - (t - v)

Separate studies of the literature on detection performance in industrial inspection tasks were conducted to identify any variables known to affect performance. These were to assist in the provision of data for inclusion in the fault tree. Human error probabilities are difficult to determine due to the unpredictable nature of human behaviour. Some attempts have been made to develop classification schemes and assign probabilities to various categories of behaviour. This approach has a tendency to result in an oversimplified view of causal mechanisms of error, an inaccurate understanding of the magnitude of error and a complacency about the need for further data.

This has resulted in either oversimplified estimates of human error probability or estimated probabilities using 'expert judgement' as a substitute for valid data.

Attempts were made in this study to eliminate the potential for such inadequacies by modelling failure paths to a level of detail where human error probability data could be applied from actual recorded detail based on field experience. Also, wherever subjective judgement became necessary, the value of an estimated human error probability was taken as a group median estimate (i.e. based on the aggregation of several individuals' estimates) rather than any single individual's estimate. Each human error probability therefore had to have an identifiable source.

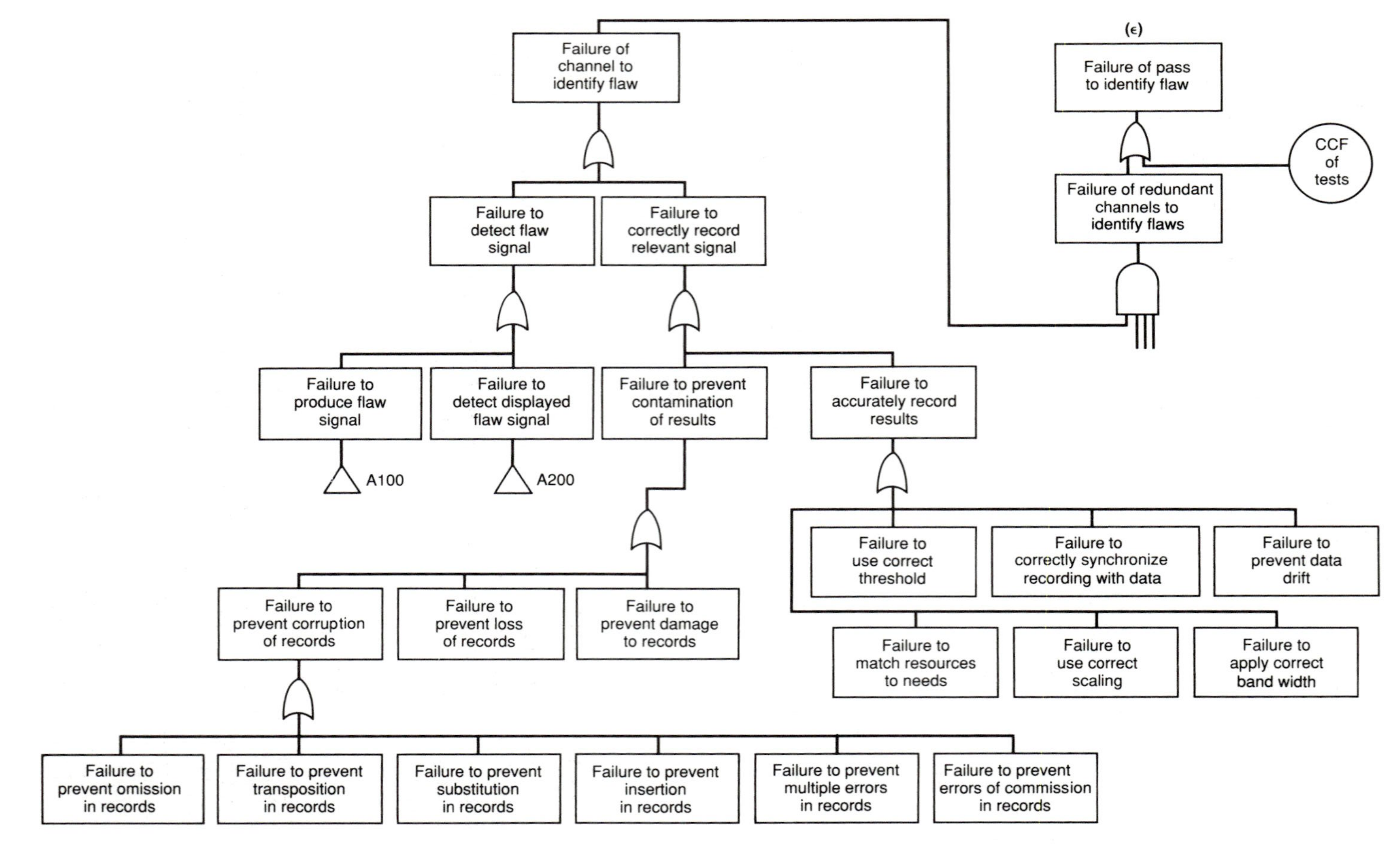

Figure 14.3 High level fault tree of identification and evaluation of flaws in welds

The human error probability estimates in Table 14.2 were entered into the bottom events of the left-hand leg of the fault tree in Figure 14.3.

Table 14.2 Human error estimates for the bottom events in Figure 14.3

Cause	Nominal human error probability
Failure to prevent omission of records	6×10^{-3}
Failure to prevent transposition of records	9×10^{-3}
Failure to prevent substitution in records	1.8×10^{-3}
Failure to prevent insertion in records	2×10^{-3}
Failure to prevent multiple errors in records	1.5×10^{-3}
Failure to prevent errors of commission in records	2.5×10^{-3}

Resources

The study took approximately eighty man-days to complete in its entirety, including 5 to 10 man-days to compare the automatic system with the manual system.

Results

Hierarchical task analysis

Figure 14.4 is a sample of the HTA completed in this study. For those readers who wish to obtain more detailed information about the order of the operator actions and the relevant issues associated with them, the detailed plans which were derived from the HTA are presented at the end of this case study as an Annex.

Event tree analysis

Figure 14.2 shows the event tree which was constructed as a result of the analysis, and the six nodes which are identified on this event tree are briefly described below. [*Editor's note: The event tree is not necessary to understand this case study, but is included to demonstrate the level of complexity that event trees may reach in real investigations*].

- NODE 1: Identifies the possibility of creating a crack which is greater than a criterion size during weld construction.

- NODE 2: Identifies the possibility of ultrasonic inspection identifying such a crack. There are two errors which are possible with this task, namely:
  — omitting to see a crack which is present (error of omission)
  — seeing a crack that is not present (error of commission)
- NODE 3: Identifies the possibility of creating a crack during the heat treatment of the weld.
- NODE 4: This is a further ultrasonic inspection stage, and as such it is similar to NODE 2.
- NODE 5: Identifies the possibility of creating a crack during the pressure testing of the welded structure.
- NODE 6: This is a further ultrasonic inspection stage, and as such it is similar to NODE 2.

Probability of a dangerous failure (F) = $F_1 + F_2 + F_3 + F_4 + F_5 + F_6 + F_7$
These Fs represent individual failure paths on Figure 14.2.

A sample of the results achieved by the fault tree analysis can be seen on Figure 14.3. Table 14.2 shows examples of the human error probabilities. These results feed into the bottom left leg of the fault tree on Figure 14.3.

General findings of the study

The analysis conducted by Shepherd (1983), showed that the overall process is governed by the pre-scan calibration. These pre-scan calibrations are ultimately dependent upon the sensitivity of the meter, which in turn is dependent upon precise following of the distance amplitude calibration procedure. Although this aspect of procedure-following is most important, the analysis showed that it is not the only way in which meter sensitivity is influenced, and furthermore, it indicated that the meter's sensitivity could be contingent solely upon the selection of a *good* area on the test piece.

Periodic correct re-calibration of the equipment is fundamental to the success of the manual inspection process. The reliability with which this task is performed, its frequency and the timeliness of such re-calibrations were areas highlighted as having the potential to cause significant deviation from successful inspection processes.

Amongst other things, pre-scan calibration was noted to be dependent upon determining the required probe range, and it was not clear at the time of the study whether this was specified in the instructions, or whether inspectors must remember the rules.

The operational sensitivity of the meter depends ultimately upon the calibration blocks chosen for the calibration exercise, and the mechanisms used to decide which block are also fundamental to the ensuing behaviour. The means by which the choices are made and their rigour were noted as needing considerable enforcement and reinforcement.

The most obvious difficulty associated with manual ultrasonic inspection is the need for the operator to perform the task of moving an ultrasonic probe

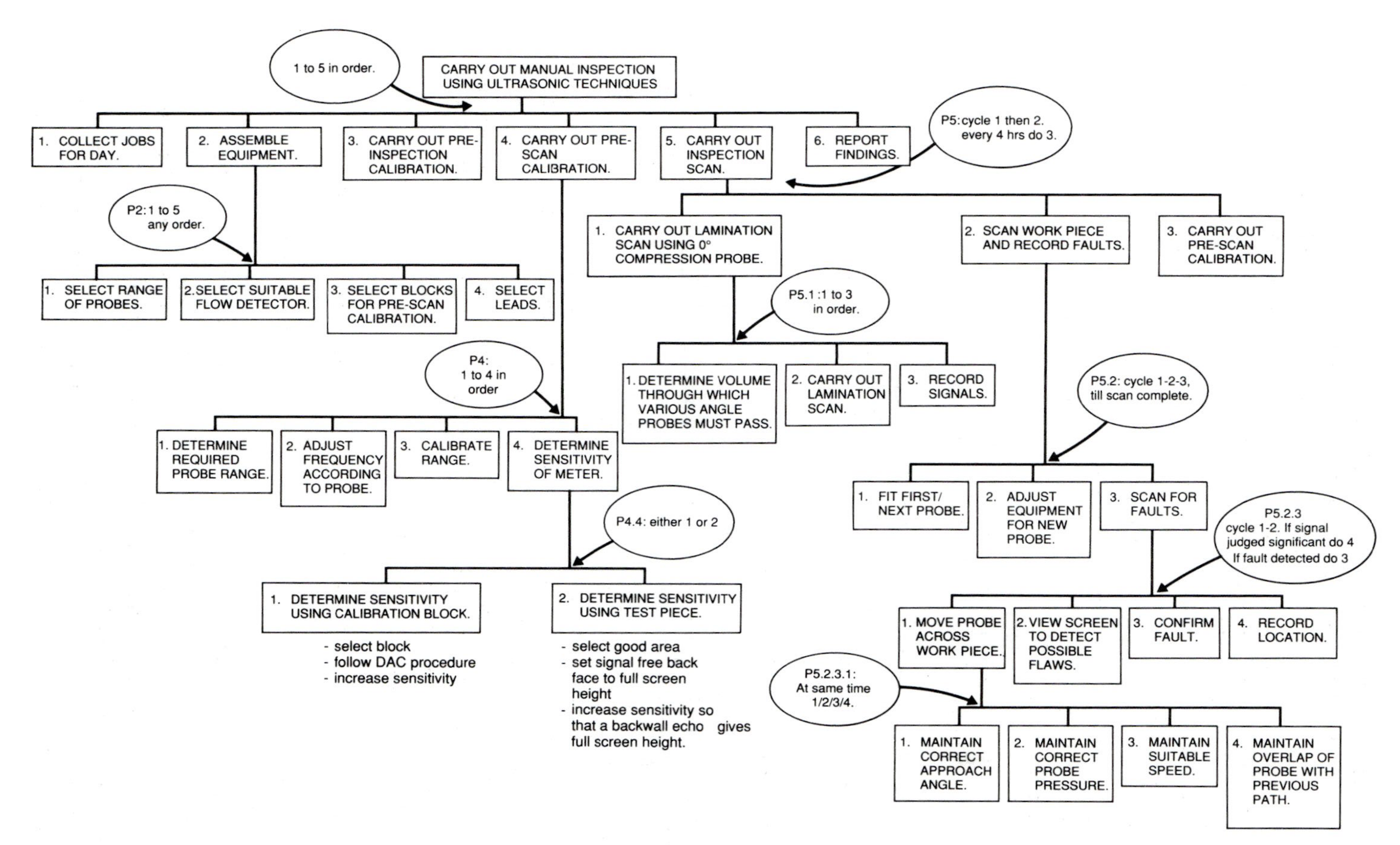

Figure 14.4 *Task analysis diagram. The diagram shows the relationship between subtasks, but reference needs to be made to the text for detailed notes.*

manually across the test-piece at an appropriate speed while maintaining the control and acoustic coupling necessary to produce meaningful echoes. At the same time the operator is required to observe and act upon the displayed acoustic information.

The way in which these tasks are performed could be seen to merit considerable further task analytical attention, because the interpretations placed upon these actions would determine the ultimate judgements to be made concerning the state of the test-piece. In an overall sense these activities were considered likely to place an exceptional burden on the operator. It was concluded that the ways in which optimal task time-sharing is achieved, and reliable performance maintained, would, as a minimum, benefit from some clarification. This would help ensure that the potential for failure is minimized, as far as possible. Maintenance of correct beam angle, probe pressure, speed of traverse and adequate overlap with previous passes, were all seen to combine to place considerable sensori-motor demands on operators, and the ways in which all these parameters are maintained within appropriate limits were thought worthy of detailed investigation. At first sight the combination of all these success-limiting factors would seem to place a very considerable burden on the operators' performance, and therefore the probability of their successful overall behaviour.

Discussion

The event/fault tree methodology proved effective for delineating the parameters likely to affect the probability of failure to identify flaws in high-integrity vessel welds. It has also facilitated the development of a means of quantifying the effectiveness of the whole weld fabrication and inspection process.

The fault tree which was created has been inspected by technical experts for accuracy and completeness, and was considered to be a fair representation of reality. Human error probabilities which have been inserted have some basis in fact and are not loose approximations. Thus it is argued that, as a first attempt, the deductions which can be made regarding the probabilities of the top event, and the sensitivity of the system which will produce such an event, provide useful insights into the design of the inspection system.

Perceived benefits to the organization

This HTA was an integral part of a much larger programme investigating the likelihood of human error in manual pressure vessel inspection. It identified areas of vulnerability and therefore provided opportunities for possible future error reduction in such processes. The work was highly cost-effective as it provided rapid, low-cost, insight into complex issues associated with man–machine systems design and operation. Its output served as a basis for considering the overall likelihood of successful task performance.

Annex to Chapter 14

Plans derived from hierarchical task analysis

(1) Carry out manual inspection

Plan 0: Carry out 1 to 6 in order:

1	Collect jobs for day	
2	Assemble equipment	Selection of items depends upon factors such as surface finish, type of material, access to inspection item. To what extent is the inspector required to know or memorize the rules for choice or can he refer to an operating instruction?
3	Carry out pre-inspection calibration	
4	Carry out pre-scan calibration	
5	Carry out inspection scan	
6	Report findings	

(2) Assemble ultrasonic inspection equipment

Plan 2: Do 1 to 4 in any order:

1	Select range of probes	According to surface finish and surface flatness of weld. What are the precise rules? Does the inspector need to memorize, or are they specified in the operating instructions?.
2	Select suitable flaw detector	
3	Select calibration blocks for pre-scan calibration	
4	Select leads for job	Choice is dependent upon access and whether or not inspection involves two men.

(4) Carry out pre-scan calibration

Plan 4: Carry out 1 to 4 in order:

1	Determine required probe range	Is this specified in instructions, or must the inspector memorize rules? There appear to be two rules.
2	Adjust frequency as per probe	
3	Calibrate range	
4	Determine sensitivity of meter	

(4.4) Determine sensitivity of meter

Plan 4.4: Do either 1 or 2: It should made clear in the operating instructions who makes this choice

1 Determine sensitivity using calibration block
- select calibration block
- follow DAC procedure
- increase sensitivity (ASME Section XI — 20 dB)

2 Determine sensitivity using test piece
- use good area of work place
- set signal from back face to full screen height
- increase sensitivity so that n backwall echoes give full screen height

(5) Carry out inspection scan

Plan 5: 1 then 2 until scan is complete. Every four hours do 4:

	Operation	Notes
1	Carry out lamination scan using 0° compression probe	
2	Scan work-piece	
3	Record faults	
4	Carry out pre-scan calibration	This is a routine repetition of operation 4. Is this done reliably, or are inspectors prone to forget, delay or ignore, in order to complete an inspection?

(5.1) Carry out lamination scan using 0° compression probe

Plan 5.1: 1 then 2 and 3:

	Operation	Notes
1	Determine volume through which the various angle probes must pass	
2	Carry out lamination scan	This is same procedure as operation 5.2, but 5.2 is a more exhaustive procedure
3	Record signals	See operation 5.3.

(5.2) Scan work-piece

Plan 5.2: Cycle through 1, then 2, 3, and repeat until the scan is complete:

	Operation	Notes
1	Fit first/next probe	Select according to operating instructions
2	Adjust equipment for new probe	Does inspector remember what he has to do and when to do it?
3	Scan work piece to locate faults	

(5.2.3) Scan for locate faults

Plan 5.2.3: 1 then 2 as a signal is judged significant, 3 then if fault detected do 4: A major source of performance error may be due to having to carry out both 1 and 2 together. In itself, 1 requires considerable monitoring of several aspects of movement simultaneously, a problem which will increase in severity as difficult geometries are scanned

1	Move probe across test-piece	
2	View screen to detect possible flaws	The signal detection problem may be substantially increased in view of the motor control being exercised in step 1
3	Confirm fault	
4	Record	

(5.2.3.1) Move probe across test-piece

Plan 5.2.3.1: At the same time 1, 2, 3, 4: As mentioned earlier the need to control four factors for a satisfactory movement is extremely demanding, especially with difficult geometries

1	Maintain correct approach angle	To ensure that angled beams are pointing in right direction
2	Maintain correct probe pressure	The feedback that the inspector must monitor to perform each of these four operations successfully is by no means obvious. Each is a considerable skill requirement. All four together may be asking a lot. Effective signal detection as well is probably unlikely
3	Maintain suitable speed	To enable signal detection. How does the inspector judge whether he is working excessively fast when he has no independent confirmation of whether he is making the correct decisions?
4	Maintain overlap of probe with previous path	How does he do this with planar geometry, let alone non-planar geometry?

References and further reading

Shepherd, A (1983) *Examination Via Task Analysis of Ultrasonic Inspection Equipment and Procedures to Explore Vulnerability to Human Error*. Rept. No. NOR 4042. Warrington: UK Atomic Energy Authority.

Williams, J.C. and Featherstone, A.M (1984) A Method for Quantifying the Effectiveness of Ultrasonic Inspection to Identify and Locate Flaws in Welds. *Proc. of the 8th. Advances in Reliability Technology Symposium*, University of Bradford, National Centre of Systems Reliability, B1/3/1-B1/3/7.

Chapter 15

Operational safety review of a solid waste storage plant

Helen Rycraft, Francis Brown, Nigel Leckey – *British Nuclear Fuels plc, Sellafield*

Introduction

During the preparation of a safety case for a solid waste storage plant at Sellafield, the importance of the operability of the inerting and ventilation equipment was highlighted. Consequently, a systematic review of the operation was proposed.

The purpose of the plant was to store solid waste underwater. The plant storage areas were ventilated to remove any hydrogen generated by a reaction between the waste and the water. Since there was a potential for enhanced hydrogen generation, facilities were available to provide an inert nitrogen atmosphere in the event of the hydrogen concentration exceeding certain levels.

There was potential for human error during the inerting process, and so this was examined more closely. During inerting, nitrogen gas is supplied and extracted through some of the storage areas by the use of fans, and then discharged through the decontamination plant and stack, which has monitoring facilities.The system has multiple routes for performing the above operation, which the operator may alter. On examination, there appeared to be some mismatches between the operating instructions and the actual operations performed, and differences between the methods of work adopted by individual supervisors. The design of some controls and displays also appeared to be in need of improvement.

A detailed task analysis was performed to identify the inadequacies and suggest how the system could be improved. This case study is an example of where task analysis has been used at the operational stage rather than the design stage of the project.

Objectives

The initial objective of this study was to establish whether there was any mismatch between performance during actual operation and testing of the inerting and ventilation system, and the reliability and assumptions made in the safety case for this store. Implicit in this objective was the aim of assessing the various aspects of the system's design and operation to identify any areas which could be subject to human error. From this analysis, practical measures for improvement could then be recommended.

The study aimed to identify the potential for human error in the following aspects of the system:

- Industrial hazards
- Procedures
- Hardware features
- Training

Information sources

In addition to discussions with operations personnel, the following sources of information were obtained:

- Operating instructions
- Safety case documentation
- Ventilation air extract and control systems operating manuals
- Inert gas facility inerting manual
- Tests log book
- Training records and schedules

Method

The investigation consisted of plant visits, interviews, examination of documents and a formal task analysis. The plant visits involved walk-throughs of each operation with an individual operator, and then a more detailed examination at the locations of the equipment. A full operating team was then observed during performance of these tasks. Examination of documents provided most of the factual detail used in plant descriptions.

The following methods of task analysis were utilized explicitly or implicitly during the study:

- Observation
- Interviews
- Walk-throughs
- Hierarchical task analysis
- Task decomposition
- Timeline analysis
- Interface surveys

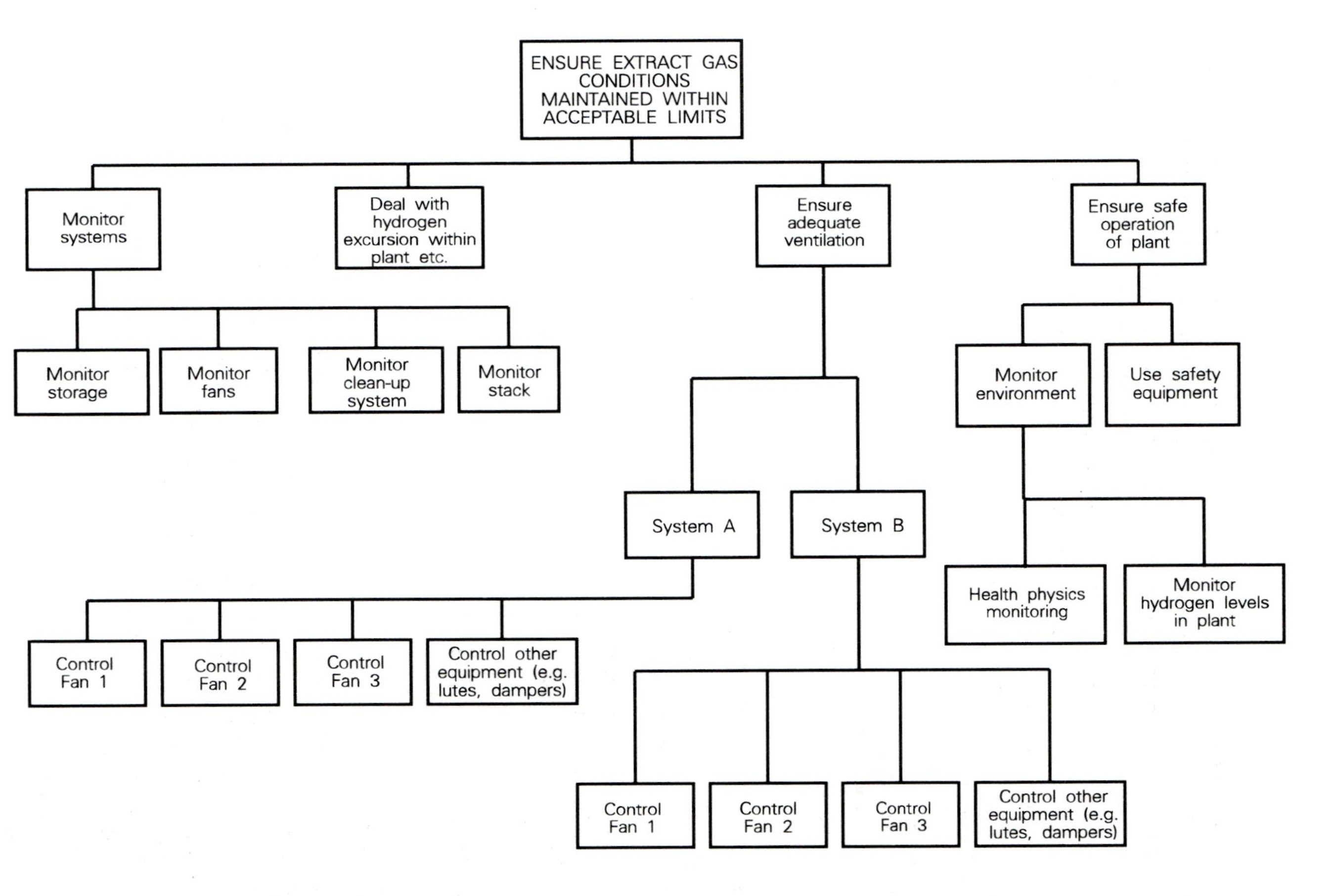

Figure 15.1 Hierarchical Task Analysis for Solid Waste Storage Plant

The major method of task analysis used was hierarchical task analysis (HTA), which was carried out for the inerting procedures (see Figure 15.1 which shows the top level of the HTA Note – details of the ventilation and inerting system are not included and therefore this figure is grossly simplified). The principal objective of the inerting process is to lower oxygen levels for each storage area to levels below those required for combustion. To do this the operators inert the plant by replacing oxygen with nitrogen in a controlled manner. There are four main components of this task:

- Dealing with increased rates of hydrogen evolution
- Ensuring safe operation of ventilation plant
- Ensuring adequate ventilation to extract gases and avoid pressurizing the storage compartments
- Monitoring the discharges

From the initial hierarchical breakdown, it appeared that the task where there was the most potential for human error was that of passing nitrogen through the storage areas. A detailed task analysis and task decomposition of these elements of the task was therefore completed.

The hierarchical diagrams were broken down into progressive levels of detail, and then each of the task elements at the lowest level of the branches were decomposed into a set of standard categories. This provided more detailed information about specific aspects which the analysts felt merited deeper investigation. These categories were as follows:

- The procedures
- The displays and controls relevant to each task step
- The sources of feedback
- The time constraints
- Potential hazards
- Possible errors

Resources

The personnel who took part in the review were as follows:

- A lead reviewer (i.e. a human factors assessor)
- Two trainee reviewers
- A specialist in HTA

The review took place over several months and the assessors spent a total of 60 person-weeks to complete the review.

Results

As a result of the task analysis, several mismatches were noted between the system design and human capabilities. These mismatches led to an increased workload upon the operators and could lead to human errors resulting in a failure to inert, and are described below.

Industrial Hazards

(a) Noise hazards were identified in particular areas in which ear defenders were not readily available. A stricter control of entry into these noisy areas was recommended.
(b) The operators experienced some problems operating particular damper controls due to poor lighting. Increased illumination was required at the damper controls and this has since been provided.
(c) Portable oxygen meters were necessary to warn of any nitrogen leaks which might affect the working areas. These were not available to all personnel, but were necessary for the operators' personal safety during the task. This piece of equipment has since been made available to all personnel during inerting.

Enforcement of the rules and regulations relating to the above hazards is the responsibility of the Management, who are required to encourage compliance with safety practices.

Procedures

Some of the operating instructions reviewed did not directly reflect the current practice for performing the task. This was partly because new equipment had recently been installed. In addition, the Management were in the process of training operators to undertake the inerting tasks which were previously only performed by supervisors.

(a) To aid the training of operators and to identify tasks which had been completed during the inerting test, the incorporation of checklists/tick sheets was recommended. This would allow the simultaneous use of more than one operator to prepare the plant for inerting.
(b) Many operations required operators to move about the plant. It was recommended that timelines should be prepared for these tasks, to allow operations performed in distant parts of the plant to be organized together. These timelines were subsequently prepared by the analysts and then used by plant management.
(c) Management had already identified some tasks which could be undertaken during the preparation for inerting which started 24 hours before the main task. The timelines assisted Management in a review of these

tasks and helped to reduce the number of tasks required to be performed during the inerting test.

(d) The inerting equipment had been installed in phases, and consequently the labelling of equipment was inconsistent. The analysis allowed management to identify a meaningful method of equipment identification and to identify access routes to equipment. This resulted in improved labels and notices being provided on plant.

(e) The nitrogen generation plant required special consideration prior to the training of operators, due to the complexity of the plant, inadequate labelling and the infrequency of testing. Management were required to modify the procedures so that the operator could be guided through the process. Greater emphasis was also placed on using the system and equipment developers to train the plant operators in system operation.

Hardware features

A number of inadequate features in the design of the controls and displays was noted. These did not prevent the operator from completing the tasks, but made their completion more difficult. In stressful conditions these features would increase the potential for operator error.

(a) Labelling could be improved, particularly for the nitrogen generation plant. Labelling groups of related instruments and control positions would help the operator in the control room. In several cases it was found that plant items were located behind trap doors: signs on plant to direct the operator to these hidden locations were recommended and subsequently installed.

(b) In some cases the grouping of instruments in the control rooms was not meaningful to the operators. Where the regrouping of instruments was not feasible, demarcation lines were recommended.

(c) There were many plant parameters which had set points, and these were recommended to be marked on the displays.

(d) A simple colour coding scheme was recommended to denote different system groupings of pipework and displays both on plant and in the training documentation.

(e) Throughout the plant there were many banks of annunciator alarms which were often extensive. While it was easy to locate a single flashing annunciator tile, it was difficult to scan them for general monitoring purposes. In addition there were many *standing alarms* (i.e. alarms which were frequently present during routine operations, but which did not require immediate action) which made it difficult to pick out those that were important to a particular task. These shortcomings were addressed by the following recommendations:

- Splitting the annunciator blocks into meaningful groups of alarms, by physically separating them and/or using demarcation lines

- Organizing the positioning of alarms within blocks to conform more closely with the operator's natural stereotype and expectations
- Standardizing the use of colour within the alarm system
- Segregating or reducing the number of standing alarms
- Highlighting important alarms, e.g. segregating process alarms from emergency alarms

Training

Several improvements in training were identified. The lack of a specific training programme and the reliance upon the senior supervisor to train his staff , were considered to be the highest priority problems. The following actions were therefore recommended:

- The development of a training programme
- A reduction of the training burden by improving the labelling and colour coding, and by the introduction of checklists and marked limits on displays
- The simplification of some of the plant layout diagrams by presenting them in a more schematic form
- The improvement of training schedules to ensure simultaneous availability of key personnel and trainers

Discussion

The testing of the ventilation and inerting systems was performed according to a schedule. Task analysis assisted the identification of the optimum way of performing these tests for the benefit of both the management and the worker. This enabled management to structure the training of plant operators and to conduct a thorough review of plant operating instructions.

The control panels located in the control rooms had been extended as the different phases of the inerting system had been installed. Thus the layout of the instrumentation occasionally hindered the operator. This task analysis suggested practical ways of improving the instrumentation layout.

Throughout the analysis, supervisor and operator input was vital to the analysis to provide an understanding of the problems and the appropriateness of the solutions. Plant visits and observation of the operators performing the task were important information sources and provided a valuable insight to plant operations that could not have been gained from a more isolated task analysis.

Perceived benefits to the organization

There are safety implications associated with the tasks performed within this plant (i.e. if the system leaked nitrogen gas or an incorrect mix of oxygen and hydrogen occurred, the safety of the personnel would be at risk). It is therefore important that the operators perform the task in an effective manner, are supported by an appropriate design and layout of instrumentation, and are trained to a high standard in order to maintain their awareness of the system state at all times. This will enable the operator to take the appropriate action if parameters are approaching unacceptable limits.

The perceived benefits to the system were not only those related to safety of the personnel on plant (and those in the surrounding areas), but also included that of increased efficiency. The system will be less prone to operator error, the operators' views of the plant and mental pictures of its operations will be consistent, and consequently training will be made easier.

Acknowledgements

The authors would like to thank Dr Les Ainsworth of Synergy for his work in completing the task analysis on which this study was based, and the operators of the plant for their co-operation and useful input to the assessment.

Chapter 16

A task analysis programme for a large nuclear chemical plant

B Kirwan, *British Nuclear Fuels plc, Risley*

Introduction

The thermal oxide reprocessing plant (THORP) is a large plant currently being constructed at Sellafield in Cumbria, England, at a cost of approximately £1.3 billion. To give an idea of the size of the plant, it has required over 800 engineering flow diagrams (EFDs), and has over 20,000 digital inputs/outputs for its distributed control system. Since the end of 1987, in line with heightened awareness in the nuclear power and reprocessing industries of the utility of human factors in enhancing safety and operability, THORP was the subject of a large and resources-intensive human factors (HF) programme (Kirwan, 1989). This programme was aimed primarily at the safety of the plant, and was run within the Safety and Technical Department at Risley, Cheshire. Task analysis was fundamental to the programme. Various forms of task analysis were utilized, creating a task analysis programme in itself, designed to maximize the usefulness of the interrelations between the task analysis methods used.

This case study defines the task analysis programme and the methods as used on THORP, lessons learned from their application and the overall impacts which the use of task analysis has achieved.

Objective

The objective of the task analysis programme were the selection and integration of task analysis techniques into the assessment programme in the detailed design phase of THORP, to evaluate the safety adequacy of THORP's man–machine interfaces from the ergonomics perspective.

Information sources

Information sources were varied in the human factors programme, but the following were usually consulted:

- Personnel – safety assessors; designers; operations personnel
- Piping and instrumentation diagrams – engineering flow diagrams; mechanical flow diagrams
- Schematics and general arrangement drawings
- Functional specifications
- Safety cases
- Contractor's documentation
- Sequence and Interlock Definitions
- Mock-ups/actual equipment (if available)

In the early and detailed phases of design, there was a heavy reliance on contact with the personnel responsible for assessing and developing the design/operational strategy for the particular part of plant in question.

Method

Before defining the actual techniques utilized, it is useful to specify the major constraints which affected the choice of techniques and their manner of implementation. These constraints were generated by the way in which the human factors programme arose and was implemented.

In late 1987, a human factors strategy was developed for the assessment of THORP (Kirwan, 1988: see Table 16.1), which included the carrying out of various task analyses. THORP was already well into its detailed design phase, and so the implementation of a significant assessment programme at this stage meant that certain human factors and task analysis related missions could not realistically be carried out (e.g. functional analysis; early HAZOP; allocation of function; early workspace layout of central and local control stations; specifications for control systems; etc.). The lateness of the human factors programme entering the project scene was thus a major constraint on the task analysis programme.

A second major constraint on the type of analyses carried out was that the human factors work occurred in the Safety Department, and was, in effect, a safety audit of the design of the plant, rather than constituting design support. This largely affected the scope and direction of the analyses, rather than the choice of which task analysis methods to use.

A third major constraint on the work was the availability of information. Although the plant was in the mid-late detailed stage of design, there were no operating instructions/procedures (at a detailed level), as such documentation is generally more effectively produced in the commissioning phase, especially for a new plant concept. As no similar plant to THORP existed, it was necessary to utilize initially task analysis techniques which would collect and create the depth of information required for human factors analysis. This constraint largely affected the sequence of task analysis methods used in the programme.

Table 16.1 Generic strategy for the Human Factors Assessment of plants (Kirwan, 1988)

Human Factors area	Concept	Detailed design	Operations support	Commissioning	Operation
Ergonomic/human factors analysis	Allocation of function • Functional analysis • Fitts List considerations • Control and other concepts • Level of automation • Other task analysis Initial staffing concepts Anthropometrics Person–machine interface philosophy • Layout Broad control and instrumentation consideration • Interlocks • Emergency shut down requirements • Environment	Detailed person-machine interface recommendations Detailed task analysis Use of CCR mock-ups Development of advanced operator aids for design support Implementation of error reduction measures (derived from HRAs) Design for maintainability Review of design specification Software ergonomics	Development of procedures and operating instructions Development of training system Simulator usage Advanced staffing concepts • Selection and qualification • Organization structure • Sociotechnical considerations Preparedness/Emergency response plans Advanced maintainability system development	Development of QA programme Certification of operators Complete entire staffing and safety management implementation Assist with changes made during commissioning Setting up incident reporting system Revision of earlier human factors data/guidelines	Implementation of QA programme (including safety management) Refresher training (annually) Implement and evaluate retrofits/ extensions as required Feedback info./data on incidents to design database
Human reliability analysis (HRA)	HAZOP Screening HRAs	HAZOP Human HAZOP Task analysis HRA Scenario anal. (inc. misdiagnosis and maintenance errors) Analysis of dependent failures Maintainability analysis Software reliability analysis Error recovery analysis	CCR audit Procedures audits Detailed HRAs of accidental event sequences Simulator experiments • Verify HRAs • Generate human error data • Evaluate training Evacuation assessments	Management audit (requires R&D) Setting up performance monitoring systems Evaluation of changes made during commissioning Revision of human reliability techniques and data	Monitor safety progress of plant operations Feedback information and data into human reliability database Carry out operational safety assessments

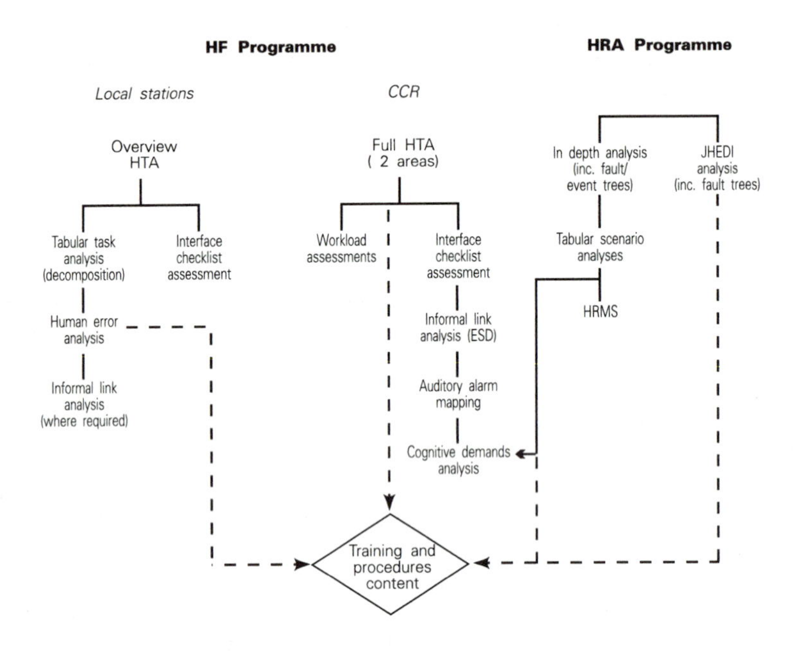

Figure 16.1 The task analysis programme for THORP

Thus, given these three major constraints of being somewhat 'late' in the design process, of being safety-driven, and of a low availability of operational documentation, a task analysis programme for THORP was generated within the larger structure of the human factors programme. The task analysis programme is represented in Figure 16.1. The task analysis methods utilized in the programme were as follows:

Human Factors
Hierarchical task analysis
Checklists
Decomposition analysis
Human error analysis
Informal link analysis
Workload/timeline analysis
Cognitive demands analysis

Human Reliability
Sequential task analysis
Fault/event tree analysis
Tabular scenario analysis
Human reliability assessment

The distinction between human factors and human reliability is not easy to make since they are highly interrelated and both deal with design of systems for humans. The distinction made above distinguishes between techniques used in the design process or during qualitative analysis (human factors), and those task analysis techniques which were used in support of quantitative assessments, such as those used in probabilistic safety analysis (PSA).

Each of these task analysis methods, some of which were developed and tailored for the THORP human factors assessment, are explained below. Comments on resources are embedded within the descriptions.

Human factors techniques

Hierarchical task analysis (HTA)

As mentioned above, one of the major constraints on all the human factors work was the lack of documentation in a form accessible for human factors assessment (although it is obviously preferable to carry out human factors analyses early to achieve design impact, rather than wait until appropriate documentation is available, when design impact will be difficult). For this reason, HTA was carried out at the beginning of each cycle of assessment whether it was for the central control room (CCR), local control rooms (LCRs), or for operation in local areas. HTA was utilized in one of two ways. The first was full HTA going into considerable detail and including all relevant plans. This was used in the assessment of the CCR and the principal LCR for THORP. The CCR comprises approximately eight operational and two supervisory 'areas' of responsibility. It was decided early on that it would not be resources-effective to carry out full detailed HTA for all process areas, as this would have taken approximately 45 person years of effort alone (the total planned human factors input for THORP, up to the commissioning phase, was 15-person years of effort). However, it was desirable from a safety perspective to demonstrate that operators could cope both with the workload and the complexity of tasks in this new plant. HTA was therefore carried out for two areas within the CCR, the (assumed) highest workload area console, and the most conceptually complex area console. An example from the HTA is shown in Figure 16.2. An *overview* HTA for the supervisor in the CCR was also carried out. The HTAs for the CCR took six person months to carry out in total.

Detailed HTAs were also carried out for the principal LCR and this work is detailed in another case study in this volume (Chapter 10).

The second approach utilized less detailed HTAs (see Chapter 9) which, with very few plans, were carried out for other locally-controlled operations on plant. This lack of explicit usage of plans in the HTAs was because these areas involved fairly straightforward sequenced operations with simple goal/sub-goal structures, often involving control of the movement of mechanical equipment in or between areas. An example of the simplified HTA utilized for these areas is shown in Figure 16.3. This level of HTA, applied to a large number of operational control stations and areas on local plant, took approximately 3 months in total.

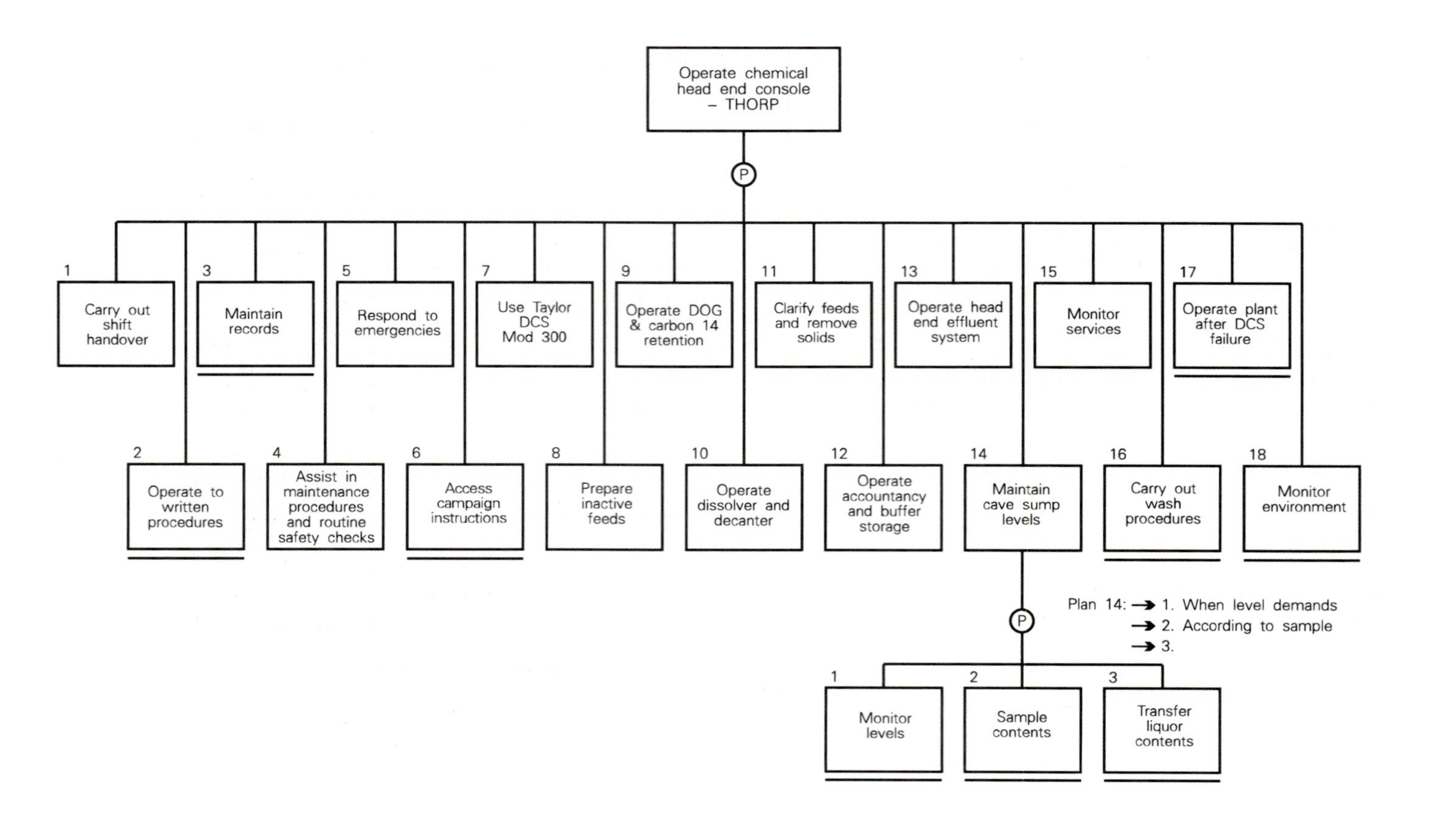

Figure 16.2 Example hierarchical task analysis for THORP CCR

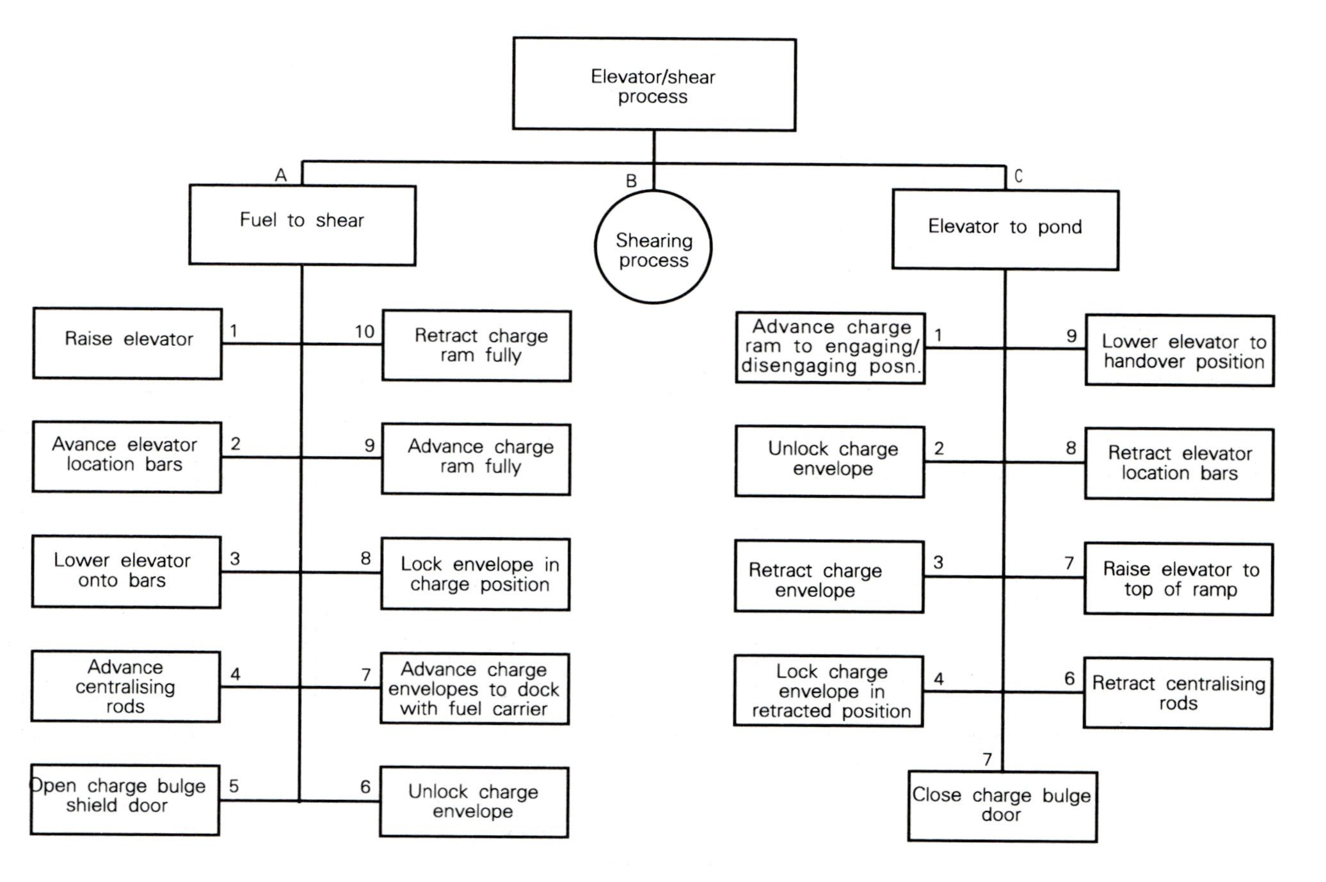

Figure 16.3 *Simplified hierarchical task analysis utilized for local plant operations*

Ergonomics checklists

Ergonomics checklists for the design of equipment were developed early on in the human factors programme. It was necessary first to generate the HTA's so that the application of checklists was not based on a superficial understanding of the process, which could have led to inaccurate assessment of equipment by checklist. The design was then assessed via a number of checklists on the following:

- Workplace layout
- Console design
- VDU system design
- Use of colour
- Alarm system design
- Environmental factors
- Communications

Later, a third checklist-type document was created for the development of procedural information and training systems and content, although these areas largely fell outside the scope of the human factors programme described here, as they were in the commissioning phase, rather than the design phase.

The design checklists contained nearly 400 assessment points derived from a literature review of relevant ergonomics texts and assessment experience. An area where there was a particular lack of standard accepted ergonomics criteria was with VDU mimic guidelines, for which ergonomics check-points were developed in British Nuclear Fuels plc. (Reed *et al.*, 1990a) and for which research is still continuing. Two examples from the checklists concerning the use of colour, and VDU systems respectively, are shown in Tables 16.2 and 16.3. One useful outcome of this work was that a design standard was developed based on these checklists, to be used on future plant design projects (Kirwan *et al.*, 1990).

Gaining the required information for a response to a checklist point varied in resources requirements. If, for example, anthropometric (body dimensional criteria) were being applied to a CCR VDU console, to ensure that at least 95% of operators could operate from it, drawings could be consulted and measurements evaluated with relative ease. Other checklist points proved more elusive (e.g. the number of raster lines on the computer screen). Application of checklists to all interfaces (100) in THORP has taken at least two-person years to date. If detailed checklists are being applied rigorously, this approach is a non-trivial assessment process, but one which yields a powerful and comprehensive assessment of the adequacy of the operator interface.

The task decomposition method described here was used to assess the design, and its format is shown in Table 16.4. This format was used in the assessment of local areas outside the CCR (the CCR itself had an intensive checklist and workload assessment). The task decomposition follows on from the simplified HTA, and for each step in the HTA notes the operations carried out and, in particular, how the operator utilizes the interface for control and feedback to achieve the task goals. Thus, the task decomposition links the HTA firmly to the actual usage of the interface. The decomposition goes further however (bridging over to the human error analysis method described below), by allowing the analyst to identify potential errors at each step in the procedure. To do this, the analyst must be familiar with human error forms (e.g. error of omission, commission, memory failure, etc.) as defined in the field of human reliability analysis.

Table 16.2 Checklist for THORP: the use of colour (on VDUs)

Colour	Process material identification	Current state	Alarm information abnormality
White	Utilities	—	—
Yellow	Gaseous feed	—	2nd level alarm, 'Caution'
Magenta	Reagent feed	—	Under maintenance calibration
Orange	Solvent	—	—
Cyan (light blue)	—	Symbol for equipment moving (if not used for low priority alarm)	Low priority alarm, abnormality notification
Red	—	Equipment stopped (symbol with red in-fill)	Emergency High priority alarm
Green	Aqueous	Equipment running (symbol with green outline)	—
Blue	Ventilation/Air	—	—

Task decomposition

Two forms of task decomposition were used for THORP, and the second form, the tabular scenario analysis, is so different in nature and purpose from its more conventional 'cousin' that it is discussed later under the human reliability heading.

The task decomposition process used in the human factors analysis phase was labour-intensive, but rendered the assessment systematic, relatively robust and highly auditable. As discussed later under perceived benefits, these were critical qualities for this type of assessment for THORP. The decomposition fed directly into the human error analysis format described below, and resources for application of both of these techniques, which are highly interrelated, are discussed within that description.

Table 16.3 VDU information presentation (extract from checklist)

- The overall screen display system should be hierarchical in nature to facilitate understanding of the process. It should be broad rather than deep, and not exceed five levels
- Each screen should have a unique and informative title
- At each level a particular title format should be used so that the operator always knows where he resides in the process
- Display pages should be consistent such that menus and titles are in the same place on each page.
- The operator should be provided with a simple means to go back a level
- The operator should have a simple means to follow process information from page to page
- Superfluous and unnecessarily long labels should not be used
- Crossing of process lines should be avoided
- Irrelevant digital detail should be avoided
- Only one console should be able to control a parameter or process items, although more than one console may view that parameter
- Size, centrality and labelling should be used to show importance of items
- The operator should be able to recover easily from an input error
- For symbols on screen, a maximum of three levels of size should be used
- Symbols should be meaningful to the operators who will operate the plant
- Target areas on screen should be depicted by a box around the target item
- Complex symbols and coloured symbols should be at least 10 mm high for normal viewing distances
- It should be possible to call up an alarm listing giving chronological and prioritized alarm information onto a single VDU screen

Human error analysis (HEA) for interface design

The human error analysis (HEA) format used for THORP is shown in Table 16.5, and bears a strong resemblance to hardware reliability techniques such as HAZOP and failure modes and effects analysis tables. Its usage is fully described elsewhere (Kirwan and Reed, 1989). The HEA considers the errors identified in the decomposition analysis and determines their likely consequences, means of error recovery (if any), and potential error reduction measures.

Table 16.4 Decomposition analysis for THORP local plant design assessment

Task/ step no.	Task goal	Information available	Required action	Feedback	Communications	Possible errors, distractions. time available, skills/knowledge
D.32	Input vessel No.	CCTV	Keyboard input (6 character code)	VDU message	Check with adjacent area of plant (intercom)	Input error May require operation of CCTV controls to obtain adequate view – this may be omitted due to time constraints, picture is of poor quality Omit communication check, or communication error occurs
F.4	Traverse equipment to co-ordinate J11	CCTV	Joystick (initiate and hold)	CCTV	Check path is clear (off-board operator – if available)	Fail to check, or partial check CCTV system poor Poor visibility of object Communication not available

Table 16.5 Human error analysis for Interface Design Assessment

Inadequacy No.	Ergonomics inadequacy	Possible errors	Possible Consequences	Remedial actions			
				Recover points	Design	Training	Procedures
1	Certain important push-buttons are not readily discriminable from each other	Actuate push - button at incorrect stage of operation	Potential risk to operator safety	None	Golour coding Perspex covers		
2	No indication to show 'barrier' status	Fail to realize barrier is closed	Collision – damage to plant and equipment, loss of throughput	CCTV check Communication	'Gate' indicator light Extra CCTV Highlighting of gate		

The HEA tables enabled prioritization of identified potential design inadequacies with respect to safety, in a qualitative sense. In discussions with design and operations departments it was then possible to determine which problems required rectification and how such rectification could be achieved. The HEA step was critical in the process of achieving design change for identified ergonomics problems. As with the decomposition analysis, it was critical that the HEA was systematically carried out and fully and accurately documented. The task decompositions and HEA for THORP together required approximately another 6 months of effort.

Informal link analysis

For certain items of equipment which involved manual operation, the sequence of use of the controls and displays was assessed to determine the adequacy of the layout (e.g. in terms of the likelihood of out-of-sequence errors). These assessments were few in number and were carried out on equipment which could, on a very infrequent basis, be used in a less automatically controlled and protected mode. No formal link diagrams were produced, but these informal link analyses led to several design changes on existing panels largely for panel layout and protection against inadvertent or out-of-sequence operation. One formal sightline analysis was also carried out, and an auditory alarm map was also developed to review the discriminability of alarms in the CCR. The alarm 'map' involved simply noting the relative positioning of the various audible alarm devices in the CCR. This was then used to assess the likely discriminability of these alarms by the operators stationed at a number of consoles in the CCR.

On emergency shut-down (ESD) panels, the layout was similarly reviewed in terms of sequences of operations, functional grouping and links between different controls and displays. This analysis led to a major revision of the layout of these CCR panels (Reed, *et al.*, 1990b).

In both the above cases no formal link analysis was carried out, but the functions of link analysis were applied with some effect. These task analyses took approximately 1 month.

Workload/timeline analysis

Three major workload assessments were carried out for THORP, two for the CCR and one LCR. The latter of these is described in detail in another case study (Chapter 10) in this volume, and so detail is spared in this section. The assessments used the HTAs, and estimated the time required for each bottom-level subordinate operation. The requirements for the operations to be carried out during a shift were estimated either by engineering judgement or via a stochastic (computer) model of THORP processes. The tasks were then plotted on a timeline, and it was calculated when workload peaks would occur. It was assumed, in the absence of more robust data, that operators would be overloaded if working for more than 75% of a shift and underloaded if busy less than 50% of the time (rest breaks and nuisance alarms, etc., were factored into the relevant

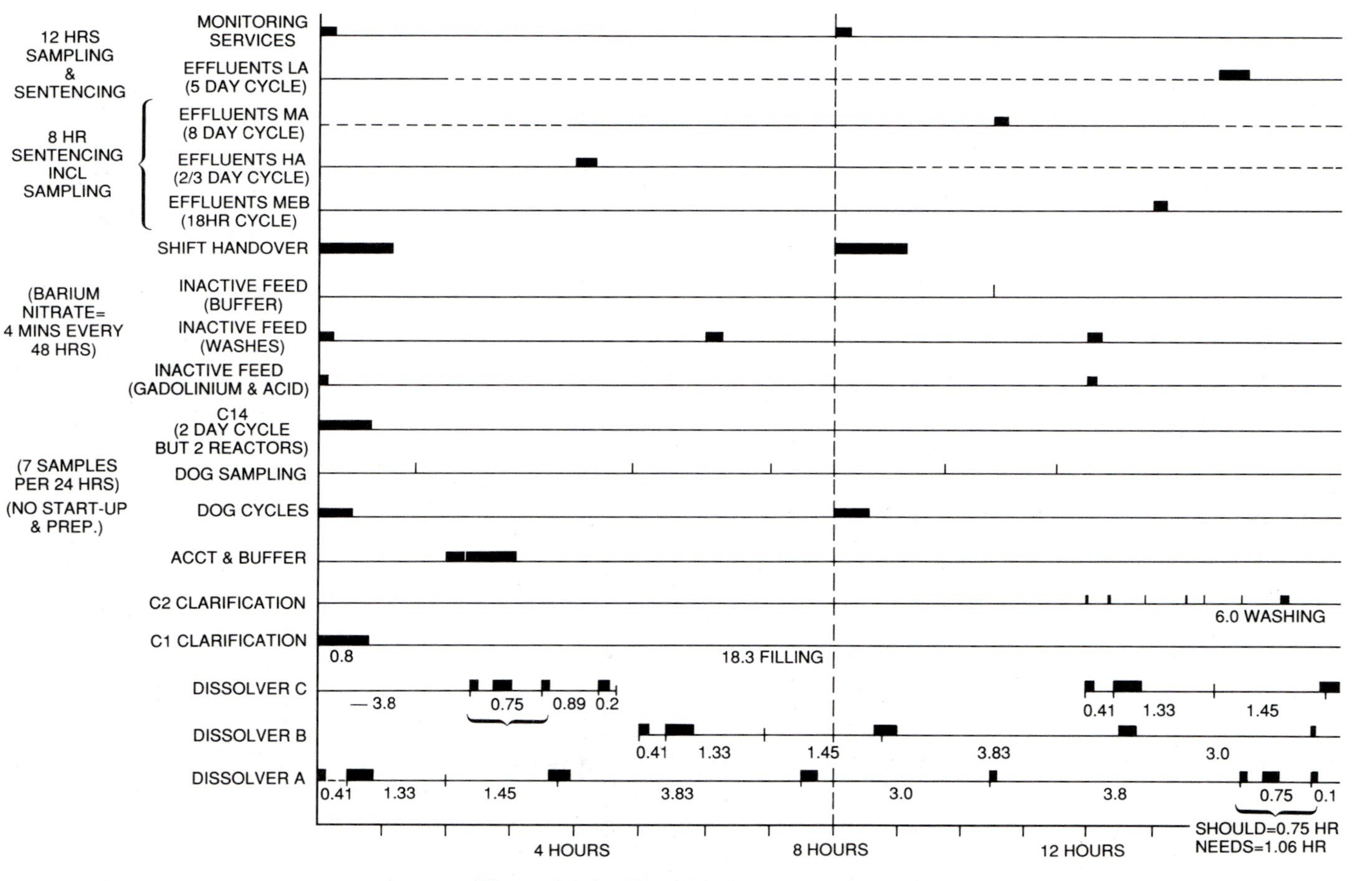

Figure 16.4 Workload assessment

equations). An extract from a *timeline* produced for the CCR is shown in Figure 16.4.

The workload assessments were difficult in terms of estimating subtask times and the occurrence of operations during a shift, and many assumptions had to be made to create a timeline. However, what was developed became a working and reasonable best estimate upon which workload forecasts could be made until better operational information could be generated (e.g. in the commissioning phase). The two workload assessments took, in total, approximately 4 months.

Cognitive demands analysis

As part of an alarm-handling performance analysis, the cognitive demands being placed on the operators during anticipated and unanticipated events were considered. This work was highly exploratory in what is a difficult technical area. The format utilized is shown in Table 16.6. This format did allow, to a limited extent, the modelling of the operator's cognitive pathway during the event sequences studied. This work, which utilized the tabular scenario analyses described later, took approximately 1 month.

Human reliability

The aforementioned techniques were largely devoted to the assessment of the interface as part of the qualitative evaluation of the human factors adequacy of THORP. In addition to these assessments, as part of the overall human factors programme to protect THORP from human error impact, human reliability assessments were also being carried out to feed into the PSA for THORP (a largely quantitative assessment). The following task analysis methods were utilized in the human reliability assessment component of the overall human factors assessment programme for THORP.

Sequential task analysis

Human reliability assessment for THORP operated at two levels: a screening approach for evaluating human error potential in safety cases, and a more in-depth method for assessing and, if required, reducing human error impact in error-sensitive scenarios. In the screening approach, a hierarchically structured sequential task analysis method was used. This documented the overall task being assessed, and its subordinate operations, and was supplemented by a verbal description of the scenario, personnel involved, and consequences of overall operator failure in the scenario. Other information on performance shaping factors was also logged on this computerized system. This approach was aimed at being auditable, justifiable and conservatively accurate. An example of the sequential task analysis portion of this system, known as the justification of human error data information (JHEDI: Kirwan, 1990) system, is shown in Figure 16.5.

Table 16.6 Cognitive demands analysis format

Step	Current top goal	Influences	SRK (skill, rule, knowledge level)	Current frame, hypothesis, action plan	Working memory	Prospective memory	Long term memory (LTM)
2	Production	Production mode prevails No perceived pressure Uncertainty as to cause (scenario very infrequent)	Activation/ Observation	Infrequent event, therefore undecided. Could be: – Fans stopped – Blockage (e.g. filters) – Failure of speed controller – Flooding of a column – Spurious *Action Plan:* – Acknowledge alarm – Study mimic - – Monitor – Continue normal operations	• Low DOG fan speeds (yellow) *Perceived causes:* – Fans stopped – Blockage – Speed controller failed – Flooding of a column – Spurious	Monitor for: • Other low flows and pressures • High pressure • Instrument failures • No other alarms (plant operating normally)	Knowledge base of causes and effects
3	Production	Increasing uncertainty and stress	Observation	Operator establishes that there is low flow and speed is falling *Action Plan:* – Acknowledge alarms – Study mimics – Check for further indications of fan failure	• Low DP across fan (yellow) *Perceived causes:* – Fans stopped - – Blockage – Speed controller failed • Low flow in DOG leg (Red) *Perceived causes:* As above	Monitor for: • Other low flows and pressures • High pressures • Instrument failures • Check levels in columns • Check status of dissolvers to find failure severity	Knowledge base of causes and effects Knowledge base of past DOG fan failures
	Safety		Interpretation		• Low DP across dissolver(s) *Perceived causes:* As above		Knowledge of consequences and time-scale
		Perceived time pressure			• Other alarms due to distractions		

FILE NAME: TEST224
ASSESSOR: A.N. OTHER
DATE: 1-8-89
SAFETY CASE: TAN P222

SCENARIO
This scenario concerns a sampling and sentencing operation. The operator must initiate a routine sample by asking a local operator to take the sample. The local operator goes to the sampling point and takes a sample, which is then dispatched for analysis. The analysis is standard and only gross analytical errors will give an incorrect result. The results are reported back to the CCR operator via the plant information computer. The operator reads the report and should realize that the contents of the tank should not be sentenced, but re-blended. The supervisor is also required to approve sentencing by reading the report independently of the operator. This scenario therefore concerns the incorrect sentencing of the tank contents due to operator error,

TASK ANALYSIS
1. Initiate sampling
1.1 CCR op recognizes need for end of batch sample
1.2 CCR op communicates with local operator
1.3 Local op goes to location and takes sample
1.4 Local op dispatches sample for test
2 Analyse sample
2.1 Analyst carries out test
2.2 Analyst records result
2.3 Analyst transmits results
3 CCR op makes sentencing decision
3.1 CCR op reads report
3.2 CCR op decides not to sentence
4 CCR supervisor QA's sentencing decision
4.1 Sup reads report
4.2 Sup decides re-blending is necessary
5 CCR op/sup re-blend tank contents

Figure 16.5 Extract of Sequential Task Analysis Portion of JHEDI System

Fault/event tree analysis

Fault and event trees (see pp. 188 and 178) were already being utilized in the PSA for THORP prior to involvement of human factors practitioners, as these are standard reliability technology-based methods. The human reliability assessment programme fed directly into these methods, although more use was made of the event tree by the human reliability assessors than the safety assessors, primarily for representing time and event-driven sequences of activities. Fault trees in the detailed human reliability phase (using an approach called scenario analysis) often went to considerable levels of detail to identify not only errors, but their root causes or psychological error mechanisms. This level of depth was utilized to devise error reduction mechanisms (Kirwan, 1990).

Examples of event tree and fault tree usage in the human reliability programme are shown in Figures 16.6 and 16.7. Both of these figures involve the same scenario, concerned with fan failure which leads to over-heating of a vessel. Cooling can be achieved by switching (via a three-way valve) from a heating circuit to a cooling circuit, or else time can be gained by switching off the pumps which are circulating the heat source. Therefore, the control room operator (CRO) and shift supervisor (SS) detect a series of alarms and either achieve recirculation by manipulating (remotely) a three-way valve within a short time frame, or else switch off the pumps to gain more time to achieve recirculation. In the latter case, however, due to the residual heat in the vessel, the pumps must be switched back on.

Tabular scenario analysis

In the detailed human reliability assessments, called scenario analyses, a tabular format of task analysis was utilized, based originally on work by Pew *et al.* (1981). This format enabled the detailed modelling of the task to the level of individual alarm occurrence. The format utilized is shown in Table 16.6, and is required for human reliability assessments of event-driven scenarios by the computerized human reliability management system (HRMS: Kirwan and James, 1989). Table 16.6 is also concerned with the dissolver off-gas (DOG) fan failure scenario utilized in Figures 16.6 and 16.7, and shows in detail the information available on the VDU, the CCR back-up panel (BUP), and the ESD panels. It also shows that immediately prior to the event, the operator would be busy on other activities, in this case the shear cave (SC) and basket handling cave (BHC) of THORP. The tabular scenario analysis (TSA) was a form of *paper simulation* of scenarios and, although it is a resources-intensive process, produces a convincingly realistic chronology of events. This level of detail is useful if trying to reduce system vulnerability to human error in error-sensitive scenarios. The tabular scenario analysis was also used for qualitative assessment of alarm-handling performance as part of the human factors programme.

Human reliability assessment

The HRMS approach already mentioned above is modular and comprises sequential task analysis (identical to that in JHEDI), tabular scenario analysis, human error identification based on psychological error mechanisms, human error screening, performance shaping factor-based quantification, and error reduction. These modules lead to a good deal of documentation of tasks in a variety of formats. As an example, Figure 16.8 shows an extract from the

Table 16.7 Tabular scenario analysis utilized by HRMS

Task goal	Time	Staff	System status	Information available	Decision/action/ communication	Equipment/ location	Feedback	Distractions/ other duties	Comments and operator expectations
Normal operations		CRO. SS	Normal operating envelope			CCR		Busy on SC/HHC activities	Single CRO. SS not busy. Assume beginning of leach sequence (worst case timescale).
Respond to first alarm	T + 10 s	CRO	Both DOG fans fail for a cause other than C&I failure	*Alarm*: (yellow) on low DOG fan shaft speed	CRO targets 'fetch alarm': Goes to level 4 mimic	VDU, DOG console	Flashing yellow alarm message	Previous tasks: Other alarms begin to occur	
Respond to further alarms	T + 20 s	CRO	Further alarms occur due to: • Low DP across fans • Low flow in DOG fan leg • Low DP across dissolver • Low pressure fall across columns	*Alarms*: Yellow Red Red Red	CRO targets next alarm: Goes to level 4 mimic. CRO goes to group alarm listing. Looks at printer	VDU, printer	Flashing, audible messages	Other alarms	
CRO identifies fan failure & calls Supervisor (accept alarms)	T + 60s	CRO/ SS	Alarms on: • Low flow DOG duct at Stack Monitoring Room • Hardwired in DOG duct manifold	*Alarms*: Red Hardwired	CRO monitors level 3 mimic CRO acknowledge alarms calls Super. SS detects alarm from own console	DOG VDU DOG VDU and BUP SS VDU console	Audible and visual Audible and visual Audible and visual	SS involved other tasks	

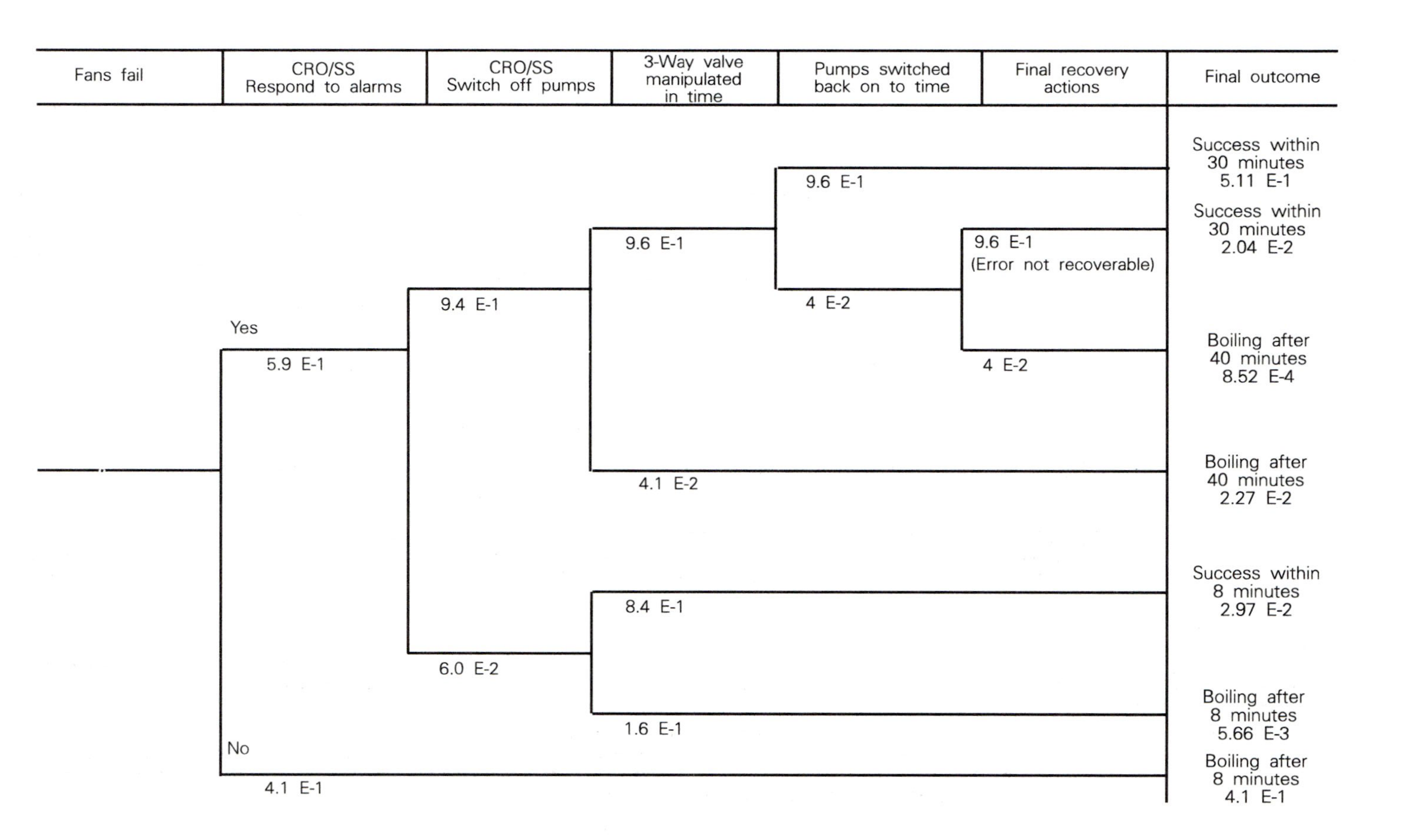

Figure 16.6 Event tree utilized in human reliability assessment

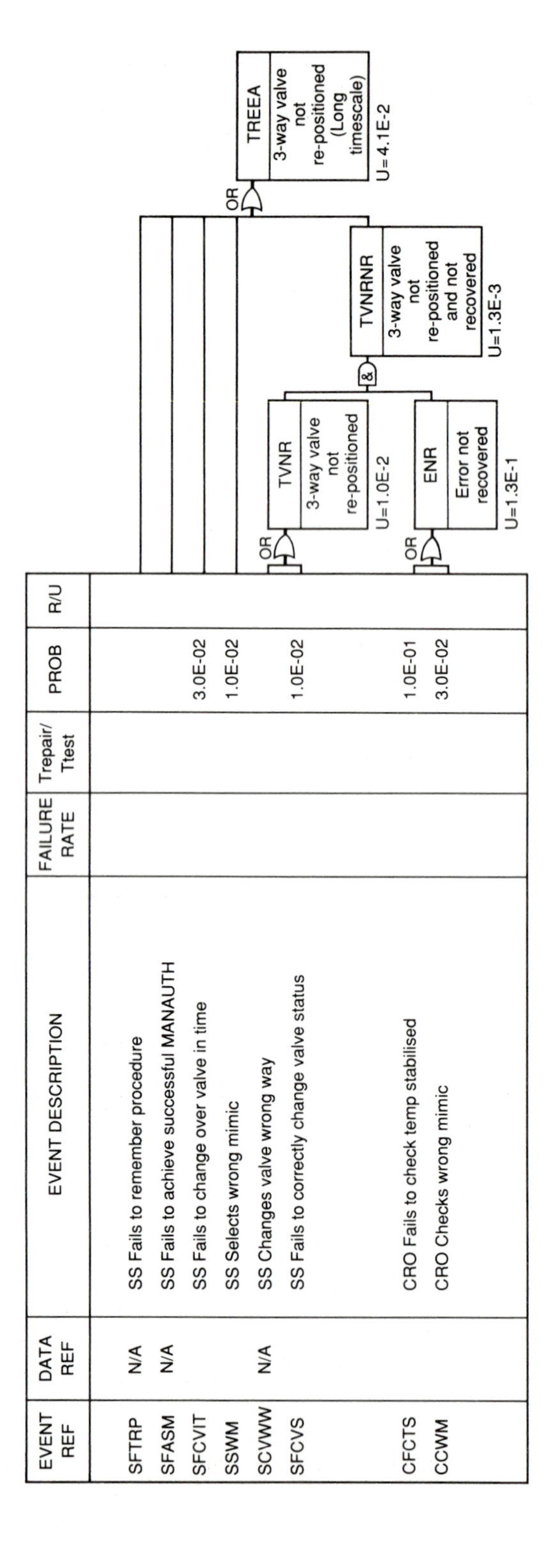

EVENT REF	DATA REF	EVENT DESCRIPTION	FAILURE RATE	Trepair/ Ttest	PROB	R/U
SFTRP	N/A	SS Fails to remember procedure				
SFASM	N/A	SS Fails to achieve successful MANAUTH				
SFCVIT		SS Fails to change over valve in time			3.0E-02	
SSWM		SS Selects wrong mimic			1.0E-02	
SCVWW	N/A	SS Changes valve wrong way				
SFCVS		SS Fails to correctly change valve status			1.0E-02	
CFCTS		CRO Fails to check temp stabilised			1.0E-01	
CCWM		CRO Checks wrong mimic			3.0E-02	

Figure 16.7 Detailed fault tree

STEP	No. 1	DETECT & ACKNOWLEDGE VDU ALARMS
ERROR	No. 01090	CRO fails to respond to all signals
ERROR MECHANISM		Cognitive/stimulus overload
RECOVERY		SS/HARDWIRED ALARM/OTHER ALARMS
DEPENDENCY/EXCLUSIVITY		
SCREENING		
COMMENTS		REC HIGHLY LIKELY TO BE QUANTIFIED AS CRO CONFUSION DUE TO HIGH DENSITY ALARMS.

STEP	No. 2	RESPOND TO HARDWIRED ALARM
ERROR	No. 07090	CRO fails to detect hardwired alarm amongst others
ERROR MECHANISM		Discrimination failure
RECOVERY		
DEPENDENCY/EXCLUSIVITY		
SCREENING		SUB
COMMENTS		HARDWIRED ALARM VERY SALIENT. 'COPIED' TO SS CONSOLE. SUB 01090: ALARM QUANTIFICATION MODULE QUANTIFIES TOTAL RESPONSE TO ALARM SET.

STEP	No. 3.2	CRO suggests fan failure as scenario
ERROR	No. 13010	CRO/SS fail to identify scenario in time
ERROR MECHANISM		Cognitive/stimulus overload
RECOVERY		SS, EMERGENCY PROCEDURES
DEPENDENCY/EXCLUSIVITY		
SCREENING		
COMMENTS		THIS ERROR IS USED TO REFER TO THE FAILURE TO IDENTIFY THE OVERALL GOAL OF PREVENTING BOILING BY ACHIEVING RECIRCULATION VIA CHANGING VALVE STATUS.

Figure 16.8 Example from a screening analysis module in HRMS

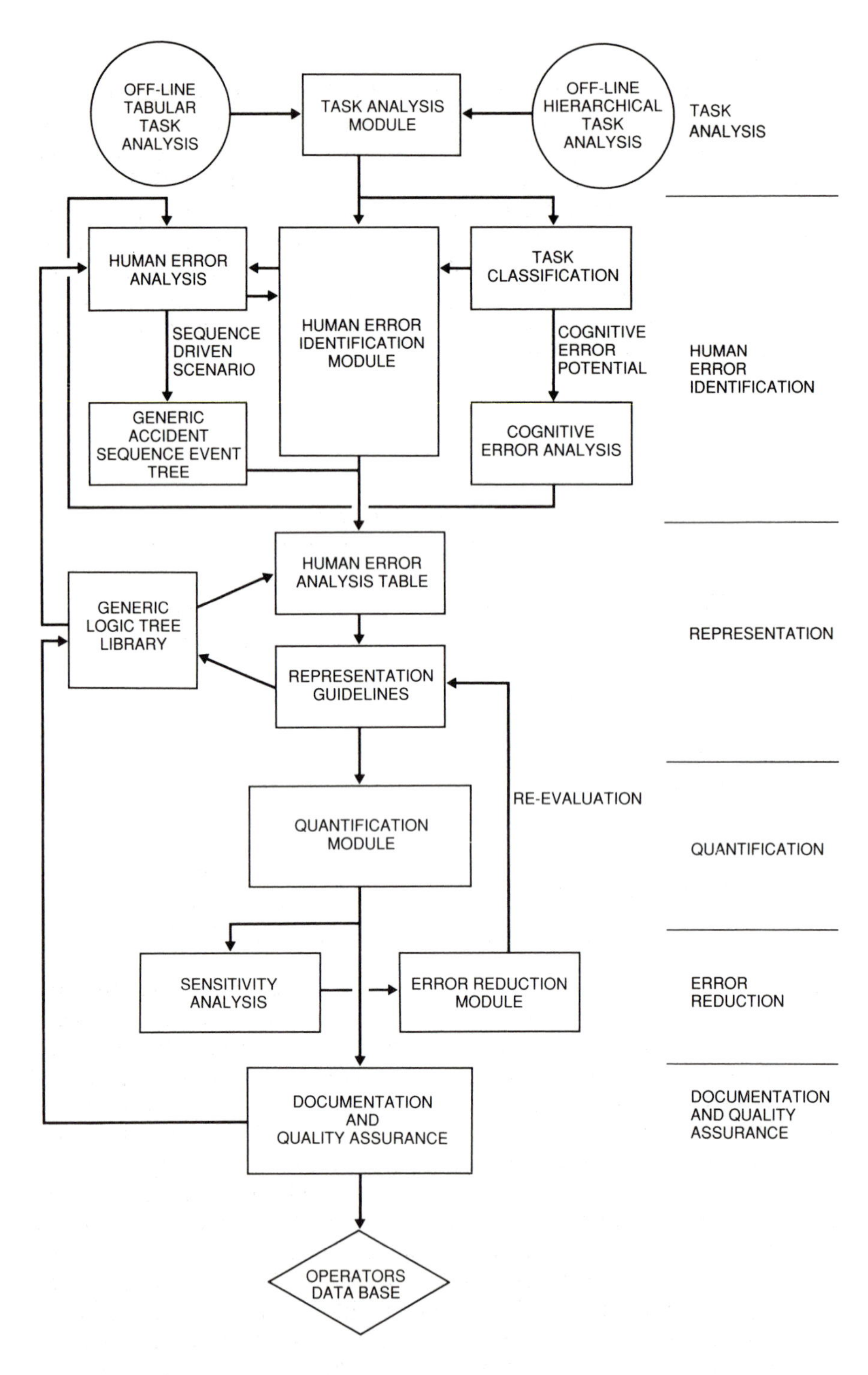

Figure 16.9 Overall sequence of assessment

screening module, which helps the analyst decide whether the error identified requires quantification or whether it can be *screened out* from further analysis (e.g. on grounds of being subsumed (SUB) by other human error probability data). Figure 16.9 shows the overall sequence of assessment in HRMS.
The task analysis component of the human reliability assessment work for THORP is difficult to quantify but is estimated as approximately one person-year of effort. The utilization of HRMS may require 1–2 weeks of task analysis per scenario, whereas JHEDI may take less than 1 hour for the task analysis component, as safety assessors are usually fully aware of the task sequence, goals and interface components.

Resources

Table 16.8 Resources required for the study

Resources utilized	
Hierarchical task analysis (full)	6 months
Hierarchical task analysis (single)	3 months
Task decomposition/human error analysis	6 months
Checklists	2 years
Link analyses	1 month
Workload analysis	4 months
Cognitive demands analysis	1 month
Sequential task analysis Tabular scenario analysis	1 year
Total	57 months

Resources utilized by each technique have been stated earlier and are summarized below for the convenience of the reader. It is worth stating that a large proportion of these resources, often as much as 50%, were often spent on information collection activities. This was because of the stage of design of THORP and the corresponding lack of detailed operational information which is required for human factors analysis. As noted earlier, this is inevitable if the programme is to be successful as a potential source of design change. This was made more difficult by the lack of an existing similar plant to THORP, and hence the corresponding lack of operational personnel with direct experience of THORP-type operations. Therefore, if doing design assessment on a new concept of plant at the design stage (which is when design assessment is often necessary), information collection resources will be intensive. Human factors personnel were also used throughout all task analysis activities except JHEDI. The JHEDI task analyses were the only ones which did not require extensive information collection because the safety assessors concerned with each safety case had already derived implicitly a task analysis for their assessments. JHEDI therefore merely formalized this implicit knowledge of the task and made it amenable to audit, justification and later re-assessment if required.

Lessons learned

The following are a summary of the major insights gained from the task analysis programme:

(a) Data collection activities are intensive for plants at the design stage (prior to commissioning).

(b) Properly executed and documented task analysis enhances justification of human factors recommendations.

(c) HTA is a fundamental form of task analysis, and is possibly the ergonomist's equivalent of the EFD or P & ID.

(d) Verification of task analyses by design/operations personnel is essential: task analyses should be auditable where possible.

(e) Human factors assessment of interfaces without having carried out a task analysis may miss more subtle problems and may identify problems which may evaporate once the operational concept is more fully understood.

(f) Prioritization of human factors problems or inadequacies (e.g. in human error analysis tables) is highly useful in attempts to achieve design change – e.g. primary (on direct safety grounds); advisable (possible indirect safety grounds); ergonomically desirable (operational grounds only).

(g) Workload assessment methods (such as timeline analysis) are relatively crude and measure the operator time involved at a console but not the cognitive demands or 'burden' actually levied on the operator in non-routine situations. More sophisticated technology needs to be developed.

(h) Some task analysis approaches produce multiple benefits (e.g. HTA enables the human factors assessor to understand the operational system, acts as the basis for human factors and human reliability assessment, and its results can be utilized directly in the development of procedures and training content).

(i) Many of the more resources-intensive techniques can be used effectively on a sub-set of representative systems/scenarios, i.e. if the scenarios are carefully chosen extrapolation from these to others is feasible.

(j) The task analysis programme used mainly human factors personnel (and one HTA developer). For the application of large parts of the checklists, and for the HTAs, it appeared that this did not necessarily require human factors professionals, although other personnel that took part required some basic training. For the task decompositions, the HEAs, the cognitive demands analyses and the link analysis, however, human factors professionals are recommended. Workload assessment requires a team effort between human factors assessors and design and operations personnel.

Perceived benefits to the organization

The THORP human factors and human reliability assessments took place on a large scale, and it was essential that efforts were systematically applied. This was necessary so that any human factors problems identified could be justified and withstand independent scrutiny, either internally or externally (by the regulatory authorities). The overall human factors programme, which has depended heavily on task analysis for its foundations, has achieved a number of significant impacts for design change (e.g. revised layouts of emergency shutdown panels and local panels on plant; changes to various VDU mimics; enhanced colour coding consistency across plant; and the identification of additional required instrumentation and control protection devices). The assessments have also influenced staffing level projections for the control rooms. In particular the HTA technique has been adopted as one of the major methods for defining training content as part of the THORP training plan, and HTA was perceived as a useful tool by many of the operations personnel who became involved in its application. Training and procedural assumptions from the THORP safety case have been fed forward to the training/procedural departments for incorporation into these systems during the commissioning phase.

It is worth re-emphasizing that for any plant in its late detailed design stage (when some design items are already constructed), many human factors recommendations for design change will have cost implications and require a good deal of justification. The task analysis work was fundamental in creating a human factors-based description of the operational design with its limitations, if any were evident (as well as its advantages), which could be evaluated independently by the designers and future operators of THORP. Often such evaluation led to changes in the task analysis as new information came to light, but it always led to a rational 'picture' of the task. Hence, any identified problems could be dealt with from a common understanding of the problem. Without such a basis for discussion, integrating human factors assessment into the design process proves difficult. Rationalization is therefore a prime benefit of carrying out task analysis.

Conclusions

A task analysis programme was carried out as part of a human factors and reliability assessment programme. A number of task analysis approaches were utilized, all of which contributed to the impact of the programme. The overall approach of task analysis was fundamental to the efficiency and effectiveness of the human factors and human reliability assessment programme for THORP.

Acknowledgements

The author acknowledges the following who took part in various task analyses: Julie Reed and Annette Verle (*British Nuclear Fuels plc.*), Michael Joy (*Diam Performance Associates*), Michael Carey and Sue Whalley (*RMC Consultants), Johanne Penington and Andy Smith (WS Atkins*) and Mike Lihou (*Lihou Loss Prevention Services*).

In addition, the designers and operators who facilitated the task analysis work and were also the recipients of its impact are sincerely and gratefully acknowledged.

References and further reading

Henley, E.J., and Kumamoto, H. (1981) *Reliability Engineering and Risk Assessment* New Jersey: Prentice-Hall.

Kirwan, B. (1988) Integrating Human Factors and Reliability into the Plant Design and Assessment Process. *Proc. Ergonomics Society Annual Conference*, 154-162.

Kirwan, B. (1989) A Human Factors and Human Reliability Programme for the Design of a Large UK Nuclear Chemical Plant. *Proc. Human Factors Soc. Annu. Meet.*, 16-20.

Kirwan, B. (1990) A Resources-flexible Approach to Human Reliability. *Proc. Safety and Reliability Society Annual Conference*. London: Elsevier Applied Science.

Kirwan, B. and James N.J. (1989) The Development of a Human Reliability Assessment System for the Management of Human Error in Complex Systems. *Paper presented at National Reliability '89*, Brighton Metropole, June.

Kirwan, B., Litherland, M. and Reed, J. (1989) The Development of an Ergonomics Standard for the Design of Operator Interfaces. *IMechE Conf. on Quality Assurance in Nuclear Power.*

Kirwan, B. and Reed, J. (1989) A Task Analytical Approach for the Derivation and Justification of Ergonomics Improvement in the Detailed Design Phase. *Ergonomics Society Annual Conference*, 36-43.

Kirwan, B., Reed, J. and Litherland, M. (1990) The Development of an Ergonomics Standard for the Design of Operator Interfaces. *IMechE Conf. on Quality Assurance in Nuclear Power*.Pew, R.W., Miller, D.C. and Feehrer, G.G. (1981) *Evaluation of Proposed Control Room, Improvements Through the Analysis of Critical Operator Decisions.* EPRI Rept. NP 1982. Palo Alto, Ca: Electric Power Research Institute.

Reed, J., Penington, J., and Kirwan, B. (1990) The Application of Human Factors to the Assessment of VDU Screen Displays in a Nuclear Chemical Plant. *Paper presented at the IMech E Conference on Operating Reliability and Maintenance of Nuclear Power Plant*, I Mech E Headquarters, London, pp. 45-59.

Reed, J., Verle, A., and Kirwan, B. (1990b) Design of Emergency Shutdown Panels. *Proc. Ergonomics Society Annual Conference*, 393-398.

APPENDIX

Summary of task analysis techniques

Summary of task analysis techniques

This section provides a quick reference guide for all the techniques referred to in Chapter 1, listed in alphabetical order. A number of these techniques 25 are described in more detail in Part II of this *Guide*, and in the annotated listing that follows these are cross-referenced to their detailed description in Part II. For the other techniques, a brief summary is given, together with a reference for further information, listed in full in the bibliography. For each technique, the earliest appropriate system life cycle stage is also specified, along with the principal relevant human factors issues which the technique can be used to address.

1. **Activity sampling** (pp. 41)
Real-time human actions are observed and recorded regularly at a selected sampling interval. Sampling must be at a frequency which is at least twice the frequency of the behaviour which is of most interest or, for routine repetitive tasks, sampling should take place at random intervals. A limiting factor is whether the data collector can keep up with the worker.

Earliest life cycle stage: Commissioning
Human factors issues: Performance assurance

2. **Barrier analysis** (pp. 169)
This approach aims to identify hazards that could lead to accidents. For each of these, any barriers that could prevent the accident should be recorded along with their method of functioning (barriers can be both physical and non-physical) and modes of failure (including human error causes). This is followed by a check on how the barrier failure could be recovered.

Earliest life cycle stage: Preliminary design
Human factors issues: Task and interface design
Performance assurance

3. **Charting and network techniques** (pp. 81)
There are a range of charting and network techniques available for representing and analysing tasks, eight of which are reviewed in Chapter 3 (pp. 82 – 93). Their aim is to represent the task concisely in a graphical form, often highlighting particular aspects of the task. Usually they are used to represent system and human task interactions in one graphical medium.

Earliest life cycle stage: Flow-sheeting
Human factors issues: All

4. **Coding consistency** (see pp. 226)
A reporting form is produced which must consider the acceptability of coding from both the coding method and the coding function points of view (one code given to one meaning, one meaning given to one code). This should enable a check for coding inconsistency or coding incompatibility and subsequent rectification.

Earliest life cycle stage: Detailed design
Human factors issues: Task and interface design
Skills and knowledge acquisition
Performance assurance

5. **Cognitive task analysis** (Diaper, 1989)
Cognitive task analysis covers a range of approaches used for looking at mental (hence cognitive) internal events or knowledge structures. One particular example of a technique in this domain is task analysis for knowledge descriptions (TAKD). TAKD uses a knowledge representation grammar to represent the knowledge required to carry out a task. This information is further represented in a task descriptive hierarchy, and so can be seen as a parallel technique to HTA. TAKD and other cognitive task analysis techniques have been used primarily in the field of human–computer interaction (e.g. for identifying training needs, evaluating computer dialogue systems, or for knowledge elicitation in expert systems development).

Earliest life cycle stage: Detailed design
Human factors issues: Task and interface design
Skills and knowledge acquisition

6. **Computer modelling and simulation** (pp. 147 and Chapter 11)
The use of computer programs to represent the workers' activities or environment. A number of simulation runs can be made aggregating activity sequences, and these will assist estimation of cycle times, operator strategies and error likelihoods. Alternatively, the workspace layout can be graphically represented on computer, allowing analysis of visual access, layout adequacy, etc.

Earliest life cycle stage: Flow sheeting
Human factors issues: Staffing and Job Organization
Performance assurance
Task and interface design

7. **Confusion matrices** (Potash *et al.*, 1981)
This is a tabular plot of a set of stimuli (e.g. displays) against a set of related responses. The frequency of actual task responses are recorded in the grid squares, with the diagonal showing the frequency of correct responses. It is then possible to look at the grid and identify which responses may be made mistakenly for a given stimulus. Confusion matrices have been used in human reliability assessments to look at potential misdiagnosis during abnormal events.

Earliest life cycle stage: Detailed design
Human factors issues: Performance assurance

8. **Critical incident technique** (pp. 47)
Collection and analysis of information about incidents. Each incident is categorized to identify factors that are associated with or responsible for an incident occurrence. The intent is to identify when, where and why critical incidents occur.

Earliest life cycle stage: Commissioning
Human factors issues: Performance assurance

9. **Decision–action diagrams** (see pp. 87)
These diagrams are information flow charts constructed in the same way as those used for computer program development. Each section is initiated by an information input, which is followed by a decision block from which there may be alternative response routes which can be developed further.

Earliest life cycle stage: Flow sheeting
Human factors issues: Skills and knowledge acquisition

10. **Decomposition Methods** (pp. 95 and Chapters 9 and 16)
This produces an expansion of a task description into a series of statements about the task. Initially a set of short descriptions of all task elements is produced, more information is then elicited for each using a pre-determined set of subheadings. These sub-headings are chosen by the analyst to obtain the necessary information to address the particular issue under consideration (e.g. cues initiating action, controls, decisions, errors, responses, feedback, etc.).

Earliest life cycle stage: Detailed design
Human factors issues: Task and interface design
Skills and knowledge acquisition
Performance assurance

11. **Ergonomics checklists** (pp. 217)
There are a range of checklists available for directly assessing tasks and their interfaces (e.g. for assessing the workplace, its environmental aspects, or VDU systems, etc.). Some checklists (e.g. Fitts list) can be used for broad allocation of function.

Earliest life cycle stage: Concept
Human factors issues: All

12. **Event trees** (pp. 178 and Chapters 14 and 16)
An event tree is a graphical logic method for identifying the various possible outcomes of a given event, known as the initiating event. The course of events from the occurrence of the initiating event until its final consequences is determined by the operation or non-operation of various systems (human and hardware). An event tree is typically used to model the reliability of safety systems, including operator intervention, designed to prevent an initiating event turning into a catastrophic accident.

Earliest life cycle stage: Preliminary design
Human factors issues: Performance assurance

13. **Failure modes and effects analysis** (pp. 184 and Chapter 9)
This technique is used in human reliability analysis. It allows the analyst to consider what errors might occur during a task and their likely consequences for the system. It can also be used to enhance error recovery and to identify error reduction measures.

Earliest life cycle stage: Preliminary design
Human factors issues: Performance assurance

14. **Fault trees** (pp. 188 and Chapters 14 and 16)
Fault trees show failures that would have to occur in order for a 'top event' (accident) to occur. These are constructed as a series of logic gates descending through subsidiary events resulting from basic events at the bottom of the tree. These 'basic events' may be human errors, or hardware/software failures, or environmental events. Sensitivity analysis to determine the most important events (i.e. which events contribute most to the likelihood of system failure) is possible once the basic event likelihoods have been quantified.

Earliest life cycle stage: Flow sheeting
Human factors issues: Performance assurance

15. **Function analysis system technique (FAST)** (see pp. 112 and Chapter 7)
FAST was originally developed to understand how systems really work and how they may be modified to save costs. As the functional analysis is undertaken, two questions must be asked: '*How*' does a sub-ordinate function achieve a super-ordinate one? and '*why*' is a super-ordinate function dependent upon a sub-ordinate function? Super-ordinate and sub-ordinate functions can be connected by **AND** and **OR** gates in order to facilitate a logical check of the overall function requirements.

Earliest life cycle stage: Concept
Human factors issues: Allocation of function
Staffing and job organization
Skills and knowledge acquisition
Performance assurance

16. **Goals, operators, methods and selection rules model (GOMS)** (Card *et al.* 1983)
Initially the tasks are described as a set of goals and sub-goals similar to a HTA. At the required level this breakdown is stopped and each sub-goal is described in terms of the operations required to achieve it. An operation (or operator) is an elementary perceptual, motor or cognitive act, whose execution is necessary to change any agent of the user's mental state or to affect the task environment. Methods describe the procedures used to achieve a goal which have a specified chance of success. Selection rules are rules used to predict which of several possible methods will be selected by the user, based on task environment characteristics. GOMS allow hierarchical representations of goals which is useful, and GOMS can provide a complete dynamic description of behaviour. If times can be estimated for individual systems, then the GOMS model predicts the time necessary to complete tasks, although it will not consider errors. GOMS has mainly been used for human–computer interaction.

Earliest life cycle stage: Detailed design
Human factors issues: Skills and knowledge acquisition
Staffing and job organization
Performance assurance

17. **Hazard and operability study (HAZOP)** (pp. 194)
Based on pipework and instrumentation diagrams (P&IDs) or operating instructions, a set of guide words is applied to each stage of a process to identify potential hazards, possible consequences and preventive mechanisms. Success of a HAZOP is dependent upon co-operation of an interdisciplinary group, sometimes including ergonomics (human factors) expertise.

Earliest life cycle stage: Flow sheeting
Human factors issues: Task and interface design
Performance assurance

18. **Hierarchical task analysis (HTA)** (pp. 104 and all of Part III apart from Chapters 7 and 11)
The best known task analysis technique, used to represent the relationship between tasks and subtasks. It records system requirements and how these can be achieved, including the order in which tasks and subtasks must take place. If recorded pictorially it resembles a tree with branches and sub-branches as required.

Earliest life cycle stage: All
Human factors issues: All

19. **Influence diagrams** (pp. 201)
Target events (e.g. system failure) are defined prior to using the technique, and an assessor then produces a directed graph representing the influences (e.g. procedure adequacy etc.) that determine the outcome of each event (i.e. success or failure), together with any dependencies between them. The effect of each influence is then evaluated, with the resulting values used to weight human error probability estimates.

Earliest life cycle stage: Preliminary design
Human factors issues: Performance assurance

20. **Information and controls analysis** (see pp. 224)
First the task description is tabulated, then each subtask is examined in order to identify any associated decision requirements plus the information needed to ensure that the decision is effective. If an active response is required, the type of control should be noted with a description of feedback requirements.

Earliest life cycle stage: Preliminary design
Human factors issues: Task and interface design
Skills and knowledge acquisition

21. **Interface surveys** (pp. 223)
These are a group of methods which can be used to identify specific ergonomics problems or deficiencies in the interfaces which are provided (e.g. the labelling of controls and displays, or the environmental conditions which are provided).

Earliest life cycle stage: Detailed design
Human factors issues: Task and interface design

22. **Link analysis** (pp. 118)
An annotated diagram showing visual/physical movement between system components, including frequency of movement and component importance. It is used to analyse the relationships between system components to optimize their arrangement by minimizing movement times/distances and by placing important items in prime positions.

Earliest life cycle stage: Preliminary design
Human factors issues: Staffing and job organization
Task and interface design
Performance assurance

23. **Management oversight risk tree (MORT)** (pp. 208)
A technique used to investigate the adequacy of safety management structures, either to ensure that these exist or, if an incident has occurred, to determine which safety management functions have failed. The MORT system, in its accident investigation role, firstly defines what barriers have failed to allow the accident to occur, and then searches for the root causes of these failures as failed safety management functions. The MORT system uses a diagrammatic (qualitative) fault tree and a highly structured accident investigation audit system to identify the causes of the accident.

Earliest life cycle stage: Commissioning
Human factors issues: Performance assurance

24. **Observation** (pp. 53)
Observation is the most fundamental data collection technique and can range from direct viewing to video recording, or to participative observation, in which the analyst is involved in carrying out the task. These methods vary in their effectiveness and their intrusiveness on the job incumbents.

Earliest life cycle stage: Detailed design
Human factors issues: All

25. **Operational sequence diagrams** (pp. 125)
A diagram that makes use of standard symbols, often in the form of a flow chart, linking operations in the order in which they are normally carried out, supported by a text description. It is used to illustrate relations between personnel, equipment and time.

Earliest life cycle stage: Preliminary design
Human factors issues: Staffing and Job Organization
Skills and knowledge acquisition
Performance assurance

26. **Operator action event tree** (see pp. 180 and Chapter 16)
A representation of success and failure routes through a sequence of 'actions', each of which is an information input, a processing activity or an output element. Any level of task detail can be used since it was originally designed as a holistic approach. Each stage in the route can be given a failure probability resulting in an overall probability of failure/success for the complete event sequence.

Earliest life cycle stage: Preliminary design
Human factors issues: Performance assurance

27. **Operator modification survey** (see pp. 228)
An assessor enters the workplace and looks for any temporary modifications made by the workers. These are recorded under three headings; memory aids, perceptual cues and instrument grouping. From this it is possible to establish user difficulties even with well-established systems.

Earliest life cycle stage: Commissioning
Human factors issues: Task and interface design
Performance assurance

28. **Petri-nets** (see pp. 91)
These are state transition networks which distinguish between states (conditions) and transitions (events). As a condition is achieved it is marked, but a transition is impossible until all previous states are marked. Progress is shown as dependent upon the system state. This forces the analyst to record the result of all activities.
Earliest life cycle stage: Operation and maintenance
Human factors issues: Performance assurance

29. **Position analysis questionnaire (PAQ)** and **Job components inventory** (see pp. 62)
These are specific questionnaires which are used for gathering job information. They are usually used to identify general job characteristics and have been used primarily for personnel purposes. They seek to identify which job elements are present or absent in an occupation. The Job Components Inventory gathers information on a total of 194 job elements grouped into six divisions.

Earliest life cycle stage: Operation and maintenance
Human factors issues: Person specification
Skills and knowledge acquisition

30. **Process/system checklists**
Used to check that a system complies with pre-determined standards. These are easy to use, and can be applied at any stage in the system life cycle. They provide a systematic baseline which can also be used to assess success. A checklist may be designed as a form for approval prior to a project moving to its next phase. The success of a checklist is dependent upon the expertise of the author and its interpretation by the assessor.

Earliest life cycle stage: Concept
Human factors issues: All

31. **Questionnaires** (pp. 58)
A formalized and standardized set of questions, which can include open-ended questions, produced to elicit a wide range of responses. In general they are used to collect systematically and in an unbiased way a variety of individuals' views of a particular task or system.

Earliest life cycle stage: Concept
Human factors issues: All

32. **Signal flow graph analysis** (see pp. 92)
This technique identifies the important variables within a system and enables their relationships to be detailed. An output variable from the system is selected and all the variables that can influence this are identified. The variables that affect these are then identified, until all the output and input variables have been found and linked together. Variables are shown as nodes connected by lines to show their causal dependencies.

Earliest life cycle stage: Flow sheeting
Human factors issues: Task and interface design
Skills and knowledge acquisition
Performance assurance

33. **Simulators/mock-ups** (pp. 150)
Development and use of equipment or information (this may be through high-or-low fidelity simulators) that is representative of what will be used during the task. Task activity while using this is then observed and recorded.

Earliest life cycle stage: Preliminary design
Human factors issues: Task and interface design
Skills and knowledge acquisition
Performance assurance

34. **Structured interviews** (pp. 66 and Chapter 8)
A systematic collection of verbal information. An interview consists of a basic question and answer session with prepared questions asked by the interviewer with the replies either written down or taped. Extra information can still be added or a relevant issue pursued.

Earliest life cycle stage: All
Human factors issues: All

35. **Table-top analysis** (pp. 155)
A group of experts meet to discuss a problem perspective of the task, using task scenarios to explore the problem and derive a solution. The technique seeks to aggregate expert opinion in a problem-solving mode.

Earliest life cycle stage: Concept
Human factors issues: Skills and knowledge acquisition
Performance assurance

36. **Timeline analysis** (pp. 135)
Timeline analysis maps operators tasks along the time dimension, taking account of task frequency and duration, and interactions with other tasks and other personnel. Timeline analysis is useful for estimating staffing requirements via workload considerations, and can also be used in human reliability analysis to consider whether an operator or operating team are likely to complete a task within a particular time.

Earliest life cycle stage: Detailed design
Human factors issues: Staffing and organization
Performance assurance

37. **Verbal protocols** (pp. 71)
Verbal protocols are the recorded verbalizations of operators as they carry out their tasks. It is important that these verbalizations should not interfere in any way with task performance and that the operators should freely report on what they are doing without any direction from the analyst. These protocols are particularly useful for gaining information about the unobservable cognitive reasons for operators actions in certain situations, which cannot be directly observed. Verbal protocols are limited by the ability of subjects to freely and honestly state why they are carrying out particular actions, without first making a conscious effort to explain these actions.

Earliest life cycle stage: Commissioning
Human factors issues: Skills and knowledge acquisition
Task and interface design
Person specification

38. **Walk-through/Talk-through analyses** (pp. 160)
Workers who know the system demonstrate associated tasks, either on the plant or a representation of it. As a full technique they physically move around the workspace while describing and explaining the required actions and work methods (doing the actions prompts the worker). The movements (only usually for walk-throughs) and comments (both techniques) must be recorded and analysed.

Earliest life cycle stage: Detailed design
Human factors issues: Skills and knowledge acquisition
Task and interface design
Performance assurance

39. **Withheld information** (Duncan and Reiersen, 1988)
Many decision-making tasks require the operator to look at several information sources, presented simultaneously on a variety of display media. This technique is used to identify how operators select and use the information. A simulated task environment is constructed and information is withheld from the operator until it is specifically requested. A set of events are presented one at a time to the operator who must identify and deal with each of them. This is achieved by asking for items of information, one at a time. As each item is requested it is recorded by the analyst within a previously constructed matrix.

Earliest life cycle stage: Detailed design
Human factors issues: Person specification
Task and interface design
Skills and knowledge acquisition

40. **Work safety analysis** (see pp. 172)
This is a systematic analysis of a chosen work situation for all possible occupational accidents plus the measures that may be adopted to reduce or eliminate their likelihood. The work for analysis is divided into major steps and placed within a tabular format, and each work step is then examined using expert judgement, experience and accident reports to determine the following: possible accidents, causative factors, relative likelihood, seriousness of consequences, a risk index, and corrective measures. By assessing the initial and reduced risk figures given by implementing each suggested corrective measure, the cost effectiveness is derived and the most suitable corrective measures recommended.

Earliest life cycle stage: Operation and maintenance
Human factors issues: Skills and knowledge acquisition
Performance assurance

41. **Work study techniques** (Barnes, 1968)
A family of techniques devised to rigorously itemize the different steps required to perform a task, to establish if the best method is being used. A recorder must observe the task and make a note of basic physical work units in the order of completion, while using a standard notation system. This can be improved by sensory motor process charts which focus on the sensory processes and decision-making. The charts are used to determine if a simpler (or more ergonomic) method can be used to action the task. This would also require reference to ergonomics data bases on anthropometry, biomechanics, repetitive strain injuries etc., if applied to manual handling or repetitive tasks.

Earliest life cycle stage:	Operation and maintenance
Human factors issues:	Skills and knowledge acquisition
	Performance assurance
	Task and interface design
	Allocation of function

References

Barnes, R.M. (1969) *Motion and Time Study*, New York: John Wiley.

Diaper, D. (1989) *Task Analysis for Human Computer Interaction*. Chichester: Ellis Horwood.

Duncan, K.D. and Reiersen, C.S. (1988) Long Term Retention of Fault Diagnosis Skills. In *Training, Human Decision Making and Control*, Patrick, J. and Duncan, K.D. (eds.). pp. 93-118. Oxford: North Holland Publishing.

Glossary

Alarm suppression

The reduction of the number of alarms following particular events, so that the operator is not 'overloaded' with information.

Allocation of function

Determination of which tasks are best performed by people and which by machines, plus the identification of tasks which should be kept together to ensure an adequate understanding of the system. These work units are then given to particular people and equipment to perform.

Audit

Formal and independent evaluation of a system.

Chunking

Division of tasks into sub-units that are complete in themselves.

Commissioning

Following construction of a plant, commissioning readies the plant for operation by testing all the equipment and ensuring it will work satisfactorily.

Critical errors

Errors which, if they occur, will have a significant impact on system goals (i.e those errors which are most potentially harmful to the system objectives of safety and/or productivity).

Data collection

Accumulating information on how operators carry out their tasks, in particular data on how failure influences performance (positively and negatively), and data on the probabilities of errors in a task per task demand.

Data representation

Placing data collected into a useful format so that human performance aspects can be recorded, further investigated and/or enhanced.

Decision aids

Support for making decisions, whether diagnosing an abnormal event or carrying out maintenance trouble-shooting. Usually in written verbal text form or provided from on-line computer 'decision support systems'.

Decommissioning

Shutting down, dismantling, and long-term storage or removal of a system.

Diagnostic tools

See 'Job aids', and 'Decision aids'.

Downtime
System unavailability (e.g. when the system is non-operational either due to system failure, maintenance, or due to planned/ unplanned shutdown).

Environmental events
Such as earthquakes, large waves, poor weather conditions etc. Events which will put systems at risk, and/or act as stressors on their integrity.

Ergonomics
The study of human behaviour in relation to work. It is concerned with the adaptation of work conditions to the physical and psychological nature and well-being of the human operator.

Expert system
A computerized system which contains a knowledge base for a particular scientific domain, and a means of using this knowledge to answer questions on that scientific domain e.g. expert systems exist in the medical field which can fairly reliably diagnose certain medical conditions if given the right information on symptoms (the expert system itself generates the questions for information on symptoms).

Fault diagnosis
Analysis of a system disturbance to identify the nature of the disturbance, either in terms of its overt symptoms and/or its underlying causes, to identify means of fault rectification or mitigation.

Function
A function is the nature of the work thing or task that something is designed to do, or that someone is asked to do.

Functional specification
See 'Task allocation'.

Goal
A goal is an overall aim which can be achieved by a varying range of tasks, based on set objectives to achieve the goal.

Hardware
Mechanical and electrical equipment and interfaces.

Human error assessment
Analysis of opportunities for error and error recovery, and identification of factors (performance shaping factors) which affect the likelihood of error/recovery.

Human factors engineering
The American term for ergonomics.

Human factors issues
The specific topics of importance to ergonomists (human factors engineers) that must be addressed to ensure that a system is adequate for the user.

Human reliability assessment
The identification of important human errors associated with a specific task or system function, the modelling of those errors and the quantification of the probability of task failure, based on data attached to, or generated by, the model. Human reliability assessment may also be able to state how best to reduce human error impact on system performance.

Information flows
The manner in which messages about the system status and changes need to be passed to and from the user, including requests to the system and feedback to confirm the results of the requests.

Interface design
The design of the equipment or medium through which the user and system interact, including the environment in which the interface exists.

Job aids
Equipment or documentation supplied to assist the worker to successfully perform the required tasks. Designing the complete set of tasks into a coherent whole for each individual worker based on the information needed and actions required, plus knowledge of human limitations and capabilities (ergonomics).

Job design and work organization
The allocation of tasks between individual team members, the co-ordination of their activities and the provision of suitable information channels to ensure everyone within the system has an adequate understanding of the system status and requirements.

Job organization
Ensuring task success by checking time constraints, workload and allocation of function for individual workers plus the co-ordination and co-operation between team members and across shifts.

Management safety structure
Established management goals and methods for ensuring the safety of their system.

Manual handling
Physical manoeuvering of objects by operators, with little or no mechanical device assistance.

Operable
Usable to the extent that successful performance can be regularly achieved.

Operation
An operation describes what is actually done in a situation, and is usually a description of the behaviour or cognitive activity carried out to achieve the task objective.

Operations and maintenance
Operations refers to the day-to-day operations required to run the plant (i.e. the tasks associated with its normal running). Maintenance refers to the off-line examination, maintenance, and repair, testing, and calibration of sub-systems either required to prevent sub-system failure, or in response to sub-system failure.

Operator support
See 'Job aids'. Also covers the provision of system information and management structures.

Performance
The way in which people complete their tasks.

Performance Assurance
Methods for checking that the joint performance of individuals within a system will be adequate to keep the system running within pre-set limits.

Performance checking
Looking to see whether people are working in the manner required and achieving the required objectives.

Person specification
Detailing the characteristics required of people who will be working within a particular system in terms of their personal, mental and physical capabilities.

Plan
A plan defines the way in which the subordinate tasks/operations must be carried out to achieve the super-ordinate goal. In particular plans define when tasks/operations should be carried out, and in what sequence etc.

Problem investigation
Checking to see why deviations from pre-set goals (production or safety) have occurred within a system, or investigating other areas of concern raised by management or the work force.

Procedures
Formal operating instructions, usually for more than one person.

Process control
A type of system which is based on the existence of a chemical and/or physical process which can be reduced to a set of formal equations (e.g. chemical or mass-energy equations), which can be controlled (manually and/or automatically) by adherence to a set of parameters (e.g. temperature, pressure) within pre-specified ranges or 'setpoints'.

Project manager
Person responsible for the timely running of a project to cost and other targets (e.g. effectiveness, safety) specified both internally (by the company) and externally (e.g. via regulatory authorities).

Quality assurance
Ensuring that safety/availability/productivity targets will be met at all levels to system specification requirements.

Quantified risk assessment
Quantitative evaluation of the risks imposed by a system design, whether those risks are from human, hardware or software failures, or environmental events, or (usually) from combinations of such failures/events. Also termed 'Probabilistic risk assessment' and 'Probabilistic safety assessment'.

Resources requirements
Resources required to carry out the task or the task analysis technique (e.g. in terms of person-hours, special tools or equipment, computer software, etc.).

Safety assessors
Those responsible for estimating the risks associated with system design, operation and maintenance.

Skills and knowledge acquisition
The means by which people can gain the capabilities required to work successfully within a specific system.

Software
Computer software used in systems.

Staffing
Personnel or team complement responsible for operating and maintaining a system.

Sub task
A part of a task that when performed with one or more additional sub-tasks will result in successful task completion.

System
The formal interaction of different items in order to produce a specific product or service.

System functions
Specified ways in which a system is expected to work.

System life cycle
The evolution of a particular system from its first suggestion through its design, building, testing, running and maintaining to the final removal of the system.

Systems of work
Work practices as specified by formal procedures/training. Usually used in the context of 'safe systems of work' (i.e. a system of carrying out required work which will avoid hazards associated with the task).

Targetting
Isolating a particular problem or problem aspect for investigation.

Task
A task is a set pattern of operations which alone, or together with other tasks, may be used to achieve a goal.

Task aids
See 'Job aids'.

Task allocation
Division of individual tasks between members of an organization and between people and machines.

Task analysis
Task analysis is a method of describing what an operator is required to do, in terms of actions and/or cognitive processes, to achieve a system goal. It is a method of describing how an operator interacts with a system, and with the personnel in that system.

Task analysis process
The task analysis process is one of collecting information on how a task is carried out, and representing such information such that the task can be analysed to see if it can be improved, or to assess its task design adequacy. The task analysis process is most useful when properly integrated into the system life cycle.

Task analysis techniques
Different ways of identifying tasks and their requirements in terms of the resources necessary for their successful accomplishment. Individual techniques may also access different information about the tasks.

Task attributes
Specific characteristics of a task that makes it unique when viewed in the context of a particular system.

Task design
Design of each separate task so that the worker can adequately receive all the information necessary to succeed and can adequately perform any action or send any information necessary to achieve the required goals.

Work needs
The information and control facilities assigned to each task which are necessary for successful functioning of the system.

Acronyms

AFWS	Auxiliary feedwater system
AOA	Advanced operator aids
ASME	American Society of Mechanical Engineers
BHC	Basket handling cave
BOP	Blowout prevention
BUP	Back-up panel
CCR	Central control room
CCTV	Closed-circuit television
CES	Cognitive environment simulation
CRO	Control room operator
CRT	Cathode ray tube
C-SAT	A commercial satellite system
CSL	Control station layout
DAC	Distance amplitude correction
DOG	Dissolver off-gas
DOP	Dropped object protection
DPS	Data processing system
EFD	Engineering flow diagram
EOAT	Enhanced operator action tree
ESD	Emergency shut down
FA	Functional analysis
FAST	Functional analysis system technique
FDSC	Fully developed safety case
FMEA	Failure modes and effects analysis
FPV	Floating production vessel
GOMS	Goals, operators, methods, selection
HAZAN	Hazard analysis
HAZOP	Hazard and operability study
HCI	Human–computer interaction
HEA	Human error analysis
HEART	Human error assessment and reduction technique
HEP	Human error probability
HF	Human factors
HFI	Human factors issue
HRA	Human reliability assessment
HRAET	Human reliability analysis event tree
HRMS	Human reliability management system
INMARSAT	A commercial satellite system
JHEDI	Justification of human error data information
HTA	Hierarchical task analysis
LCR	Local control room
LOCA	Loss of coolant accident

LTM	Long term memory
MCR	Main control room
Micro-SAINT	(workload analysis computer package – see SAINT)
MMI	Man–machine interface
MoD	Ministry of Defence
MORT	Management oversight risk tree
NDT	Non-destructive testing
NGT	Nominal group technique
OA	Operator aids
OAT	Operator action tree
OIM	Offshore Installation Manager
OSA	Operational safety appraisal
OSD	Operational sequence diagrams
P & ID	Piping and instrumentation diagram
PA	Public address
PABX	Private automatic branch exchange
PAQ	Position analysis questionnaire
PCR	Production control room
PCSR	Pre-construction safety report
PMI	Person–machine interface
POI	Plant operating instructions
PORV	Power operated relief valve
PRA	Probabilistic risk assessment
PSA	Probabilistic safety assessment
PSF	Performance shaping factor
PSSCR	Production and sub-sea control room
PWR	Pressurized water reactor
QA	Quality assurance
RCS	Reactor coolant system
RHRS	Residual heat removal system
SAINT	System analysis of integrated networks of tasks
SAMMIE	System for aiding man–machine interface evaluation
SC	Shear cave
SG	Steam generator
SGTR	Steam generator tube rupture
SI	Safety injection
SNUPPS	Standard nuclear unit power plant system
SOI	Station operating instructions
SRK	Skill, rule, knowledge (model)
SS	Shift supervisor
STA	Sequential task analysis
SUB	(*Error probability*) subsumed by another HEP
TA	Task analysis
TAKD	Task analysis for knowledge description
THERP	Technique for human error rate prediction

THORP	Thermal oxide reprocessing plant
TSA	Tabular scenario analysis
TTA	Tabular task analysis
UHF	Ultra high frequency
VDU	Visual display unit
VHF	Very high frequency
WS	Work station
WSA	Work safety analysis

Subject Index